T0186308

Introduction to

Wireless Communication Circuits

2nd Edition

Forouhar Farzaneh, Ali Fotowat, Mahmoud Kamarei

Ali Nikoofard and Mohammad Elmi

RIVER PUBLISHERS SERIES IN CIRCUITS AND SYSTEMS

Series Editors

MASSIMO ALIOTO
National University of Singapore
Singapore

KOFI MAKINWA
Delft University of Technology
The Netherlands

DENNIS SYLVESTER
University of Michigan
USA

Indexing: All books published in this series are submitted to Thomson Reuters Book Citation Index (BkCI), CrossRef and to Google Scholar.

The "River Publishers Series in Circuits & Systems" is a series of comprehensive academic and professional books which focus on theory and applications of Circuit and Systems. This includes analog and digital integrated circuits, memory technologies, system-on-chip and processor design. The series also includes books on electronic design automation and design methodology, as well as computer aided design tools.

Books published in the series include research monographs, edited volumes, handbooks and textbooks. The books provide professionals, researchers, educators, and advanced students in the field with an invaluable insight into the latest research and developments.

Topics covered in the series include, but are by no means restricted to the following:

- Analog Integrated Circuits
- Digital Integrated Circuits
- Data Converters
- Processor Architecures
- System-on-Chip
- Memory Design
- Electronic Design Automation

For a list of other books in this series, visit www.riverpublishers.com

Introduction to
Wireless Communication Circuits
2nd Edition

Forouhar Farzaneh
Professor
Sharif University of Technology, Iran

Ali Fotowat
Associate Professor
Sharif University of Technology, Iran

Mahmoud Kamarei
Professor
University of Tehran, Iran

Ali Nikoofard
Research Engineer
University of California at San Diego, USA

Mohammad Elmi
Research Engineer
KavoshCom Asia Co., Iran

River Publishers

Published, sold and distributed by:
River Publishers
Alsbjergvej 10
9260 Gistrup
Denmark

River Publishers
Lange Geer 44
2611 PW Delft
The Netherlands

Tel.: +45369953197
www.riverpublishers.com

ISBN: 978-87-7022-140-5 (Hardback)
978-87-7022-139-9 (Ebook)

©2020 River Publishers

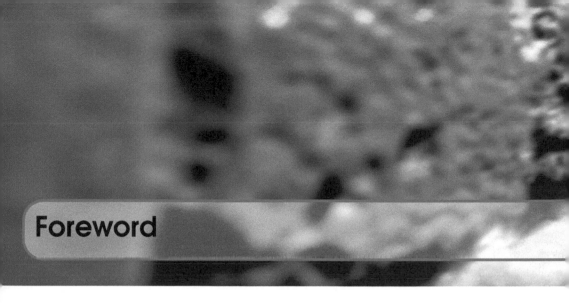

Foreword

As wireless technology takes over every aspect of our lives, the university curricula must keep up with the developments and impart proper skills to their graduates so as to prepare them for this rapidly-evolving industry. In particular, the vast body of undergraduate students must be trained in this domain, but efficiently, as course proliferation is undesirable in most universities.

"Introduction to Wireless Communication Circuits" addresses this need by selecting the most relevant topics and teaching them in a language that appeals to undergraduate students. The textbook methodically guides the reader through the concepts and, using numerous detailed examples, reenforces these concepts. The reader is then invited to exercise his/her understanding by solving problems at the end of each chapter.

The contents of the book have been chosen carefully to allow coverage in one semester or quarter. That is, the book can serve as a self-contained text that the students can read "cover to cover" in one term without skipping any major sections. These pedagogical aspects of the book facilitate its use for both students and instructors.

Behzad Razavi
Professor
University of California, Los Angeles
February 2018

Preface to the Second Edition

During the past couple of years where we used this book as our teaching reference in wireless communication circuits, we encountered a number of points to be clarified or improved in the text. To this effect, we have prepared the material for the second edition. Scores of equations and a few figures were added; therefore, the second edition contains 1161 equations and 505 figures which help more in the analysis and understanding of the text. We have changed the text in hundreds of instances to make the material straightforward and more comprehensible. We hope that this new edition will be more useful for the students and practicing engineers.

F. Farzaneh, A. Fotowat, M. Kamarei, A. Nikoofard, and M. Elmi
Tehran, November 2019

Preface to the First Edition

The tremendous development of the wireless communications during the past twenty years has had such a great impact on the human life and the society that no one can ignore its overwhelming presence in today's life, let alone the today's engineering. Students in Electrical Engineering, nowadays, are normally exposed to courses in wireless communications, wireless circuits, and the electromagnetic-wave propagation. The authors have been involved in teaching and research in the field of wireless circuits during the past thirty years in the Iranian universities and institutions namely Sharif University of Technology, University of Tehran, and KavoshCom Asia R&D Company. During these years we have felt a lack of a comprehensive book which would cover the needed material for a course coverage at the B.Sc. level of Electrical Engineering. Furthermore, as engineers at a research institution, we observed the necessity of a comprehensive text which would help the RF engineers in their RF circuit design and implementation. As a matter of fact, a number of circuits and ideas presented in this book were obtained during the development of new RF transceiver circuits, GPS receivers and wireless communication systems intended for fleet control in ground transportation at KavoshCom Co.

We have tested or verified the most of the presented circuits in this book by standard RF simulation tools to be sure of their proper operation. For materializing this book, we have used most of our course materials especially intended for the communication circuits course. It took us a long three years of intensive work to realize this book. Our intention was to make it accessible to the Electrical Engineering community worldwide, as a result of our efforts in the field of RF circuits and wireless communication.

This book is divided into three parts. In Part I, chapter 1 is dedicated to the wireless communication systems and the building blocks of a modern radio transceiver. Chapter 2 describes the major operation and configuration of the RF oscillators where the major topologies of the modern electronic oscillators are presented as well as the large signal modeling and evaluation of these circuits.

Part II of the book is dedicated to the major building circuit blocks of modern transceivers. Chapter 3 presents PLLs, different PLL topologies, FM modulators and FM demodulators. Chapter 4 deals with the RF mixer circuits where different type of mixers from the switching circuits to analog multipliers are presented. The major concepts of nonlinearity in RF circuits namely the compression, the intermodulation products, and the intercept point are introduced in this chapter. Chapter 5 is dedicated to Amplitude and Phase Modulation. This chapter begins with analog amplitude modulation techniques and then goes through present day multilevel amplitude and phase digital modulations. Chapter 6 describes the Limiters and the Automatic Gain Control circuits. Offset compensation circuits are presented in this chapter followed by different Automatic Gain Control methods. The amplitude detectors and methods for increasing the Gain Bandwidth of amplifiers are also described in this chapter as well.

Part III of the book is dedicated to Transmission Lines, Microwave circuit modeling, and Microwave Amplifiers using Scattering Parameters as well as the Power Amplifier description. Chapter 7 describes the fundamentals of RF Transmission Lines and Impedance Matching Circuits. Chapter 8 is intended to the introduction of Scattering Parameters as a modern tool for amplifier circuit design in the microwave range. Chapter 9 presents the analysis and design of microwave amplifiers using S-parameters. The problem of stability of two-ports using S-parameters is studied in detail. The simultaneous conjugate matching of an amplifier using S-parameters is presented alongside the design of Low Noise Amplifiers using noise parameters. The design of two-stage amplifiers including the noise and the gain parameters is described at last. In chapter 10 the Power Amplifiers are presented. Different classes of power amplifiers, their mode of operation, their efficiencies, and their power capabilities are studied in this chapter. The linearization methods for the power amplifiers as a modern tool for the present day transmitters are presented in this chapter as well.

We would like to thank all the colleagues and the students for their helpful discussions and encouragements which was necessary for the materialization of this book.

F. Farzaneh, A. Fotowat Ahmady, M. Kamarei, A. Nikoofard, M. Elmi
Tehran, Iran
December 2017

Contents

I   Part 1

II   Part 2

III   Part 3

List of Figures

List of Tables

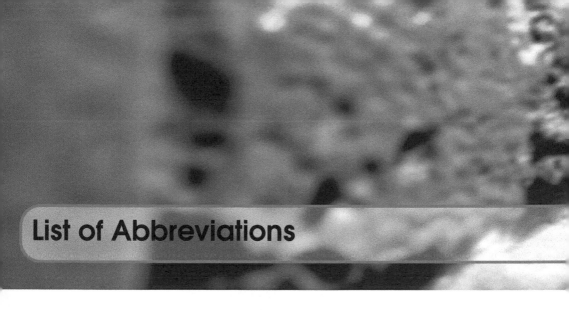

List of Abbreviations

2G (Mobile Network)	Second Generation
4G LTE (Mobile Network)	Fourth Generation, Long Term Evolution
A/D	Analog to Digital (converter)
AC	Alternating Current
ACPR	Adjacent Channel Power Ratio
ADC	Analog to Digital Converter
ADS	Advanced Design System (Software Tool)
AGC	Automatic Gain Control
AM	Amplitude Modulation
AMPS	Advanced Mobile Phone System
APG	Available Power Gain
BiCMOS	Bipolar Complementary Metal Oxide Semiconductor
BJT	Bipolar Junction Transistor
BPF	Band-Pass Filter
BPSK	Binary Phase Shift Keying
BW	Bandwidth
C/I	Carrier to Intermodulation (Ratio)
CB (Amplifier)	Common-Base
CC (Amplifier)	Common-Collector
CD (Amplifier)	Common-Drain
CDMA	Code-Division Multiple Access
CE (Amplifier)	Common-Emitter
CG (Amplifier)	Common-Gate
CMOS	Complementary Metal Oxide Semiconductor
CS (Amplifier)	Common-Source
DAMPS	Digital Advanced Mobile Phone System
DC	Direct Current

DCS	Digital Cellular System
DDS	Direct Digital Synthesis
DFF	D Flip-Flop
DSBSC	Double-SideBand Suppressed Carrier
DSP	Digital Signal Processing
DUT	Device Under Test
EER	Envelope Elimination and Restoration
EIRP	Effective Isotropic Radiated Power
ERP	Effective Radiated Power
EVM	Error Vector Magnitude
FFT	Fast Fourier Transform
FLL	Frequency-Locked Loop
FM	Frequency Modulation
FSK	Frequency-Shift Keying
GBW (GBWP)	Gain Bandwidth Product
GF	Gaussian Filter
GMSK	Gaussian Minimum Shift Keying
GPS	Global Positioning System
GSM	Global System for Mobile communication, *originally* Groupe Spécial Mobile
HPF	High-Pass Filter
IF	Intermediate Frequency
IIP3	Input third-order Intercept Point
IM	Intermodulation
IM3	Third-order Intermodulation
IM5	Fifth-order Intermodulation
IMD	Intermodulation Distortion
IP3	third-order Intercept Point
IS-95	Interim Standard 95
KCL	Kirchhoff's Current Law
KVL	Kirchhoff's Voltage Law
LINC	Linear Amplification with Nonlinear Components
LNA	Low Noise Amplifier
LO	Local Oscillator
LPF	Low-Pass Filter
LSB	Lower SideBand
MOS	Metal Oxide Semiconductor
MOSFET	Metal Oxide Semiconductor Field Effect Transistor
NF	Noise Figure
NMOS	N-channel Metal Oxide Semiconductor
NPR	Noise Power Ratio
OIP3	Output third-order Intercept Point
OPG	Operational Power Gain
OQPSK	Offset Quadrature Phase-Shift Keying
PA	Power Amplifier
PAE	Power Added Efficiency
PAR	Peak to Average Ratio

PCB	Printed Circuit Board
PCS	Personal Communications Service
PD	Phase Detector
PDF	Probability Density Function
PDM	Pulse Deletion Modulation
PFD	Phase Frequency Detector
PLL	Phase-Locked Loop
PM	Phase Modulation
PMOS	P-channel Metal Oxide Semiconductor
PSD	Power Spectral Density
PSK	Phase-Shift Keying
PWM	Pulse Width Modulation
QAM	Quadrature Amplitude Modulation
QPSK	Quadrature Phase-Shift Keying
RBSG	Random Bit Sequence Generator
RF	Radio Frequency
RFC	Radio Frequency Choke
RSSI	Received Signal Strength Indicator
RX	Receiver
S/N	Signal to Noise (Ratio)
S/P	Serial to Parallel (Converter)
SMA	SubMiniature version A (Connector)
SNR	Signal-to-Noise Ratio
S-parameter	Scattering Parameter
SSB	Single-SideBand
SSBSC	Single-SideBand Suppressed Carrier
SWR	Standing Wave Ratio
TCXO	Temperature Compensated Crystal Oscillator
TDMA	Time Division Multiple Access
TEM	Transverse ElectroMagnetic
THD	Total Harmonic Distortion
T-line	Transmission line
TPG	Transducer Power Gain
TRX	Transceiver
TX	Transmitter
UHF	Ultra High Frequency
USB	Upper SideBand
VCO	Voltage Controlled Oscillator
VGA	Variable Gain Amplifier
VSWR	Voltage Standing Wave Ratio
WCDMA	Wideband Code Division Multiple Access
WiFi	Wireless Fidelity
XOR	Exclusive-OR

Part 1

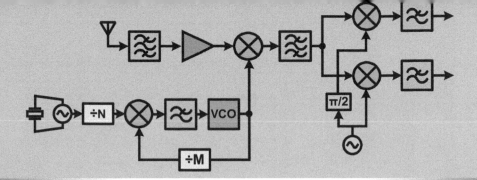

1. The Amazing World of Wireless Systems

Wireless system and circuit design is one of the most interesting fields in electrical engineering. From the economic point of view, wireless applications can be categorized into cellular/smart phones, cordless phones, wireless data networks, sensor networks, global positioning systems, and digital television broadcasting (terrestrial or satellite based). A huge investment has already been made in this sector and experts project further growth in the years to come. From the engineering point of view, the design of wireless systems has different levels of abstraction which are relevant to radio frequency (RF) antennas, wave propagation phenomena, RF and microwave circuit design, evaluation of noise and intermodulation phenomena, digital modulation, coding, and digital signal processing.

1.1 Introduction to Communication Circuits

Communication circuits is a comprehensive course which is normally taught for senior-level undergraduate students. This course is an aggregation of a number of materials including analog circuits, digital circuits, and digital signal processors. One of the most important technological developments which have thoroughly changed lifestyle of the people in the past two decades has been undoubtedly the inventions pertaining to wireless systems. In this section, we discuss the design of a basic radio transceiver (transmitter plus receiver) and analyze its system-level behavior. Furthermore, we focus on the behavior of each building block of a reciever/transmitter chain and investigate the mutual interactions of these building blocks on the overall signal performance.

Figure 1.1 shows a general transceiver block diagram. Here we briefly describe the function of each block in the receiver and the transmitter. Through the following chapters, we describe more precisely the functions and the analysis of each of the blocks. It should be noted that as the frequency spectrum is crowded with many transceivers for wireless applications ranging from AM/FM radios to TV transmitters, cellular phones, air transport communications, police and fire stations, emergency

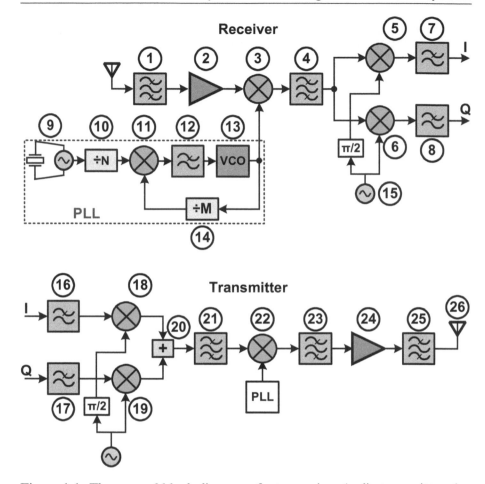

Figure 1.1: The general block diagram of a transceiver (radio transmitter plus receiver).

aid radios, and WiFi networks, we need to select the frequency channel of interest properly (i.e., reception) while interferences from all the other systems may be present. Similarly, we must transmit a channel in the frequency allocated to an application without causing excessive interference for other applications (i.e., transmission).

To start after the receiving antenna, we normally use a band-pass filter, block (1), to preselect the spectrum of our application (e.g., the full 25 MHz bandwidth in the 869.2–893.8 MHz receive band of GSM). In block (2), the weakly received signal will be amplified, usually by about 5 to 20 dB. This block normally consumes several milli-amps of current because it is normally operating in class A and at the highest frequency. In block (3), the amplified signal is downconverted through a mixer which brings the signal to a lower frequency for further processing. Block (3) is symbolized by a multiplication sign because the multiplication of two sinusoids is known to result in two new frequency components at the sum and the difference frequencies. Block (4) is a band-pass filter that usually eliminates the undesired frequencies and selects one of the two output signals. The frequency at block (4) is called the intermediate frequency (IF). The channel selection in radios is performed by changing the local oscillator (LO) frequency applied to block (3). The frequency from block (4) onward is fixed, making its processing more simple. Blocks (5) and (6) downconvert the IF to the baseband (in modern radios, where blocks (3) and (4) are suppressed and the RF frequency is directly converted to the baseband, they are called zero-IF downconverter). Blocks (7) and (8) are low-pass filters that are narrow enough to select the desired information. The I and Q outputs go to a DSP (digital signal processing) block for further digital processing intended for the display or the speaker for example. The difference between filters (1), (4), (7), and (8) is that as we go through the receiving chain, the filters become narrower, eliminating undesired frequency components. To generate the LO signal used to derive block (3), we start with a crystal oscillator in (9) whose frequency is usually between 5 MHz and 50 MHz. Block (10) divides the crystal oscillator frequency by integer N to provide a lower stable frequency. Block (13) is a voltage-controlled oscillator (VCO) whose output frequency is divided by integer M in block (14). The resulting two frequencies out of block (10) and (14) are compared in block (11). The output of block (11) is low-pass filtered by block (12) which provides an error voltage to drive the VCO. The ratio M is digitally controlled. When the loop is settled, the frequency of the VCO will be set to M/N of the frequency of the crystal oscillator. We describe these blocks in more detail in the upcoming chapters. It is important to note at this point that the first LO generates the signal required by block (3) to select the desired channel. Block (15) is a fixed oscillator that supplies the second LO for blocks (5) and (6).

For the transmitter portion in modern receivers, the baseband signals (either voice, video, or data), after analog-to-digital conversion, form the I and Q signals which are low-pass filtered by blocks (16) and (17). The outputs are upconverted by blocks (18) and (19) mixers and summed in block (20). The resulting signal is band-pass filtered in block (21) which is called the IF of the transmitter, then applied to a second mixer of block (22), and is upconverted to the desired RF channel. Filter (23) is a band-pass filter that selects the desired radio frequency and leaves out the undesired components. Block (24) is a power amplifier that may amplify the output to the desired wattage. Block (25) is the final stage filtering that will guarantee proper compliance with the regulatory standard preventing undesired frequency components (here, the harmonics or the intermodulation) for other systems or subscribers. The LO frequencies needed for the transmit path might be generated by the same scheme as the receiver. The difference between filters (16), (17), (21), (23), and (25) is that as we move forward in the transmit path, they become wider to allow the transmission of the full spectrum of the application to be used. For example, in GSM 850, the final filter (25) is a band-pass filter in the range of 824.2 MHz–849.2 MHz.

As an example for a transceiver, we investigate the block diagram of the second generation (2G) Digital AMPS system (DAMPS[1]). We have deliberately chosen this system because it includes both analog and digital modulations. In this system, the channel spacing is 30 kHz. The AMPS standard was fully analog, but evolved to

---

[1]Digital advanced mobile phone system.

contain digital modulation in DAMPS. The analog modulation in this system is based on frequency modulation with a maximum frequency deviation of 12 kHz (and given the 3 kHz baseband, consequently a total bandwidth of 30 KHz). The digital modulation is based on $\pi/4$ QPSK. Considering a possible bit rate of about 60 kb/s in each channel with a bandwidth of 30 KHz, three users can be present. The increased number of subscribers is due to digital modulation and the time division among them, and consequently sequential transmission of digital information. The procedure of sharing one channel between three users is based on three time slots (time division multiple access or TDMA). Each subscriber's speech data are recorded and transmitted in its time slot. This is accomplished at the cost of a maximum of three time slots delay. The speech data are also compressed with advanced algorithms to reduce its bit rate (to less than 20 kb/s). It is possible that the subscribers are in different geographical positions, and as a result, we need a base station for management and control of the three time slots allocated to different subscribers in different places. In this system, the frequencies of the receive and the transmit have 45 MHz difference. The frequency allocation for this system is in the range of 824 MHz−849 MHz which is used for transmission of the mobile set and is called uplink. Similarly, the downlink for this system (the reception frequency of the mobile set) is defined in the range of 869 MHz−894 MHz that is used by the base station. As it is evident, the difference between the center frequencies of the downlink and the uplink bands is 45 MHz and each has a 25 MHz bandwidth. In Figure 1.2, the spectrum usage and the frequency allocation of this system are shown.

In Figure 1.2, two simultaneous subscribers are shown. In Figure 1.3, it is shown that there is a free 30 kHz channel between two adjacent channels.

As illustrated in Figure 1.3 even though the bandwidth of each channel is 30 kHz, in the same cell, 60 kHz channel spacing is considered. This is due to maintaining an interference-free reception that is discussed in the following chapters. What was described earlier for DAMPS can be similarly applied for GSM[2] assuming a 200 kHz bandwidth and digital performance. It is recommended that in mobile networks there be always an empty channel between two adjacent channels.

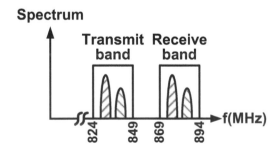

Figure 1.2: The spectrum of DAMPS showing two simultaneous subscribers.

---

[2]Groupe Speciale Mobile.

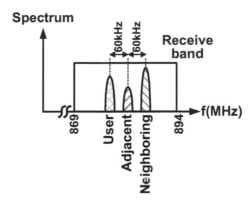

Figure 1.3: The DAMPS channel spacing stipulates one empty channel between two adjacent channels in every cell (30 kHz guard band is considered between two adjacent channels).

While receiving a subscriber channel, it is possible that the adjacent or neighboring channels might be stronger up to 60 dB. The adjacent, neighboring, and other channels which lie in the receiving band are called in-band interferes. As the total receive bandwidth is 25 MHz, with a channel bandwidth of 30 kHz, we have 833 channels. Assuming a frequency reuse pattern of seven (see Figure 1.4), there will be about 119 channels ($833/7 = 119$) available for allocation in one cell. With the ever increasing number of mobile phone usage, it should be clearly obvious that 119 simultaneous calls do not satisfy the requirements of a dense urban area. So digital TDMA is provided in DAMPS to increase the capacity by a factor of three with respect to AMPS.

It is usually common to compare the voltage or the power of signals with logarithmic ratio as follows

$$Ratio(dB) = 10\log\left(|\frac{P_1}{P_2}|\right) = 10\log\left(\frac{\frac{V_1^2}{2Z_0}}{\frac{V_2^2}{2Z_0}}\right) = 20\log\left(|\frac{V_1}{V_2}|\right) \qquad (1.1)$$

where Z_0 is the reference impedance. It is observed that once both signals have the

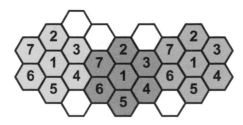

Figure 1.4: Three clusters of seven cell frequency distribution for DAMPS or GSM.

same reference impedance, the power ratio and the voltage ratio in dB (decibels) would have the same value. Another popularly used definition in radio engineering is dBm which is used for describing the absolute power of the signals and is defined as the ratio of the power in milli-watts to a 1 mW reference power and defined as

$$P(\text{dBm}) = 10\log\left(\frac{P_{\text{mW}}}{1\,\text{mW}}\right) \tag{1.2}$$

We now explain the difference between dB and dBm. When we use the term dB, we are expressing the logarithmic ratio of two signal amplitudes; once we are using dBm, we are expressing the logarithmic power ratio of the signal with respect to a 1 milli-watt reference or describing the power in dBm. Here are a few conversion examples in Equation 1.3.

$$5\,\text{dBm} = 3\,\text{mW} \tag{1.3a}$$

$$0\,\text{dBm} = 1\,\text{mW} \tag{1.3b}$$

$$-10\,\text{dBm} = 0.1\,\text{mW} \tag{1.3c}$$

$$-100\,\text{dBm} = 0.1\,\text{pW} \tag{1.3d}$$

With the above definitions, the sensitivity in the GSM system implying the minimum signal which could be properly detected is about $-103\,\text{dBm}$, and for DAMPS, the sensitivity is $-114\,\text{dBm}$.

1.2 Signal Levels and Rayleigh Fading

The signal levels from the base station transmitter to the mobile phone receiver may experience an attenuation of several tens of dBs up to 100 dB. Now consider two users who receive mobile signals, due to the fact that these signals come from different paths, the received signal strength may vary from one user to another because of constructive or destructive interference of rays coming from different paths. As an example, the wavelength of a 1 GHz carrier signal is just 30 cm. Thus, one by moving 7.5 cm (quarter wavelength) may go through the signal deep point from its peak point in the space, as such the signal level might change dramatically. One may ask despite the fact that one particular channel can go through multipaths and cause variation in the signal level, how does a receiver detect the desired signal level with all the interferes that come from other sources. Before answering the aforementioned question, we introduce the concept of Rayleigh fading. This phenomenon is a statistical model for radio wave propagation in a multipath medium. As the waves have multiple reflections on hills, buildings, and trees, calculating the exact signal level is extremely complex. Figure 1.5 exhibits the typical level of the received signal propagating from a base station to the mobile set receiver. It is clear that when the subscribers' distance to the base station antenna increases, the received signal becomes generally weaker. On the other hand, the signal power level seems to peak and dip as the user increases his/her distance with the base station. These variations come about from the superposition of different waves propagating and reflecting through different paths. In radio engineering, it is convenient to use a statistical model instead of an exact model for estimation of the

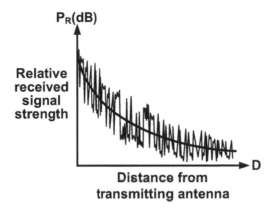

Figure 1.5: Concept of Rayleigh fading.

field strength. This model predicts the amplitude of the signal which experiences changes or fading by passing through the multipath media. The distribution of this statistical model is based on Rayleigh distribution. The received signal strength in general is dependent on the base station's transmitted power, its antenna gain, and the fading phenomenon in the propagation medium. In general, the received signal strength diminishes with distance plus or minus some local variations. In large cities, the above-mentioned problem becomes more acute because of the reflections from the ground, and reflections or diffractions from the multiple buildings on the propagation path. Due to higher traffic in large cities and higher population density, the number of cells is increased which in turn results in lower cells' radii. In suburban or rural areas, where there are not too many base stations, the radius of the cell may increase and the RF power coming from the base station may be as high as 43 dBm and the minimum detectable signal may be as low as −103 dBm. The difference between the transmitted and the received signal being as high as 146 dB implies the complexity of design at radio frequencies. Handling the large dynamic range required in the rural or suburban areas and the significant Rayleigh fading phenomenon encountered in urban areas is one of the main challenges encountered in radio systems.

1.3 Calculation of the Sensitivity in Different Standards

In this section, we intend to calculate the sensitivity of the each of the radio transmission standards defined for AMPS, GSM, and WiFi 802.11a/g. In general, the sensitivity of each standard is determined by its bandwidth (bit rate and modulation scheme), noise figure or noise temperature, and the minimum required signal-to-noise ratio:

$$S_{\min} = K(T_e + T_0)B(S/N)_{\text{minimum}} \tag{1.4}$$

where K is the Boltzmann constant, T_e is the equivalent receiver's noise temperature ($T_e = (F - 1)T_0$), B is the bandwidth in Hertz, and $S_{\min}$ is the minimum acceptable

Table 1.1: Specifications of AMPS, GSM, and WiFi 802.11a/g.

	Bandwidth	Min. Acceptable Signal-to-Noise Ratio	Max. Acceptable Noise Figure	Sensitivity
Wi-Fi (802.11a/g, BPSK)	20 MHz	14 dB	5 dB	−82 dBm
GSM	200 kHz	9 dB	10 dB	−102 dBm
AMPS	30 kHz	12 dB	5 dB	−112 dBm

signal-to-noise ratio. Now rewriting Equation 1.4 in decibels, one (at the standard temperature of 290°k) obtains

$$S_{min}(dBm) = -174 + 10\log(F) + 10\log(BW) + (S/N) \tag{1.5}$$

As such, the sensitivity in these three standards will become as what is demonstrated in Table 1.1.

1.4 Considerations in RF System Design

Let's start by studying, as an example, the operation of the DAMPS receiver in Figure 1.6. The antenna is the first element in the front-end of a receiver that absorbs the electromagnetic energy propagating through the air and converts it to electrical signals, voltage, and current. Radio receivers usually are designed for extremely weak signal reception. In many applications, these weak signals may be accompanied by strong interference from nearby transmitters. In order to attenuate the unwanted signals, a band-pass filter is placed after the antenna to pass the whole 25 MHz bandwidth of the DAMPS receiver with a center frequency of 882 MHz. Note that the other signals like UHF band television or microwave links will be eliminated to some extent. It is a common practice in a receiver design to make the bandwidth of the filters narrower as we move forward to the back-end to select the desired signal bandwidth. Right after the band-pass filter, an amplifier is placed to provide a *RF* gain for the whole bandwidth. The received signal might be about 0.5 μV rms, and goes through a gain of 12 dB (or the voltage will be multiplied by four). In many cases, the interferer might be 60 dB larger than the desired signal (for instance, consider the desired signal with power of −113 dBm and the interferer level of −53 dBm at other UHF frequencies). With the mentioned condition, the first filter may introduce a 40 dB loss for the out-of-band signal which may not be adequate and therefore the second band-pass filter may attenuate the interferer signal for another 40 dB such that the interferer signal is attenuated about 80 dB in total. These filters are made of passive components and based on their materials, these resonators may vary in size from near a couple of millimeters to one to two centimeters. Next, the amplified band enters a mixer circuit. Here the *RF* frequency of 882 MHz is downconverted to an *IF* of 90 MHz. It is common in most of the receivers, to have the *IF* frequency much lower than the input *RF*. This entails

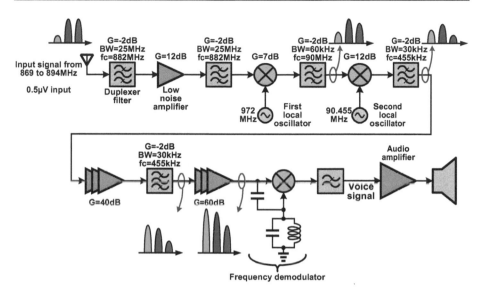

Figure 1.6: System-level schematic of the DAMPS receiver (the analog portion).

a few advantages as follows. First of all, achieving high gain is much easier at low frequencies rather than high frequencies. Secondly, in most receivers the *IF* frequency is held constant allowing for more accurate narrowband filter/amplifier design. Up to the mixer, the signal has experienced 4 dB loss in the passband of the filters and 12 dB gain of the *RF* amplifier (8 dB total gain) which for an input of 0.5μ Vrms brings the signal to 1.25 μ Vrms level. In a mixer circuit, the analog multiplication occurs for which we assume an ideal multiplier. Therefore, the sum and difference components of $882 + 972 = 1854$ MHz and $972 - 882 = 90$ MHz are generated. The former component is eliminated in the first band-pass filter, while the latter reaches the next stage. At this point, the desired signal is further amplified by 7 dB in the active mixer and then injected into the filter with 60 kHz bandwidth.

The desired channel passes alongside adjacent channels that are now attenuated. Finally, the second mixer with an *LO* frequency of 90.455 MHz brings the 90 MHz signal to 455 kHz with, say, 12 dB gain. The 455 kHz signal goes through two 30 kHz filters plus a chain of amplifiers with 40 dB and 60 dB gain. It is instructive to note that most of the gain is obtained at low frequency and with a small current consumption. In addition, the design of narrowband band-pass filters is generally much easier at lower frequencies rather than high frequencies.

A receiver which exploits one downconversion is traditionally called heterodyne receiver and if more than one downconversion occurs, it is called superheterodyne receiver. The words heterodyne and superheterodyne, while having historical significance, imply one mixing stage, and two or more mixing stages, respectively, and the prefix hetero- stands for mixing of different frequencies and the word dyne stands for analog multiplication or mixing. In the full receiver chain of Figure 1.6, the overall gain is of the order of 121 dB (the point in the superheterodyne receiver is that the total amplification is performed in three different frequency ranges, and therefore the probability of instability is reduced). If the effective input signal is of the order of $0.5\,\mu$Vrms, the overall chain gain of 121 dB brings the low signal level up to 562 mVrms. The detector shown is a frequency demodulator which detects the frequency deviation and extracts the voice signal which is applied to the speaker after audio amplification. The values shown in this example are typical values in the receiver. Although the lowest gain was placed at the front-end in the low-noise amplifier, it draws the highest current from the supply voltage (typically near 3 mA). However, the high gain of 100 dB in the second IF can be achieved with only $750\,\mu$A bias current. This point indicates one of the challenges of high-frequency amplifier design.

It should be noted, however, that the last audio amplifier stage draws a high current of several tens of milliamperes for the power amplification of the audio signal as well.

Example 1.1 Is it possible to add more low-noise amplifier stages at the front-end instead of filtering in order to have a better noise performance?
Answer:
The answer is no. In fact, placing more low-noise amplifiers at the front-end results in a better noise performance if there were no strong out-of-band interferers. However, the strong blocker signals without filtering will bring the last low-noise amplifier stages into saturation which results in decreased effective gain and possibly the mixing of the desired channel signal with the amplified blocker ones (this is discussed in more detail in Chapter 4). In addition, the power consumption cost of high-frequency amplification and the possibility of parasitic feedback may jeopardize the stability of the front-end (in case of high gain at a single frequency).
∎

The first band-pass filters attenuate out-of-band blocker signals and the first image signal $(2f_0 - f_s)$, while f_s is the *RF* signal frequency and f_0 is the first LO frequency. As such, the frequency of the first image is 1062 MHz, i.e., 90 MHz above the first LO frequency. Thus, the importance of those front-end band-pass filters is now obvious. The third filter after the mixer passes the desired channel, but will attenuate the adjacent and neighboring channels to some extent. Figure 1.7 shows a possible condition of the received signals. The two adjacent channels shown in Figure 1.7 will produce another signal which is due to the mixer nonlinearity and is called the third-order intermodulation product (IM_3) which is thoroughly discussed in Chapter 4. The IM_3 component will fall on the desired signal; if the IM_3 signal is larger than the desired signal, signal detection will not be possible. Therefore, the third filter mitigates this issue by attenuating the adjacent and neighboring channels.

Example 1.2 Is it possible to insert a band-pass filter with a bandwidth of 60 kHz at the front-end of the receiver? Then we will receive the desired channel much more easily without the blocking signals.
Answer:
The answer is no. The design of such a filter with the carrier frequency of the order of Gigahertz and such narrow bandwidth needs a very high quality factor of the order of 15000. Available passive elements, inductors, and capacitors, at the Gigahertz frequency range do not have such quality factors. ∎

Now, we discuss the spectral behavior of the receiver described earlier. As stated earlier, the importance of the third filter in the receiver is the attenuation of the two adjacent channels that lie in 60 kHz and 120 kHz away from the desired channel. This issue is demonstrated in Figure 1.7 in which the weak desired signal lies at the 882 MHz frequency.

Now, consider the frequency response of the third band-pass filter which is shown in Figure 1.8. Our goal is to calculate the attenuation of the adjacent and the neighboring channels' signals if the filter has a second-order behavior.

One may remember that with two poles in the transfer function, the magnitude of the signal will decrease by 40 dB/decade or 12 dB/octave, or in other words, the magnitude response will fall by 12 dB once the relative frequency is doubled. Actually, the attenuation of the filter is calculated based on the relative offset from the center frequency. For instance, in Figure 1.8, the adjacent channel is just one octave above the 3 dB cut-off frequency; thus, it experiences 12 dB attenuation. Similarly, the neighboring channel is 120 kHz offset from the desired channel frequency; thus, it experiences

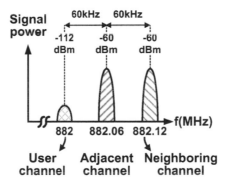

Figure 1.7: The desired channel, the adjacent channel, and the neighboring channel in a typical DAMPS radio signal.

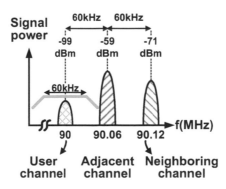

Figure 1.8: Typical received DAMPS signal levels after the third band-pass filter at 90 MHz .

24 dB attenuation. It can be shown that the attenuation for an nth order filter can be calculated as

$$L = 20n\log\left|\frac{2(f-f_0)}{BW}\right| \tag{1.6}$$

In Equation 1.6, f_0 is the center frequency of the filter, and $BW = f_{U,3dB} - f_{L,3dB}$ is the frequency that the magnitude response of the filter will experience 3 dB attenuation. In fact, this filter acts as a first channel selection filter. It also attenuates the adjacent and the neighboring channels; nonetheless, those unwanted channels could be still stronger than our desired channel. The fourth band-pass filter bandwidth is precisely equal to one channel bandwidth. Thus, by the second downconversion, both channel selection and amplification are realized at the second IF. The effect of the fourth filter is illustrated in Figure 1.9.

Now, consider the attenuation of the fourth filter. The frequency content residing at 515 kHz and its counterpart residing at 575 kHz will experience 24 dB and 36 dB attenuation, respectively. Thus, the unwanted adjacent and neighboring channels are further attenuated. Then, the linear power amplification at low frequency can be performed by a small current (e.g., 300 μA for 40 dB gain). The fifth filter has the same behavior as the fourth one. The signal after the fifth filter is demonstrated in Figure 1.10.

Finally, by the fifth filter, the adjacent channel will be attenuated by 24 dB and the neighboring channel experiences by another 36 dB attenuation. The spectra of the wanted and unwanted signals at the FM detector input are depicted in Figure 1.11.

As it is obvious from Figure 1.11, the power of the adjacent channel is below the desired channel and the neighboring channel has been practically suppressed. Note that the effect of the additional gains of the second IF amplifiers has been included in the computation of the final signal levels and the amplifiers are considered to be linear. In Figure 1.11, we have assumed the total IF gain (90 MHz and 455 kHz) as $G_2 = -2 + 12 - 2 + 40 - 2 + 60 = 106$ dB which is the sum of gains starting from the 90 MHz IF all the way to the end of the receiver's second IF. If we add $G_1 = -2 + 12 - 2 + 7 = 15$ dB which is the gain of the RF front-end, a total gain of 121 dB for the desired channel is achieved.

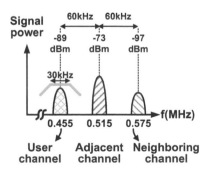

Figure 1.9: Typical received DAMPS signal levels after the fourth band-pass filter at 455 kHz.

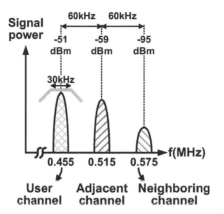

Figure 1.10: Typical received DAMPS signal levels after the fifth band-pass filter at 455 kHz.

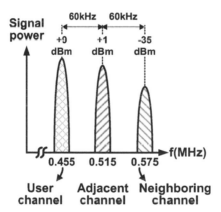

Figure 1.11: Typical received DAMPS signal levels at the input of the frequency demodulator.

One of the most important parameters in RF reactive components is the quality factor (Q). The concept of the quality factor for a reactive element or a resonant circuit, is the ratio of the stored energy to the dissipated energy. This dimensionless parameter describes how much lossy a component is. The general definition of the quality factor is as follows

$$Q \triangleq 2\pi \times \frac{\text{Average energy stored}}{\text{Energy dissipated per cycle}} = 2\pi f_0 \times \frac{\text{Average energy stored}}{\text{Power loss}} \quad (1.7)$$

where f_0 is the operating frequency. For a reactive element or a resonant circuit, it can be shown that the quality factor could be calculated as

$$Q = \frac{\text{Center frequency}}{\text{Bandwidth}} \quad (1.8)$$

By bandwidth, we mean 3 dB bandwidth of a circuit. Now with this definition of quality factor, we calculate them for the band-pass filters in our receiver chain as follows

$$Q_1 = \frac{882\,\text{MHz}}{25\,\text{MHz}} = 35.28 \rightarrow \text{for the RF filter} \tag{1.9a}$$

$$Q_2 = \frac{90\,\text{MHz}}{60\,\text{KHz}} = 1500 \rightarrow \text{for the first IF filter} \tag{1.9b}$$

$$Q_3 = \frac{455\,\text{KHz}}{30\,\text{KHz}} = 15.1 \rightarrow \text{for the second IF filter} \tag{1.9c}$$

The first one is a ceramic front-end filter, the second one is a crystal filter, and the third one is a ceramic IF filter. The high quality factor is a measure of the sharpness or the selectivity of a band-pass filter. Depending on the material used in a filter, its quality factor may change. The quality factor of crystal filters is better than other devices. When we reach the second IF, we put most of the gain of the receiver at the second IF. If we consider that the manufacturing cost of a filter is proportional to its Q, we would have all the interest for the cost and power consumption reduction to put the gain at lower frequencies. As such, we have succeeded to realize most of the filtering for the channel selection at the second IF frequency at a lower cost.

In this superheterodyne receiver that has two downconverting mixers and two IF frequencies, we observe the use of three types of filters. The filters become narrower as we proceed in the receiver chain. The first set of filters at the RF input frequency pass the full DAMPS 25 MHz band. The second filter's bandwidth (at the first IF frequency) is wide as to pass two or three DAMPS channels, in this example 60 kHz. The third set of filters (at the second IF frequency) are wide enough just for a single channel, here, 30 kHz.

On the other hand, the gain distribution is such that the RF front-end has 10 to 20 dB gain and about the same amount in the first IF, and most of the gain realize at the second IF frequency, here, of the order of 100 dB.

In our receiver example, the first filter selects the whole 25 MHz bandwidth and rejects the image signal to some extent. Actually, the image frequency which is at $2f_{\text{LO}} - f_{\text{RF}}$ is a signal which would be equally transferred to the IF frequency by the mixer (if it is present at the RF input) and therefore, it will appear as an interference at the first IF output. To avoid this inconvenience, it is mandatory that the image signal to be rejected at the RF input. This function should be materialized by the first set of filters. As such, while the input RF filter passes the whole DAMPS service band, it rejects the unwanted image frequency which is at 1062 MHz in this example.

There are two problems in high-frequency filters, the first one is their loss and the second one is their tunability. How can we switch from one channel to another in the front end of the receiver? The simplest way is to change the frequency of the first LO. After the first LO, the frequency of the first IF does not change. Therefore, the front-end filter must be able to accommodate all the frequencies of the desired spectrum (e.g.,, the whole GSM spectrum), if it is not subject to tuning.

Many years ago, the tuning of the oscillator frequency was manual. Nowadays using standard frequency synthesizers, this process is performed precisely and automatically upon a digital command word. In the next section, we briefly discuss the basic concept of the frequency synthesizers.

1.5 A Basic Understanding of Frequency Synthesizers

To understand the behavior of frequency synthesizers, we must first introduce the concept of voltage-controlled oscillators (VCO). The input of a VCO is a voltage that makes the output frequency change monotonically or possibly linearly with respect to it. The frequency range of a VCO in our receiver example is $959.5\,\text{MHz} \le f_{\text{OSC}} \le 984.5\,\text{MHz}$ which covers the whole 25 MHz input band. The channel spacing in the DAMPS standard is 30 kHz. The integer N frequency synthesizer is shown in Figure 1.12.

The frequency of the output signal is divided by a counter M which generates a pulse after receiving M pulses at its input. This process can also be done by cascading a few digital dividers. The divided signal is then injected to the phase detector to be compared with the reference signal coming from the crystal oscillator divider. The phase detector measures the phase difference between the two incoming signals. The resonant frequency of the crystal is usually between 5 MHz and 50 MHz. The frequency of the crystal oscillator in Figure 1.12 is 14.4 MHz. For 30 kHz channel spacing, the input signal to the phase detector must be 30 kHz (that is N should be 480). It can be shown that the VCO frequency in this structure can be derived as

$$f_{\text{VCO}} = \frac{M}{N} f_{\text{crystal}} \tag{1.10}$$

The circuit in Figure 1.12 is a negative feedback loop. If the input of the phase detector is a signal with 30 kHz fundamental, the other input must have the same frequency component at the steady state. By allowing enough loop gain, the error signal will tend to zero. This system is called a phase-locked loop frequency synthesizer. We will go over the concept of the PLL in Chapter 3. It is noteworthy that in a PLL at locked state, not only will the two input frequencies be the same, but also the phases of the two signals will track each other with a constant phase offset. Using a programmable counter, one may change the oscillation frequency to receive the desired channels. For instance, with a VCO frequency of 972 MHz, we calculate the division ratio as

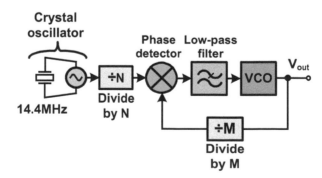

Figure 1.12: Typical integer N frequency synthesizer for the DAMPS receiver.

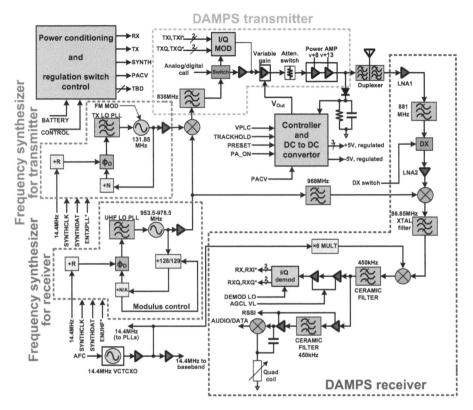

Figure 1.13: A complete block diagram of a DAMPS transceiver.

$$N = \frac{14.4 \times 10^6}{30 \times 10^3} = 480 \qquad (1.11)$$

$$M = \frac{972 \times 10^6}{30 \times 10^3} = 32400 \qquad (1.12)$$

To tune to a desired frequency, usually the divider N is constant and the desired channel is obtained by changing the value of M.

Let's now proceed to investigate a more complete DAMPS radio as shown in Figure 1.13.

As depicted in the red portion of Figure 1.13, the signal is received via an antenna. It is amplified by a low-noise amplifier and the whole band goes through the band-pass filter. It also attenuates the first image signal. Next, the signal is downconverted by a mixer to the first IF. Channel selection filter attenuates side-band channels, and by a second mixer, the signal is translated to a second IF. The LO frequency for a second mixer comes from a multiplier circuit which makes multiple 6 of crystal frequency. In the second IF, the signal again is filtered and via two paths goes for digital and analog demodulation. In the upper path, the signal is demodulated digitally through an AGC and an I/Q demodulator, and the other path demodulates signal by an FM quadrature detector. Frequency synthesizer of the receiver has a frequency range of 953.5 MHz–978.5 MHz, and by dividing by 128 or 129 followed by more division, a 30 kHz signal is obtained and the loop will be locked. As it is evident in frequency synthesizer of the receiver, the output signal of the oscillator is buffered and then by passing through a band-pass filter drives the first mixer.

A similar procedure is realized at the phase-locked loop of the transmitter. Its PLL generates proper LO frequency for driving the upconverter mixer. The high-frequency signal then passes through the band-pass filter and reaches the power amplifier. Finally, the amplified signal is coupled to the antenna for radiation into the air. The signal modulation is frequency modulation (FM); however, one may use quadrature digital phase modulation.

Nowadays, the radio receivers employ frequency translation to zero-IF which is called zero-IF receiver where there is no image signal.

Example 1.3 A radio receiver has the block diagram shown in Figure 1.14 and its specifications are denoted on the figure.

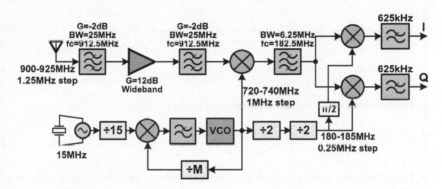

Figure 1.14: A typical heterodyne digital radio receiver.

(a) The input filters' specifications are the same. If the desired signal has a power level of -100 dBm and an image signal accompanies it with a power of -45 dBm (both at the input of the first RF filter), calculate the out of band attenuation of the RF filters, for the image frequency, such that the image signal goes 10 dB below the desired signal at the first mixer input.

(b) Calculate M for channel spacing of 1.25 MHz in the desired band. Note that the second LO frequency is not fixed.

(c) If at the receiving channel of 912.5 MHz there exists an adjacent channel signal at 915 MHz, determine the order of the low-pass filters such that the adjacent channel is rejected by 40 dB. Assume that $f_{3\,\mathrm{dB}}$ for the low-pass filter is 625 kHz.

Solution:

(a) The image frequency can be obtained as follows

$$f_{\mathrm{Im}} = 2f_{\mathrm{LO}} - f_{\mathrm{RF}} \tag{1.13}$$

As the RF frequency can be varied between 900 and 925 MHz, then the lower edge of the receive band is downconverted by 720 MHz. The image frequency would be

$$f_{\text{Im}1} = 2 \times 720 - 900 = 540 \, \text{MHz} \tag{1.14}$$

Now, consider the upper edge of the receive band. In this case, 925 MHz signal is downconverted by 740 MHz LO. Thus, the image frequency can be written as

$$f_{\text{Im}2} = 2 \times 740 - 920 = 560 \, \text{MHz} \tag{1.15}$$

As we desire that the image frequency to be 10 dB lower than the desired signal, it must be attenuated by 65 dB ($55 + 10 = 65$ dB). As a result for the identical RF filters, each one must have an out-of-band attenuation of 32.5 dB at least. The normalized frequency difference of the image signal will be

$$\frac{f_{\text{RF}} - f_{\text{Im}}}{\frac{BW}{2}} = \frac{912.5 - 560}{12.5} = 28.2 \tag{1.16}$$

The normalized frequency difference in octaves becomes

$$D = \frac{\log 28.2}{\log 2} = 4.81 \ \text{octaves} \tag{1.17}$$

The bandpass filter order becomes

$$n = \frac{32.5}{6 \times D} = 1.12 \tag{1.18}$$

Therefore, we choose $n = 2$ for the RF filters. This will satisfy the required attenuation for the other image frequency (540 MHz) as well.
(b) The channel spacing at the RF frequency in this receiver is 1.25 MHz ; however, in this architecture 1 MHz spacing is realized by the first mixer (because the crystal frequency is divided by 15) and 250 kHz is realized by the second mixer (because the VCO frequency is divided by 4). Thus, the frequency synthesizer will have a 1 MHz frequency step. As a result, the minimum value of M is 720 and its maximum value is 740.
(c) As the adjacent channel is $915 - 912.5 = 2.5$ MHz above the desired channel, the attenuation for the nth order low-pass filter can be written as

$$20n \log \left(\frac{2.5 \, \text{MHz}}{625 \, \text{KHz}} \right) = 40 \tag{1.19}$$

Then, $n = 3.32$. So we choose $n = 4$ as an integer and the filter will be a 4th order one.   ∎

1.6 Conclusion

The world of wireless communications has conquered many aspects of the modern human life. The technical aspects of this field are of great importance for an electrical/electronic engineer. In this chapter, we made a general presentation for the RF communication systems. We surveyed the general architecture of an RF transmitter and an RF receiver. Specifically, we briefly studied the architecture of a superheterodyne receiver which consists of two frequency conversion (mixer) stages as well as a zero-IF receiver which consists of a receiver with the same LO and RF frequencies. Furthermore, we observed how in a superheterodyne receiver the large interfering signals in the adjacent and the neighboring channels are suppressed (or attenuated) with respect to the desired signal along the receiver chain. In addition, we saw how using a phase-locked loop and frequency dividers we can synthesize the desired local frequencies in a receiver. A DAMPS transceiver block diagram was studied as an example. We deliberately ignored some more complex issues such as the nonlinearity of the mixers, VCOs, or amplifier circuits here. But as the reader goes forward through the text, he/she would gain more understanding about the nonlinearity issues. Then the reader is urged to return back to this chapter to gain more understanding of the related problems.

1.7 References and Further Reading

1. T.S. Rappaport, *Wireless Communications: Principles and Practice*, second edition, New York, NY: Prentice-Hall, 2002.
2. B. Razavi, *RF Microelectronics*, second edition, Castleton, NY: Prentice-Hall, 2011.
3. L.W. Couch, *Digital and Analog Communication systems*, eighth edition, New Jersey: Prentice-Hall, 2013.

1.8 Problems

Problem 1.1 Figure 1.15 shows a triple downconversion receiver in which the input signal range at the antenna is from 0.02 GHz to 5 GHz. If the first VCO with its initial control voltage is oscillating at 7.5 GHz,
1. Find the values of M and N in such a way that desired frequencies are provided for the corresponding mixers.
2. What is the frequency of the image signal at the first mixer's input? How this component is eliminated in this structure? What is the frequency of the second image at the second mixer's input, and what is the corresponding values at the input of the antenna. Moreover, find the third image frequency at the third mixer's input and its corresponding frequencies alongside the structure.
3. Suppose that the input frequency is 3.5 GHz, then find the first VCO frequency. In this situation, if the input low-pass filter is a second-order one with 3 dB frequency of 5 GHz, and if the three subsequent band-pass filters' frequency response are as those depicted in Figure 1.16, and if a strong blocker signal emerges at 100 MHz above the input signal, how much it will be attenuated through the receiver chain?

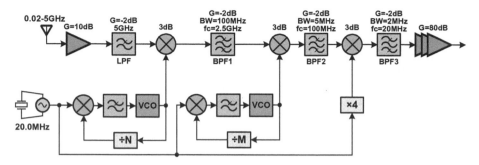

Figure 1.15: Triple downconversion wideband receiver.

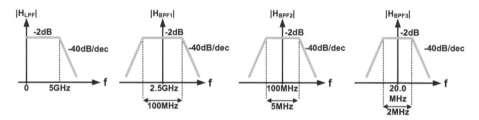

Figure 1.16: Frequency response of filters.

Problem 1.2 In the transceiver depicted in Figure 1.17, first *IF* frequency resides at 90.1 MHz and the second *IF* is at 455 kHz. If $f_{VCO_1} = 966.3$ MHz, find the receiving channel frequency. In this situation, find the frequency of the second VCO such that the transmitted carrier signal is 45 MHz lower than the received signal.

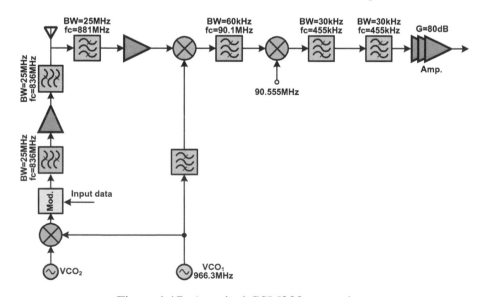

Figure 1.17: A typical GSM900 transceiver.

Problem 1.3 In the FM transceiver depicted in Figure 1.18, determine the unknown VCO frequencies alongside with division ratio M_2. If f_{VCO_2} and M_2 have two different possible values each, discuss the advantages and disadvantages of either of the values if the transceiver has a tuning bandwidth of 15 MHz.

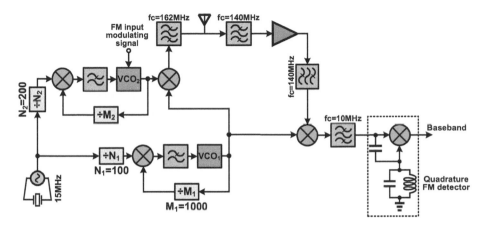

Figure 1.18: A VHF FM transceiver block diagram.

Problem 1.4 In the wideband receiver depicted in Figure 1.19, the first VCO is tuned at 3.5 GHz,
(a) Find the received signal frequency.
(b) What is the first image frequency of the first mixer? Is it in the receive band? In this situation, the image frequency in the second mixer can be emanated from two RF components. Determine those components' frequencies at the antenna RF input.
(c) If the input low-pass filter is a first-order one with a 3 dB corner frequency of 1 GHz and the other two bandpass filters have a frequency response given in Figure 1.19, determine the attenuation values of the first image and those two RF components which could result in the second image in this circuit.

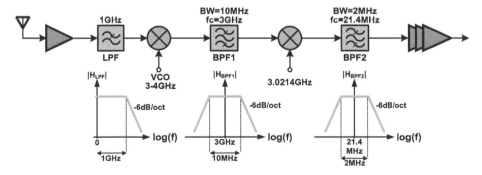

Figure 1.19: A dual conversion wideband receiver.

Problem 1.5 In the WiMax receiver depicted in Figure 1.20, the input frequency range is between 3.4 GHz and 3.6 GHz with 20 MHz channel spacing. The input band-pass filter is of second order with the bandwidth of 400 MHz and the center frequency of 3.5 GHz. The first IF signal is at $f_{in}/5$ where f_{in} is the input signal frequency. The second IF frequency is zero. The output low-pass filters are of third order with a corner frequency of 10 MHz. Moreover, the frequency response of the amplifier A and the mixers are constant.

(a) What is the image frequency at the receiver input which experiences the maximum attenuation, in dB?

(b) Find the division ratio M for the receiver for the input frequency range.

(c) If the input frequency is 3.5 GHz, what is the attenuation of the unwanted adjacent and the unwanted neighboring channels? (channel spacing is about 20 MHz)

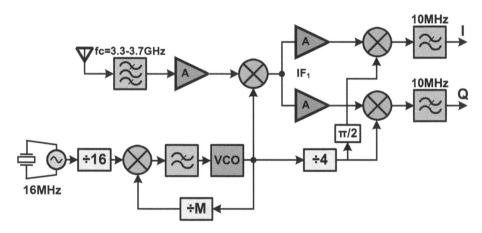

Figure 1.20: An I/Q double conversion WiMax receiver.

Problem 1.6 In a radio receiver depicted in Figure 1.21, we have cascaded the blocks with the given specifications.

(a) If at the input two signals (the main channel and the adjacent channel, respectively) with a power of -60 dBm each at frequencies of 900.060 MHz and 900.120 MHz are present, what is the power of IM_3 components at the output of the mixer?

(b) If the 45 MHz band-pass filter has a passband of 60 kHz and is of second order, then what is the output of the 45 MHz filter emanating from the IM_3 component at 900 MHz. (See IM_3 concept in Chapter 4.)

Block	LNA	Filter	Mixer	Filter	VGA
Gain(dB)	15	−3	10	−3	43
IIP₃(dBm)	−8.5	∞	5.5	∞	10

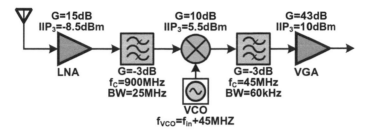

Figure 1.21: A simplified GSM900 receiver.

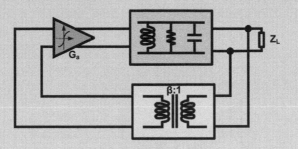

2. Oscillators

2.1 An Introduction to Oscillators

Oscillators are of most important blocks in RF circuit design because they generate the fundamental RF signals required in the transceiver systems. As we have seen a complete radio in Chapter 1, oscillators are integral parts of each system. These circuits which generally behave in a large-signal regime mandate specific considerations for proper analysis and design. Furthermore, they are one of the most power hungry blocks whose figure of merit depends on the signal purity and stability. The basic behavior of every oscillator is due to interaction of the noise with the circuit nonlinearity that should be examined in detail; however, in this chapter, we introduce simple methods which are based on linear systems to better understand the oscillator behavior. This block is one of the first blocks in the receiver which provides the driving signal for a downconverting mixer. In this chapter, we introduce two different methods for analysis of an oscillator, namely, positive feedback in a loop and negative resistance in a resonant circuit.

2.2 First Approach: Positive Feedback

Consider Figure 2.1 where a tuned amplifier output is fed back to its input through a divider circuit.

The important point in oscillator's behavior is that there is no external excitation except for the bias. The input of these circuits is noise which is fed at the input of the loop amplifier. Different blocks as depicted in Figure 2.1 are the amplifier, the resonant circuit or the frequency selection tank, and the divider. As one might recall from the feedback control systems, one may write the transfer function for a positive feedback loop as

$$H(s) = \frac{G_a(s)}{1 - \beta G_a(s)} \qquad (2.1)$$

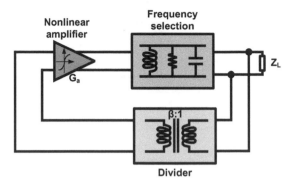

Figure 2.1: Block diagram of a typical oscillator.

The loop will be unstable (the oscillation will occur) once $\beta G_a(s) = 1$ which is called the Barkhausen condition. Consider the noise at the input of the amplifier which is amplified and passed through the tank circuit and is fed back through the divider to the input of the amplifier (in this example, we have the resonant frequency of the tank which is $f_0 = 1/(2\pi\sqrt{LC})$), and by means of a nonlinear amplifier, the returning signal would be in phase at the input. Typical white noise samples in the frequency domain and in the time domain are shown in Figure 2.2.

Thermal noise has a white spectrum; in other words, it has a fixed power spectral density at least in the RF range. The power spectral density (in W/Hz) can be expressed as $N_0 = kT$ and the RMS voltage across a resistance R can be expressed as

$$\overline{V_n^2} = 4kTBR \tag{2.2}$$

The available power of this noisy resistance could be described as $N = kTB$ in watts where in Equation 2.2, V_n is the RMS noise voltage, k is the Boltzmann constant equal to $1.38 \times 10^{-23} \text{J/K}$, T is temperature in Kelvin, and B is the noise bandwidth in hertz. Noise is a random process and it is not possible to determine its instantaneous value

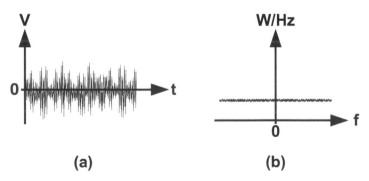

Figure 2.2: Typical white noise samples, (a) in time and (b) in frequency domain.

at a specific time. Thus, the noise level is usually specified by its power spectral density or its RMS value. Now considering the noise shown in Figure 2.2 passes through a frequency selective circuit, the noise spectrum will be changed according to the resonant circuit response. This will result in rejection of the noise in frequencies out of the passband of the resonant circuit. White noise is transformed to colored noise with a selective skirt shaped spectral density. Its time-domain waveform would resemble approximately a sinusoid, if the resonant circuit has a high quality factor. Then, the shaped noise passes through the divider and the amplifier, and the amplifier amplifies it and once again it is applied to the amplifier input after the division. If the loop gain of the system is larger than unity, the fed back noise would build up until the nonlinearity of the amplifier compresses the gain and the loop gain of the system approaches unity. It should be noted that the larger the fed back signal becomes through this process, the more it will approach a pure sinusoid. Thus, when looking at the output, there is approximately an amplified noisy sinusoid whose spectrum approaches a pseudo-impulse-shaped spectrum in the frequency domain. The larger the quality factor of the frequency selective circuit, the better the spectral purity of the output signal. That is, the output signal would approach the sinusoidal form. That is because of better rejection of the noise sidebands by a sharper filter. This process is shown in Figure 2.3.

As depicted in Figure 2.3, for high quality factor tank circuit, the output signal is more similar to a sinusoid; however, degrading the quality factor will result in more phase/amplitude noise and higher harmonics level.

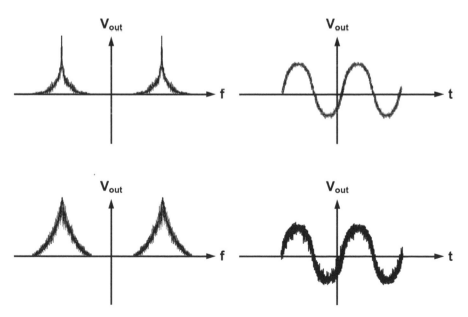

Figure 2.3: The output signal of the resonance circuit for different quality factor values.

Now let's consider the amplifier nonlinearity. We would now investigate the definition of the gain in a large-signal/limiting regime. Table 2.1 shows the output voltage versus the input voltage of a saturating amplifier.

The large-signal gain of the amplifier is defined as the ratio of the output fundamental to the input fundamental voltage of the circuit. As Table 2.1 suggests, increasing the signal level will result in decrease in the large-signal gain. This limiting behavior of this circuit will stabilize the oscillation amplitude eventually. Furthermore, characteristic of a typical nonlinear amplifier is illustrated in Figure 2.4 and Figure 2.5 which demonstrates the limiting behavior of the amplifier for large signals.

We have assumed an implicit approximation in our above descriptions. By entering the large-signal regime due to the nonlinear behavior of the circuit, a number of harmonic frequencies of the fundamental signal will be generated as well. We can define a more precise definition for an effective large-signal gain of the amplifier as the ratio of the first harmonic amplitude at the output to the fundamental input voltage amplitude, while neglecting the other harmonics at the output.

How does a positive feedback loop for an oscillator stabilize? With respect to the definition of the effective large-signal gain of an amplifier, the Barkhausen's criterion suggests that the loop should have a unity gain with zero phase at the oscillation frequency ($G_a\beta = 1\angle 0$), as such the loop will be stabilized. It should be noted that the feedback in any oscillator is positive which results in initial noise amplification and consequent oscillation. There are a number of important parameters in oscillator

Table 2.1: Effective gain with input and output signal of a saturating amplifier.

Input Voltage(V)	Effective Gain	Output Voltage(V)
0.25	14	3.5
0.5	10	5
1	5.2	5.2

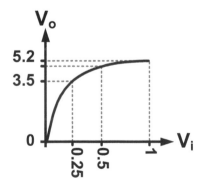

Figure 2.4: Input/output phasor voltage characteristic of a typical limiting amplifier.

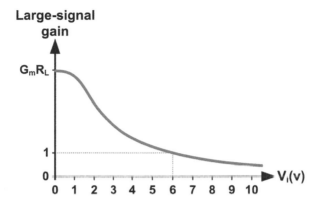

Figure 2.5: Typical variations of the large-signal gain of a nonlinear amplifier as a function of input voltage.

design, such as topology, resonant circuit, small-signal loop gain, large-signal loop gain, signal amplitude, and phase noise. Phase noise in the oscillators is one of the most interesting and challenging issues. We briefly discuss about the phase noise here. Phase noise is the result of the interaction of baseband white noise signal with the sinusoidal oscillation signal in the nonlinear circuit of the oscillator, in the sense that the amplitude and the phase of the sinusoidal oscillation signal are modulated by the random noise signal. As such, we can describe the general output of a sinusoidal oscillator as $V(t) = (A + a_n(t))\cos(\omega_0 t + \phi_0 + \phi_n(t))$, where $\frac{d\phi_n}{dt} \ll \omega_0$ and $\frac{a_n}{A} \ll 1$. Here A is the amplitude of the oscillations, $a_n(t)$ represents the random amplitude modulation, ω_0 is the radian frequency of oscillation, ϕ_0 is the phase of the oscillation, and $\phi_n(t)$ represents the random phase modulation. In Figure 2.6, the oscillations' start-up of a typical oscillator as a function of time is depicted.

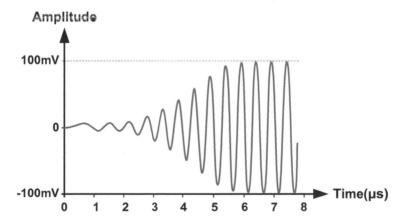

Figure 2.6: Typical oscillations' start-up of an oscillator as a function of time (oscillation frequency is about 2 MHz).

If one computes the auto correlation function of this signal, and takes the Fourier transform of it, he/she will obtain the spectral density of the oscillator signal. This signal would be in the form of a narrow skirt around the sinusoidal carrier. The importance of the phase noise is in the coherent receivers. In the sense that one intends to detect, for example, different phase modulating levels on the carrier signal, the random phase noise of the carrier induces a random phase shift at the output of the detector. This exacerbates the signal detection process. We refer the interested reader to more advanced texts regarding sinusoidal oscillators and the phase noise for further investigation [5].

2.3 Second Approach: Negative Resistance/Conductance

The concept of negative resistance is useful for oscillators using two-terminal devices and/or two-terminal resonators. It is obvious that the positive resistor (passive resistor) dissipates energy. In contrast to it, one can imagine a negative resistor which generates electrical energy. Now, consider the circuit depicted in Figure 2.7.

In Figure 2.7, positive resistors dissipate energy; however, the source and the negative resistors generate electrical energy. In other words, the negative resistor acts as an active load in generating energy. As an example, consider the output voltage in this circuit

$$V_O = \frac{R_L}{R_L + R_S + R_1} V_S = -10 V_S \tag{2.3}$$

As it is obvious, the output voltage is larger than the input voltage and excess power is generated indeed. A similar concept is shown in Figure 2.8.

As it is obvious, there is no external excitation in the circuit in Figure 2.8 except the noise current. The existing thermal noise in the resistor (random movement of electrons) may be amplified in the circuit by means of positive feedback.

$$\frac{I_L}{I_n} = \frac{G_L}{G_L + G_S + jC\omega - \frac{j}{L\omega}} \tag{2.4}$$

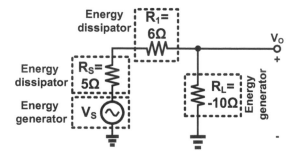

Figure 2.7: Negative and positive resistors in a simple circuit.

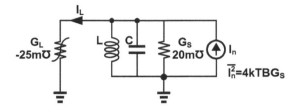

Figure 2.8: Energy generation in resonant circuit with no external excitation.

At the resonant frequency, the imaginary part vanishes, and consequently, the above equation reduces to

$$\frac{I_L}{I_n} = \frac{-25}{-25+20} = 5 \tag{2.5}$$

As such, the noise current will be obviously amplified. If one calculates the overall tank, conductance will reach to the total $G_{eq} = -5\,m\mho$, and thus there will be a net energy generator in the circuit. This negative conductance will amplify the noise and the frequency response of the tank shapes its spectrum in a way as described in the previous section. Once the large-signal oscillation is established, one can neglect the noise current source and writing the KCL at the common node of the circuit, one would obtain

$$Y_L(V)V + Y(j\omega)V = 0 \tag{2.6}$$

As V has a nonzero value, therefore,

$$Y_T(V,\omega) = Y_L(V) + Y(j\omega) = 0 \tag{2.7}$$

This is the oscillation condition of a two-terminal device oscillator. The process of amplification of noise by the negative resistance and its frequency selection by the tank circuit continues till the nonlinear behavior of negative conductance decreases the absolute value of the negative conductance down to $20\,m\mho$. In other words, increasing the amplitude of oscillations results in decreasing of the negative conductance's absolute value till the dissipated energy in the load resistance and the generated energy of the negative conductance are equal. This is the point of stable oscillations. The negative conductance is normally dependent on the oscillation voltage amplitude. When its conductance decreases to $-20\,m\mho$, the amplitude of oscillation can be calculated precisely from the nonlinear characteristics of the negative conductance. The oscillator based on negative resistance/conductance can be demonstrated in Figure 2.9 where a selective two-terminal resonator is connected to a negative resistance device. This configuration can be a potential sinusoidal oscillator depending on the value of the negative resistance and the resonator loss resistance.

Writing the KVL in the circuit, one obtains

$$\left(-R(A) + jL\omega - j\frac{1}{C\omega} + R_L\right)I = 0 \tag{2.8}$$

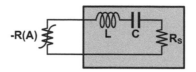

Figure 2.9: Negative resistance model of an oscillator.

where A is the amplitude of the current phasor. I having a nonzero value, we would have

$$-R(A) + R_L + j \left(L\omega - \frac{1}{C\omega} \right) = 0 \qquad (2.9)$$

As such, the oscillation condition simplifies to

$$Z_T(j\omega, A) = 0 \qquad (2.10)$$

where $Z_T(j\omega, A)$ stands for the total loop impedance.

Example 2.1 Is it possible to replace the negative conductance in Figure 2.8 equal to -20 m℧ at first? Will the circuit stabilize immediately, then?

Solution:
No, because the oscillator will never start. We have stated that for starting the oscillations the net negative conductance must be less than zero and the circuit by its nonlinear behavior will decrease the net negative conductance to zero, and the oscillation will be stabilized. ∎

It can be asserted that most types of oscillators can be analyzed by either of the two methods presented in the previous sections.

2.4 Oscillator Topologies

In this section, we describe the basic oscillator circuit topologies. Here the active device could be either a bipolar transistor or a MOSFET, with proper bias with the same topologies. We start with a general oscillator topology as depicted in Figure 2.10 with a fictive ideal amplifier (with infinite input impedance and zero output impedance) which has a real gain A_v.

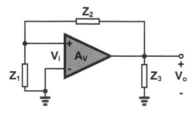

Figure 2.10: General oscillator topology with three reactive elements.

In this topology, the positive feedback is realized by the voltage division through Z_1 and Z_2. Normally, the three external elements in the oscillator circuit are purely reactive elements. Now, consider $Z_1 \approx jX_1$, $Z_2 \approx jX_2$, and $Z_3 \approx jX_3$. The three reactive elements should resonate at the oscillation frequency. For the oscillation condition, one can write

$$A_v \frac{X_1}{X_1 + X_2} = 1 \quad \text{Unity loop gain condition} \tag{2.11}$$

$$X_1 + X_2 + X_3 = 0 \quad \text{Resonance condition} \tag{2.12}$$

Since $X_1 + X_2 = -X_3$, then for satisfying the Barkhausen oscillation condition, we should have

$$-A_v \frac{X_1}{X_3} = 1 \tag{2.13}$$

Now, we consider three distinct cases depending on the value of A_v.

Case 1 (common source/common-emitter amplifiers): $A_v < 0$ then given Equation 2.13, it is imposed that $\frac{X_1}{X_3} > 0$. In this case, if X_1 is chosen to be inductive, then X_3 should be inductive as well, and considering the resonance condition X_2 should be necessarily capacitive, (Figure 2.11(a)). On the other hand, if X_1 is chosen to be capacitive, then X_3 should be capacitive as well, and considering the resonance condition X_2 should be necessarily inductive (Figure 2.11(b)).

Case 2 (common gate/common-base amplifiers): $A_v > 1$ then given Equation 2.13, it is imposed that $\frac{X_1}{X_3} < 0$. In this case, if X_1 is chosen to be inductive, then X_3 should be capacitive, and since $A_v > 1$, then it is imposed that $\frac{X_1}{X_1 + X_2} < 1$, therefore, X_2 should be necessarily inductive (Figure 2.12(a)). On the other hand, if X_1 is chosen to be capacitive, then X_3 should be inductive, and since $A_v > 1$, then it is imposed that $\frac{X_1}{X_1 + X_2} < 1$, therefore, X_2 should be necessarily capacitive (Figure 2.12(b)).

Case 3 (common drain/common-collector amplifiers): $0 < A_v < 1$ then given Equation 2.13, it is imposed that $\frac{X_1}{X_3} < 0$. In this case, if X_1 is chosen to be inductive, then X_3 should be capacitive, and since $A_v < 1$, then it is imposed that $\frac{X_1}{X_1 + X_2} > 1$, therefore, X_2 should be necessarily capacitive as well (Figure 2.13(a)). On the other

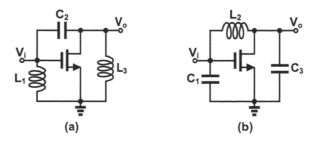

(a)   (b)

Figure 2.11: Two possible oscillator topologies with negative voltage gain ($A_v < 0$).

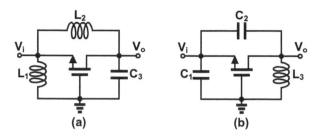

Figure 2.12: Two possible oscillator topologies with positive voltage gain greater than unity ($A_v > 1$).

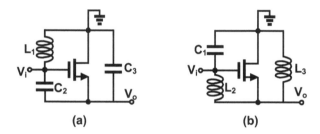

Figure 2.13: Two possible oscillator topologies with positive voltage gain less than unity ($0 < A_v < 1$).

hand, if X_1 is chosen to be capacitive then X_3 should be inductive, and since $A_v < 1$, then it is imposed that $\frac{X_1}{X_1+X_2} > 1$, therefore, X_2 should be necessarily inductive as well (Figure 2.13(b)). In the above-mentioned, circuits and the corresponding figures, we used MOSFET transistors as the active element to show the implementation of different oscillator topologies. In those figures, all the MOSFET transistors can be literally replaced by bipolar transistors without any alteration. To demonstrate these implementations, in the following sections, we use the bipolar transistor as the active element, including the bias circuitry.

2.4.1 Common-Emitter Oscillator Circuit

The oscillator with grounded emitter is shown in Figure 2.14.

As depicted in Figure 2.14, if one considers the noise voltage across the base–emitter of the transistor, the signal is amplified through the base to the collector by a small-signal gain of $A_{ss} = -g_m R_L$. At the oscillation frequency where another 180° phase shift is materialized by the passive LC circuit (note that C_B is AC short-circuit), the criterion of oscillation which is the positive feedback is realized. The other criterion which is the unity closed-loop gain will be materialized after the amplitude growth of the oscillator, through the nonlinearity of the transistor, the gain is compressed to its large-signal value $A_{ls} = -G_m R_L$.

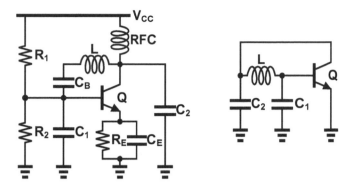

Figure 2.14: The common-emitter oscillator circuit with 180° phase shift through the LC circuit.

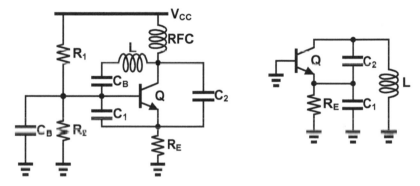

Figure 2.15: The common-base oscillator circuit.

2.4.2 Common-Base Oscillator Circuit

The oscillator with grounded base is shown in Figure 2.15.

Here both C_B capacitors have large capacitances which are approximately short-circuit at the operating frequency and the RFC is a large inductor which is considered as open-circuit at the operating frequency. As depicted in Figure 2.15, if one considers the noise voltage at the base–emitter junction of the transistor, this signal will be amplified by a small-signal gain of $A_{ss} = g_m R_L$ at the collector in phase with the input. Then, the signal is fed back to the base by the capacitive divider and as such provides the necessary positive feedback at the oscillation frequency. Again due to the nonlinearity of the transistor, the stabilization of the oscillation amplitude will occur by the gain compression through the feedback loop.

2.4.3 Common-Collector Oscillator Circuit

The oscillator with grounded collector is shown in Figure 2.16.

Here C_B is a large capacitor whose RF impedance is considered to be as a short-circuit. As depicted in Figure 2.16, the noise voltage at the transistor's base appears at

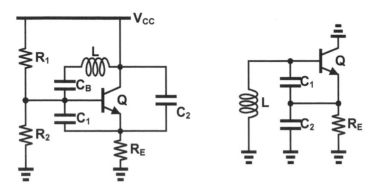

Figure 2.16: The common-collector oscillator circuit.

the emitter by the common-collector voltage gain. This gain is slightly less than but near to unity and the output is in phase with the input. The output voltage is fed back to the base by the capacitive step-up transformer. At the frequency where the positive feedback is realized and once the loop gain is compressed to unity, the oscillations will materialize. Note that here the transistor amplifier has a voltage gain of less than unity but a current gain larger than unity and consequently a power gain greater than unity.

Note that in all the above oscillators, there is a resonant and dividing circuit which consists of an inductor and two capacitors which is called the Colpitts oscillator. The dual of these circuits could be used as an oscillator as well, that is, with a resonant circuit of a single capacitor and two inductors which is called the Hartley oscillator.

2.4.4 Colpitts versus Hartley Oscillators, a New Insight

The oscillators shown in Figs. 2.14, 2.15, and 2.16 are called Colpitts family of oscillators. If one omits the bias circuitry from those oscillators and draws an AC model for them (considering R_E is sufficiently large in the common-base and the common-collector configurations), he/she would arrive at a common core circuit which is depicted in Figure 2.17(a). If one replaces all the RF capacitors by inductors, and all the RF inductors by capacitors, he/she has made a dual circuit of Colpitts oscillators which is called Hartley family of oscillators (Figure 2.17(b)). It is instructive to observe that all these oscillators have the same concept (considering R_E is sufficiently large in the common-base and the common-collector configurations). It is noteworthy that all the oscillator topologies described in section 2.4 can be reduced to either of these two core oscillators.

If one computes the closed-loop gain in either of Colpitts or Hartley oscillators, neglecting the transistor parasitic reactances, he/she would reach the following resonance criteria for the oscillations:

$$L\omega - \frac{C_1 + C_2}{C_1 C_2 \omega} = 0 \quad \text{for Colpitts oscillators} \tag{2.14a}$$

$$C\omega - \frac{1}{(L_1 + L_2)\omega} = 0 \quad \text{for Hartley oscillators} \tag{2.14b}$$

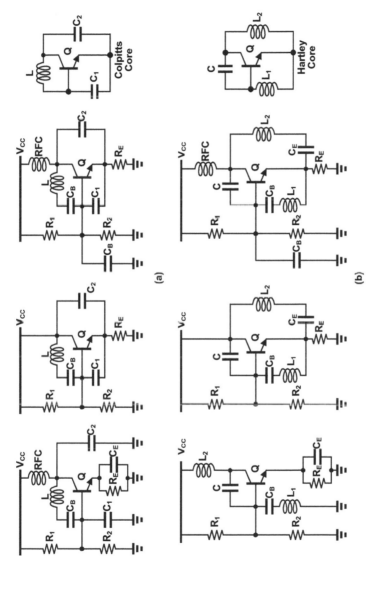

Figure 2.17: The Colpitts and the Hartley oscillators (in common-emitter, common-collector, and common-base configurations) and their corresponding main core circuits.

Either of these two equations show a parallel resonance condition (why?) in the reactive elements surrounding the transistors. Furthermore, there is a positive feedback from the output to the input (180° phase of the voltage gain and 180° phase of the voltage division for CE case, and zero-degree phase of the voltage gain and zero-degree phase of the voltage division for the CB and CC cases).

Example 2.2 Determine the oscillation condition in the following common-emitter transistor oscillator considering the parasitic elements of the transistor.

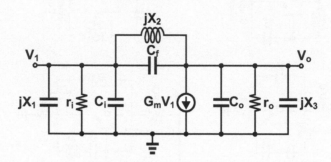

Figure 2.18: A Colpitts oscillator core including the parasitic elements of the transistor.

Solution:
For determining the oscillation condition, we just absorb the parasitic elements of the transistor into the surrounding reactances as follows

$$Z_1 = \left(\frac{1}{jX_1} + \frac{1}{r_i} + jC_i\omega \right)^{-1} \tag{2.15}$$

$$Z_2 = \left(\frac{1}{jX_2} + j\omega C_f \right)^{-1} \tag{2.16}$$

$$Z_3 = \left(\frac{1}{r_o} + j\omega C_o + \frac{1}{jX_3} \right)^{-1} \tag{2.17}$$

As such, we can consider the following equivalent circuit for this oscillator.

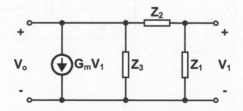

Figure 2.19: The equivalent circuit of the Colpitts oscillator.

The output voltage can be easily computed as

$$V_o = \frac{-G_m V_1}{Y_3 + \frac{1}{Z_2 + Z_1}} \tag{2.18}$$

Then, the input feedback voltage can be written as

$$V_1 = \frac{-G_m V_1}{Y_3 + \frac{1}{Z_2 + Z_1}} \times \frac{Z_1}{Z_1 + Z_2} \tag{2.19}$$

As such, the complex oscillation condition becomes

$$\frac{-G_m Z_1}{1 + Y_3 (Z_2 + Z_1)} = 1 \tag{2.20}$$

Or

$$\frac{-G_m Z_1 Z_3}{Z_1 + Z_2 + Z_3} = 1 \tag{2.21}$$

2.5 Crystal Oscillators

Quartz crystal is a piece of a wafer of crystalline silicon dioxide mineral which is cut at certain angles to give electro-acoustic resonances at the required frequency range. Two metallic surfaces cover the two faces of the quartz wafer. These two surfaces constitute the electric terminals of the crystal. The quartz crystal has a piezoelectric property. In piezoelectric materials, mechanical pressure is transduced to the electric voltage and electric voltage is transduced to mechanical pressure. As such, the mechanical vibrations in the crystal are transformed to electric vibrations by the virtue of its piezoelectric property. A typical crystal's electric model is shown in Figure 2.20. It is noteworthy that the natural resonances of a piece of quartz crystal are of a very high quality factor and a very good temperature stability, the points which are the advantages of the use of quartz crystals in the oscillators.

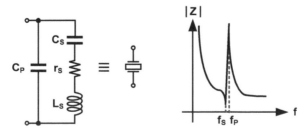

Figure 2.20: Electrical model of a crystal and its impedance behavior.

As it is obvious from the crystal model in Figure 2.20, there exists a series and a parallel resonance for the crystal. That is the crystal impedance, at first, falls to a very small value at its series resonance frequency (f_S) and then changes to a very high impedance at its parallel resonance frequency (f_P). Then, its impedance due to the parallel capacitor falls gradually to a low impedance value at frequencies much higher than f_P. The values of the equivalent circuit element of the crystal are such that the difference between the series and the parallel resonance frequencies is quite small while the difference between the impedances at these two frequencies is quite large. The input impedance of the crystal can be easily computed as

$$Z(j\omega) = \frac{1 - \omega^2 L_s C_s + j\omega r_s C_s}{j\omega C_p \left(1 + C_s/C_p - \omega^2 L_s C_s + j\omega r_s C_s\right)} \tag{2.22}$$

Given the fact that $r_s \ll \frac{1}{\omega_s C_p}$, the crystal impedance value at ω_s with a very good approximation would be

$$Z(j\omega_s) \approx r_s \tag{2.23}$$

This is a quite small value for the crystal impedance. On the other hand by the fact that $Q = \frac{L_s \omega_s}{r_s}$ has a very large value and $\frac{C_s}{C_p} \ll 1$, then the crystal impedance value at ω_p with a very good approximation would be

$$Z(j\omega_p) \approx \frac{QC_s}{\omega_p C_p^2} \tag{2.24}$$

This impedance value is normally quite large. Interestingly, the crystal impedance above f_s and under f_p is inductive (with a large inductive reactance derivative, $\frac{\partial X}{\partial \omega}$) and its impedance for frequencies under f_s is capacitive (with a large capacitive reactance derivative, $-\frac{\partial X}{\partial \omega}$). This is the phenomenon which stabilizes the oscillation frequency in the crystal oscillators. Figure 2.21 shows the computational results of a typical 10 MHz crystal impedance with $C_s = 9.1$ fF, $r_s = 35\,\Omega$, $L_s = 27$ mH, and $C_p = 2$ pF. The series resonant frequency of the crystal would be 10.153542 MHz and the parallel resonant frequency would be 10.176615 MHz. The quality factor of the crystal would be 49200. As such, the crystal impedance goes from $35\,\Omega$ to $1.75\,M\Omega$ within a frequency span of 23 kHz only. At low frequencies, the capacitors are open circuit, and therefore the impedance will be high. The first resonance frequency in the circuit is due to the

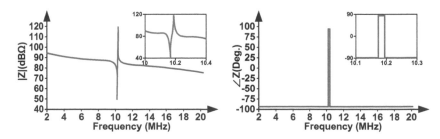

Figure 2.21: Simulation of crystal impedance.

resonance of the series inductor L_s and the series capacitor C_s which is called the series resonance f_s. At this frequency, the inductor and the capacitor will tune out and make approximately a short circuit (very low impedance) at the right-hand branch. Thus, the overall impedance of the crystal will be equal to $r_s \parallel j\omega C_p$. By increasing the frequency, C_s goes low impedance and the right-hand branch becomes inductive, the next resonance will occur approximately due to the resonance of L_s and C_p which we call the parallel resonance and show it by f_p. Finally, if we continue increasing the frequency, the parallel capacitor will short out the whole crystal impedance at high frequencies. The series resonance frequency can be calculated from Equation 2.25:

$$f_s = \frac{1}{2\pi\sqrt{L_sC_s}} \tag{2.25}$$

The parallel resonance frequency can be calculated from Equation 2.26:

$$f_p = f_s\sqrt{1 + C_s/C_p} \tag{2.26}$$

The crystal's quality factor can be computed as

$$Q = L_s\omega_s/r_s = 1/\omega_sC_sr_s \tag{2.27}$$

Crystal is generally used in oscillators where the acoustic vibration occurs in the body of the crystal, and by the virtue of the piezoelectricity of the crystal, those oscillations are transformed into electrical oscillations. A crystal has a very high quality factor which will result in the purity of the oscillator signal spectrum. This quality factor for crystal in Figure 2.20 is equal to 49200. As stated in Equation 2.24, the maximum impedance of the crystal occurs at the parallel resonant frequency and it has a very high value described by this equation.

A number of crystal oscillator topologies are shown in Figure 2.22.

In Figure 2.22(a) which is a common-collector Colpitts-like oscillator, the crystal acts as an inductor. That is the inductive reactance of the crystal resonates with the capacitances of C_1 and C_2. The feedback voltage across R_E appears through the step-up capacitive transformer across the crystal terminals. The oscillation frequency would be slightly above f_s.

In Figure 2.22(b) which is a common-base Colpitts-like oscillator, C_B is a large (short-circuit) capacitor. The crystal is in series within the feedback loop. The resonant

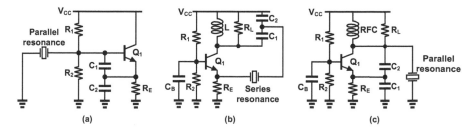

Figure 2.22: Three different Colpitts-like crystal oscillator configurations with a bipolar transistor.

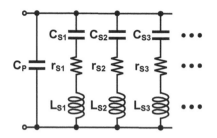

Figure 2.23: A complete model of a crystal.

frequency of the LC circuit should be approximately the same as the series resonance frequency of the crystal. the oscillation frequency would be approximately f_s.

In Figure 2.22(c) which is a common-base Colpitts-like oscillator, C_B is a large (short-circuit) capacitor. The crystal is put in parallel with the capacitive divider and it resonates with these capacitors at the oscillation frequency. The inductive reactance of the crystal is tuned out by the capacitors. The oscillation frequency would be slightly above f_s.

Normally, a piece of quartz crystal has several electro-acoustic resonant modes. These modes are called overtones which occur approximately at the odd multiples of the fundamental resonant frequency. A more generalized circuit model of a crystal is shown in Figure 2.22. In this model, the higher order resonances (overtones) of the crystal are shown by the additional parallel RLC branches in the circuit. Normally, the higher order modes resonances have a lower Q than the fundamental mode resonance.

As depicted in Figure 2.23, a crystal may have a number of higher order resonant frequencies. Thus, by choosing the main oscillation frequency in the crystal circuit meticulously, one may use it for higher desired overtone. For instance, in Figure 2.22(c) where there is no inductor, the circuit is forced to oscillate at the fundamental frequency of the crystal. That is to say the crystal will become purely inductive near the fundamental resonant frequency. It is possible to design a frequency selection circuit whose frequency is a multiple of the fundamental frequency of the crystal. For example, if one designs an LC tank with 75 MHz resonance frequency, with the fundamental frequency of 15 MHz of the crystal, the oscillator finally will oscillate at 75 MHz. However, the price of higher oscillation frequency is injecting more energy and as a result more power loss and therefore lower quality factor. In reality, one may order the manufacturer to make a crystal with a specific parallel resonance frequency by realizing specific parallel capacitor. As a rule of thumb, a parallel capacitor is usually near a few pFs or a few tenths of pFs (this value can be adjusted by the crystal manufacturer to have a precise oscillation frequency using the crystal). Moreover, the resonance frequency variations of a crystal with temperature for a range of $0 - 70°$ Celsius is just about 50 ppm. In other words, if the oscillation frequency is 1 MHz, by those temperature variations, the oscillator may have about 50 Hz frequency drift.

Example 2.3 Is it possible to choose the fundamental frequency of a crystal resonance frequency at a very high value? Therefore, surpassing the need for overtone driving.

Answer:

No, because there is a technical manufacturing problem in the design of very high frequency crystals. As a matter of fact, the crystal disc will become too thin to fabricate, the package capacitance would increase, microphonic issues would arise, and unwanted frequency modulation might happen. (Microphonic effect is an acoustic frequency modulation effect once the crystal is physically shaken). ∎

Now, assume that we have a transmitter with a carrier frequency of 900 MHz. With 50 ppm frequency variations in the reference crystal, there will be 45 kHz frequency variation at the carrier, which in GSM with 200 kHz channel bandwidth is too much. Thus, we need a more precise frequency for this kind of application. A temperature-compensated crystal oscillator (TCXO) might be a solution, where by microtuning and using temperature-dependent biasing, the temperature variations of the oscillation frequency is compensated. Therefore, the frequency will be stabilized within a range of few *ppm*'s, e.g., a frequency variation of 3 ppm here will result in a variation of 2.7 kHz in the carrier frequency. If this range of variation is not yet acceptable, a PLL-based synthesizer carrier might be employed. We discuss more about this subject in the next chapter.

2.5.1 Datasheet of a Family of Crystals

For better understanding of a crystal performance, the datasheet of a family of crystals, namely, Unit HC-49/U, is presented in Table 2.2.

The variations of series resistance (impedance of the crystals at the resonance frequency) of the family HC-49/U crystals are depicted in Figure 2.24. As it is seen in this figure, the series resistance of the crystal might change between 28 dBΩ and 70 dBΩ (25 Ω to 3000 Ω) depending on the chosen resonance frequency of the crystal. Notably, the crystals with a resonance frequency in the range of few tens of MHz have a relatively low series resistance (in the range of few tens of ohms) while the crystals

Table 2.2: The specifications of HC-49/U family of crystals.

Nominal Frequency Range	1.8 to 32 MHz-24 to 75 MHz-75 to 200 MHz
Vibration Mode	Fundamental-3rd Overtone-5th Overtone
Frequency Tolerance@25°C	±20 or ±30 ppm
Temperature Stability	±30 or ±50 ppm
Operating Temperature Range	−10°C to +60°C (Option: −20°C to +70°C)
Storage Temperature Range	−20°C to +70°C (Option: −30°C to +80°C)
Load Capacitance	8 pF to 32 pF or series
Equivalent Series Resistance	See Figure 2.24
Shunt Capacitance	5 pF max(≤18 MHz) or 7 pF max(>18 MHz)
Drive Level	200 μW max(≤5 MHz) 100 μW max (>5 MHz)
Insulation Resistance	500 MΩ min @ 100 V DC
Aging	±5 ppm per year

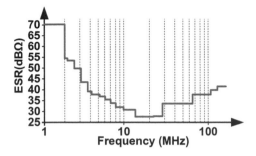

Figure 2.24: Equivalent series resistance of HC-49/U.

with a resonance frequency in the range of few MHz have a relatively higher series resistance (about few hundreds of ohms to few thousands ohms), and the crystals with a resonance frequency of few hundred MHz have a series resistance of the order of a hundred ohms.

2.6 Calculation of the Oscillation Frequency Including the Device Parasitics

As an example, consider Figure 2.25 where a Colpitts-like oscillator is shown.

The oscillator with its parasitic capacitances is shown in Figure 2.25. The gain can be computed by calculating the impedance seen at the collector times the transconductance. As it is obvious in Figure 2.25, the parasitic capacitances can be absorbed in the oscillator's circuit capacitors. Thus, the resonance frequency of the oscillator can be calculated as

$$f_0 = \frac{1}{2\pi\sqrt{L\left(\frac{(C_1+C_{be})C_2}{C_1+C_{be}+C_2} + C_{cs} + C_{bc}\right)}} \tag{2.28}$$

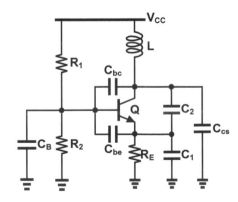

Figure 2.25: Colpitts oscillator with parasitic capacitances.

By absorbtion of the parasitic capacitances in the surrounding circuit capacitors, one can always simplify the oscillator circuit into the common core oscillators as described earlier. This will eventually result in a precise calculation of the oscillation frequency.

2.7 Quality Factor of Reactive Elements

The quality factor has an important role in the RF components applications. Very roughly one can consider it as a ratio of stored energy per cycle to the power loss in the component, or one can consider it as a ratio of a reactance to the resistance loss, or as a ratio of a susceptance to the conductance loss. All of them have the same meaning. A transfer function of a second-order band-pass filter can be written as

$$\frac{V_o}{V_i} = \frac{1}{1 + jQ\left(\frac{\omega}{\omega_0} - \frac{\omega_0}{\omega}\right)} \tag{2.29}$$

Or for the amplitude response, one can write

$$\left|\frac{V_o}{V_i}\right| = \frac{1}{\sqrt{1 + Q^2\left(\frac{\omega}{\omega_0} - \frac{\omega_0}{\omega}\right)^2}} \tag{2.30}$$

where ω_0 is the resonant frequency of the LC tank, and Q is the quality factor of the circuit. Q is a parameter which describes the ratio of the energy stored per cycle to the power dissipation, i.e., higher Q means lower power dissipation compared to the stored energy. The magnitude of the fraction in Equation 2.29 becomes equal to unity at the resonance frequency ω_0. Moreover, in the magnitude response of the filter, the same parameter Q shows the sharpness of the frequency response near the center frequency. Now consider the frequency response of the band-pass filter as shown in Figure 2.26.

By computing the upper and the lower frequency 3 dB points in the frequency response of the filter, it can be shown that the quality factor for Figure 2.26 can be written as

$$Q = \frac{\omega_0}{BW} = \frac{f_0}{\Delta f} \tag{2.31}$$

Here, the bandwidth is described in radian frequency where $BW = 2\pi\Delta f$. Here, Q is the quality factor of the bandpass filter.

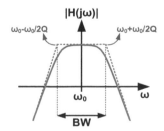

Figure 2.26: Magnitude response of a band-pass filter.

The quality factor can also be defined for lossy elements such as inductors and capacitors. For a lossless inductor, Q will be infinite; however, in reality, due to different sources of loss in the inductors (series resistance, skin effect, and magnetic core loss), they will have a finite Q. An inductor and its equivalent circuit are shown in Figure 2.27. The quality factor of an inductor is defined as in 2.32:

$$Q = 2\pi \frac{\frac{1}{2}LI^2}{\frac{1}{2}r_s I^2 T} = \frac{L\omega}{r_s} = \frac{R_p}{L\omega} \tag{2.32}$$

Note that for a specified inductor, the values of r_s and R_p are quite different. As a rule of thumb, the quality factor of a discrete inductor is between 50 and 100 and for an on-chip inductor due to its two-dimensional structure is about 3 to 5. The other reactive lossy element in circuits is a capacitor. For an ideal capacitor, the quality factor is infinite. Nonetheless, for a real capacitor due to dielectric losses or its series resistance, the quality factor is finite. The quality factor for a capacitor can be given by Equation 2.33:

$$Q = 2\pi \frac{\frac{1}{2}CV^2}{\frac{1}{2}\frac{V^2}{R_p}T} = R_p C\omega = \frac{1}{r_s C\omega} \tag{2.33}$$

Note that for a specified capacitor, the values of r_s and R_p are also quite different. The equivalent circuit of a capacitor is shown in Figure 2.28.

As a rule of thumb, for a discrete capacitor, the quality factor is between 50 and 200 and for integrated capacitors, this value is roughly between 50 and 100. The parameter Q first defined as the ratio of stored energy per cycle to the dissipated power

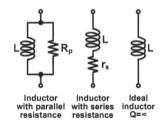

Figure 2.27: Equivalent circuit of an inductor.

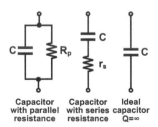

Figure 2.28: Equivalent circuit of a capacitor.

is simply related to the resistive loss in the inductors and the capacitors. As such, a resonator realized by a pair of elements like an inductor and a capacitor will have a quality factor which will be less than the quality factor of either of the elements. Therefore, realizing an LC filter with quality factors in excess of one hundred will not be possible.

Normally, the skin effect increases the quality factor of the inductors with increasing frequency, but the Q factor is reduced as the frequency passes a certain maximum value. As a matter of fact, the series resistance of inductors increases by the square root of frequency due to the skin effect. In practice, in discrete implementations, multiple inductors are placed in parallel to mitigate the skin effect, and therefore, achieve a better quality factor. The difficulty of placing a band-pass filter at the receiver's front-end is more clear now. Because of the low quality factor of the passive reactive elements, high Q band-pass circuits are barely realizable in the receiver sections. This is the reason for which to have very sharp filters with high quality factors, i.e., normally ceramic filters or crystal filters are used in the receiver chain.

2.8 Nonlinear Behavior in Amplifiers

Consider the transistor amplifier circuit in Figure 2.29.

Also, assume that the collector bias current is 1 mA, The gain of the amplifier can be obtained as in Equation 2.34:

$$A_V = -\frac{r_{in}}{R_S + r_{in}} \times g_m R_L \approx -g_m R_L \tag{2.34}$$

where g_m is the transistor's transconductance and r_{in} is the base dynamic resistance as described in Equation 2.35.

$$r_{in} = \beta \frac{KT}{qI_C} = (\beta + 1)\frac{KT}{qI_E} = (\beta + 1)\frac{V_t}{I_E} \tag{2.35}$$

Now, if the collector load resistance is 1 kΩ, for the collector bias current of 1 mA, the gain of the amplifier becomes $A_V = -g_m R_L = -38.5$. The derived equation is merely

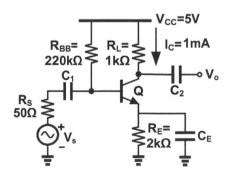

Figure 2.29: Common-emitter amplifier circuit.

valid for the linear behavior of an amplifier; however, when a large signal is imposed at the input, the gain equation should be modified. The transfer characteristics of the bipolar transistor can be assumed as in Equation 2.36:

$$i_e = I_{ES} e^{\frac{qV_{BE}}{kT}} \tag{2.36}$$

For instance, with this exponential $I - V$ characteristic, with an increase of one thermal voltage ($\frac{kT}{q} \approx 26\,\text{mV}$) at the input, the output current will be multiplied by a Neper. With a supply voltage of 5 V in Figure 2.29, the corresponding output voltage of the amplifier for different input levels is depicted in Figure 2.30. Here, β is assumed to be equal to 100.

As it is obvious from Figure 2.29, the output DC of the collector is 4 V. First, consider the input amplitude is 12.5 mV. Thus, output signal swing will be about 1 V. If we continue increasing the input voltage to 25 mV, the signal will be limited from the top to supply voltage and from the bottom to 2.4 V. Further entering large-signal input regime will result in limiting the signal from the top to 5 V and from the bottom to about 2.2 V while having a significant distortion with respect to the sinusoidal form. Consider Figure 2.31.

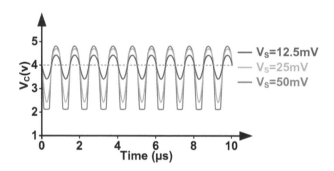

Figure 2.30: The collector voltage, for different values of input voltage, as a function of time.

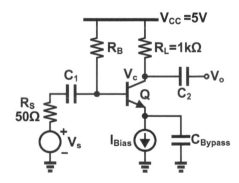

Figure 2.31: Common-emitter amplifier with a current source.

Similarly, the collector current is assumed to be 1 mA here and we gradually increase the input voltage as before. The difference between Figures 2.29 and 2.31 is the current source at the emitter of Figure 2.31. The currents' source fixes the DC current at the output. Increasing the input voltage results in exponential increase of the output current. However, the current source maintains the DC current at a fixed level. As a matter of fact, the DC voltage of the emitter is increased in such a way to maintain the DC current of the emitter at the specified bias current, I_{bias}. Therefore, the growing current of the amplifier is controlled. Simulation results of this phenomenon are shown in Figure 2.32.

If we replace the load resistance in Figure 2.31 with a parallel RLC circuit with a high quality factor, indeed we arrive at a tuned amplifier structure as in Figure 2.33.

The ideal gain of this circuit in Figure 2.33 can be obtained as

$$\frac{V_{\text{out}}}{V_{\text{s}}} = -G_{\text{m}}R_{\text{L}} \qquad (2.37)$$

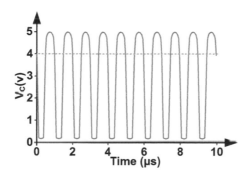

Figure 2.32: Output collector voltage with 100 mV input at 1 MHz frequency.

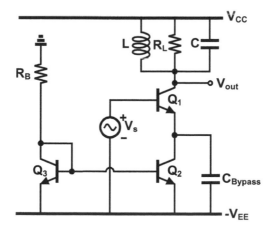

Figure 2.33: Tuned amplifier with implemented emitter current source.

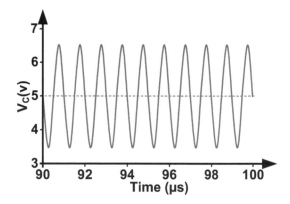

Figure 2.34: Output voltage waveform of the tuned amplifier for a 50 mV sinusoidal input ($R_L = 1\,\text{k}\Omega$, $C = 3.2$ nF, and $L = 7.93\,\mu\text{H}$).

At the large-signal regime, it is possible to choose a desired harmonic of the input signal by tuning the output tank circuit, as depicted in Figure 2.34.

2.9 A Note on the Modified Bessel Functions of the First Kind

Modified Bessel functions of the first kind are frequently encountered in the nonlinear analysis of the P-N junction diodes and the bipolar transistors due to their exponential characteristics. Here, we briefly introduce these functions and their characteristics. The harmonic expansion of an exponential sinusoidal function can be described as follows

$$e^{x\cos(\omega_0(t))} = I_0(x) + 2I_1(x)\cos(\omega_0(t)) + 2I_2(x)\cos(2\omega_0(t)) + \cdots + 2I_n(x)\cos(n\omega_0(t))$$
(2.38)

Here the functions $I_n(x)$ are called the modified Bessel functions of the first kind. These functions are the solutions of the Bessel's differential equation at certain conditions. The evolution of these functions with respect to their argument has generally an exponentially increasing form. Furthermore, the higher order function is generally smaller than the lower order function for the same argument. That is

$$\frac{I_{n+1}(x)}{I_n(x)} < 1 \quad \text{for all } x$$
(2.39)

Additionally

$$\lim_{x \to \infty} \left(\frac{I_{n+1}(x)}{I_n(x)} \right) = 1$$
(2.40)

and

$$\lim_{x \to \infty} \left(\frac{I_1(x)}{I_0(x)} \right) = 1$$
(2.41)

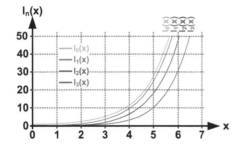 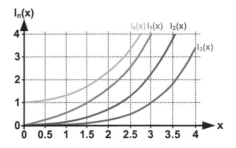

Figure 2.35: Typical variations of the modified Bessel functions with respect to their arguments.

Table 2.3: Numerical values of the first four modified Bessel functions of the first kind as a function of their argument.

x	0	1	2	3	4	5	6	7	8	9
$I_0(x)$	1	1.27	2.28	4.88	11.30	27.24	67.23	168.59	427.56	1093.59
$I_1(x)$	0	0.57	1.59	3.95	9.76	24.34	61.34	156.04	399.87	1030.91
$I_2(x)$	0	0.14	0.69	2.25	6.42	17.51	46.79	124.01	327.60	864.50
$I_3(x)$	0	0.02	0.21	0.96	3.34	10.33	30.15	85.18	236.08	646.69

and

$$\lim_{x \to 0} \left(\frac{I_1(x)}{I_0(x)} \right) = \frac{x}{2} \tag{2.42}$$

Furthermore, $I_0(0) = 1$ and $I_n(0) = 0$ for $n > 1$. Figure 2.35 shows the variations of the modified Bessel functions of different orders with respect to their argument. Table 2.3 shows the numerical values of the modified Bessel functions of different orders as a function of their arguments.

2.10 Large-Signal Transconductance and Harmonic Tuned Amplifiers

In this section, we derive relations for the amplitude of the output signal using modified Bessel functions and the concept of large-signal transconductance. The small-signal transconductance in Figure 2.33 can be obtained as

$$g_m = \frac{\partial i_C}{\partial v_{be}} = \frac{I_C}{V_t} \tag{2.43}$$

where the direct current can be calculated as in Equation 2.44:

$$I_{E_{bias}} = \frac{V_{EE} - V_{BE_{bias}}}{R_{BB}} \tag{2.44}$$

The above equation is valid for the small-signal behavior of an amplifier. One may write a general equation for large-signal bipolar transistor current as 2.45:

$$i_C = \alpha I_{ES} e^{\frac{V_{BE_{bias}} + V_i \cos(\omega_0 t)}{V_t}} \tag{2.45}$$

Equation 2.45 can be rewritten as 2.46

$$i_C = \alpha I_{ES} e^{\frac{V_{BE_{bias}}}{V_t}} \left(e^{\frac{V_i}{V_t} \cos(\omega_0 t)} \right) \tag{2.46}$$

If we define $x = V_i/V_t$ and expanding in terms of modified Bessel functions of the first kind, that will result in Equation 2.47:

$$i_C = \alpha I_{ES} e^{\frac{V_{BE_{bias}}}{V_t}} \left(I_0(x) + 2I_1(x) \cos(\omega_0 t) + 2I_2(x) \cos(2\omega_0 t) + \cdots \right) \tag{2.47}$$

If one factors out the DC component of 2.47, he/she reaches to 2.48.

$$i_C = \alpha I_{ES} e^{\frac{V_{BE_{bias}}}{V_t}} I_0(x) \left(1 + \frac{2I_1(x)}{I_0(x)} \cos(\omega_0 t) + \frac{2I_2(x)}{I_0(x)} \cos(2\omega_0 t) + \cdots \right) \tag{2.48}$$

As it is obvious from 2.48, the DC component can be obtained as

$$I_{DC} = \alpha I_{ES} e^{\frac{V_{BE_{bias}}}{V_t}} I_0(x) \tag{2.49}$$

While employing the current source, the DC component of current will be constant (how?). When x increases, $I_0(x)$ will increase similarly but $V_{BE_{bias}}$ will decrease indeed. The only mechanism that maintains the DC constant is V_{BE} depreciation as stated. The output signal for a large-signal input can be written as

$$V_{out} = V_{CC} - \alpha I_{E_{bias}} \left(Z_L(0) + Z_L(j\omega_0) \frac{2I_1(x)}{I_0(x)} \cos(\omega_0 t) + \right.$$
$$\left. Z_L(2j\omega_0) \frac{2I_2(x)}{I_0(x)} \cos(2\omega_0 t) + \cdots \right) \tag{2.50}$$

where the input signal is $V_S = V_i \cos(\omega t)$ and $\alpha \approx 1$. The typical frequency response of the amplifier is shown in Figure 2.36. Here, it is assumed that the output is tuned to the first harmonic of the input.

As it is obvious from Figure 2.36, the output current contains all harmonics of the input signal. However, by tuning the band-pass filter, any output harmonic can be selected at the output. In tuned amplifier and oscillator applications, it is normally assumed that the output circuit is tuned to the fundamental harmonic of the input. While band-pass filters are of interest in narrowband applications, we investigate their impedance behavior a little bit more. A simple parallel RLC band-pass filter fed by a current source is shown in Figure 2.37.

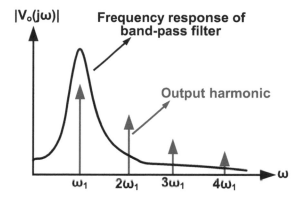

Figure 2.36: Harmonics of output and selecting behavior of the band-pass filter.

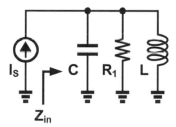

Figure 2.37: Resonant circuit.

One may obtain impedance of the band-pass filter as 2.51:

$$Z_{in} = \frac{R_T}{1 + jQ(\frac{\omega}{\omega_0} - \frac{\omega_0}{\omega})} \tag{2.51}$$

where

$$\omega_0 = \frac{1}{\sqrt{LC}} \quad R_T = R_1 \parallel R_{PC} \parallel R_{PL} \tag{2.52}$$

where R_{PC} and R_{PL} are the equivalent parallel loss resistances of the capacitor and the inductor, respectively. Q is the overall quality factor of the circuit and is expressed as 2.53:

$$Q = \frac{R_1 \parallel R_{PC} \parallel R_{PL}}{L\omega_0} = R_T C \omega_0 \tag{2.53}$$

The total quality factor of the band-pass filter will be

$$\frac{1}{Q} = \frac{1}{Q_L} + \frac{1}{Q_C} + \frac{L\omega_0}{R_1} \tag{2.54}$$

The important point in Equation 2.54 is the dominance of low Q element which is usually an inductor. Typical frequency response of the band-pass filter is shown in Figure 2.38.

It can be shown that the input impedance at $-3\,\text{dB}$ point of the circuit is as 2.55:

$$Z_{\text{in}} = \frac{R_T}{1 + j(1)} \tag{2.55}$$

By comparing Equations 2.51 and 2.55, we reach to 2.56:

$$Q\left(\frac{\omega_0 + \Delta\omega}{\omega_0} - \frac{\omega_0}{\omega_0 + \Delta\omega}\right) = 1 \quad \text{or} \quad \Delta\omega \approx \frac{\omega_0}{2Q} \tag{2.56}$$

One may derive equation for bandwidth of the circuit as

$$BW = (\omega_0 + \Delta\omega) - (\omega_0 - \Delta\omega) = 2\Delta\omega(\text{radian/s}) \tag{2.57}$$

Finally, the relation for the bandwidth is as 2.58:

$$BW = \frac{\omega_0}{Q}(\text{radian/s}) \tag{2.58}$$

Equation 2.58 is of great importance. It suggests that we can calculate the quality factor of resonant circuits by finding the ratio of center frequency to its 3 dB bandwidth. Moreover, one may obtain the bandwidth of the circuit by the division of the resonant frequency by the quality factor.

For the nth harmonic of the input, the load impedance described in Equation 2.57 can be expressed as

$$Z_L(jn\omega_0) = \frac{R_T}{1 + jQ\left(n - \frac{1}{n}\right)} \simeq \frac{nR_T}{jQ(n^2 - 1)} \tag{2.59}$$

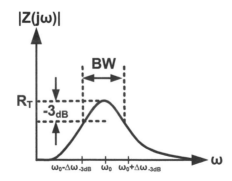

Figure 2.38: Typical frequency response of a parallel resonant circuit.

With respect to the aforementioned derivation of Q, one may obtain output voltage of the circuit depicted in Figure 2.33 as

$$V_{out} = V_{CC} - \alpha I_{Ebias} \left(\frac{2R_T I_1(x)}{I_0(x)} \cos(\omega_0 t) + \frac{4R_T I_2(x)}{3Q I_0(x)} \cos\left(2\omega_0 t - \frac{\pi}{2}\right) \right.$$
$$\left. + \frac{3R_T I_3(x)}{4Q I_0(x)} \cos\left(3\omega_0 t - \frac{\pi}{2}\right) + \cdots \right) \quad (2.60)$$

Neglecting the smaller valued harmonic terms, we can simplify the above equation to the following

$$V_{out} \approx V_{CC} - \alpha I_{Ebias} \frac{2R_T I_1(x)}{I_0(x)} \cos(\omega_0 t) \quad (2.61)$$

Or

$$V_{out} \approx V_{CC} - x\alpha I_{Ebias} \frac{2R_T I_1(x)}{x I_0(x)} \cos(\omega_0 t) \quad (2.62)$$

Or

$$V_{out} \approx V_{CC} - g_m \frac{2I_1(x)}{x I_0(x)} V_i R_T \cos(\omega_0 t) \quad (2.63)$$

Here we can define the large signal transconductance as the following

$$G_m = \frac{I_{C_1}}{V_i} = g_m \frac{2I_1(x)}{x I_0(x)} \quad (2.64)$$

In the above-mentioned equation, it is assumed that the load impedance is tuned to the first harmonic of the input. As such, it is observed that the higher order harmonics amplitudes are decreasing monotonically as a function of amplitude and frequency. If the load quality factor Q is sufficiently large, the higher order harmonics could be neglected compared to the fundamental harmonic.

Furthermore, regarding Equation 2.57, once the load is tuned to the mth harmonic of the input, the output voltage will become

$$V_{out} = V_{CC} - \alpha I_{Ebias} \left(\frac{mR_T}{Q(m^2-1)} \times \frac{2I_1(x)}{I_0(x)} \cos\left(\omega_0 t + \frac{\pi}{2}\right) + \frac{2mR_T}{Q(m^2-4)} \times \frac{2I_2(x)}{I_0(x)} \cos\left(2\omega_0 t + \frac{\pi}{2}\right) + \cdots \right.$$
$$\left. + \frac{2R_T I_m(x)}{I_0(x)} \cos(m\omega_0 t) + \cdots + \frac{nmR_T}{Q(n^2-m^2)} \times \frac{2I_n(x)}{I_0(x)} \cos\left(n\omega_0 t - \frac{\pi}{2}\right) + \cdots \right) \quad (2.65)$$

Note that, in Equation 2.65, in the developed series $n \neq m$ and here it is assumed that $m > 2$. Neglecting the smaller-valued terms, we can approximate the above equation by the following

$$V_{out} \approx V_{CC} - \alpha I_{Ebias} \frac{2R_T I_m(x)}{I_0(x)} \cos(m\omega_0 t) \quad (2.66)$$

To clarify more what is described in Equation 2.65, two special cases are considered in the following sections.

2.10.1 **Case I: Resonant circuit is tuned to the first harmonic of the input frequency (tuned amplifier case)**

For this special case, as it was already derived in Equation 2.60 for the first three harmonics, in other words for $n = 1, 2, 3$. We define nth harmonic at the output as $H_n(x)$, we will have

$$H_1(x) = \frac{2I_1(x)}{I_0(x)} R_T I_{Ebias} \tag{2.67a}$$

$$H_2(x) = \frac{2I_2(x)}{I_0(x)} \left| \frac{R_T}{1 + jQ(2 - \frac{1}{2})} \right| I_{Ebias} \approx \frac{2I_2(x)}{I_0(x)} \frac{2R_T}{3Q} I_{Ebias} \tag{2.67b}$$

$$H_3(x) = \frac{2I_3(x)}{I_0(x)} \left| \frac{R_T}{1 + jQ(3 - \frac{1}{3})} \right| I_{Ebias} \approx \frac{2I_3(x)}{I_0(x)} \frac{3R_T}{8Q} I_{Ebias} \tag{2.67c}$$

2.10.2 **Case II: Resonant circuit is tuned to the second harmonic of the input frequency (frequency multiplier case)**

Here, $m = 2$. For this case, we derive the equations for first to third output harmonics, in other words for $n = 1, 2, 3$. The equation for the first three harmonics can be written as

$$H_1(x) = \frac{2I_1(x)}{I_0(x)} \left| \frac{R_T}{1 + jQ(\frac{1}{2} - 2)} \right| I_{Ebias} \approx \frac{2I_1(x)}{I_0(x)} \frac{2R_T}{3Q} I_{Ebias} \tag{2.68a}$$

$$H_2(x) = \frac{2I_2(x)}{I_0(x)} R_T I_{Ebias} \tag{2.68b}$$

$$H_3(x) = \frac{2I_3(x)}{I_0(x)} \left| \frac{R_T}{1 + jQ(\frac{3}{2} - \frac{2}{3})} \right| I_{Ebias} \approx \frac{2I_3(x)}{I_0(x)} \frac{6R_T}{5Q} I_{Ebias} \tag{2.68c}$$

Typical output harmonic currents and the load impedance variations are shown in Figure 2.39 for both cases (I and II).

As it is obvious from Figure 2.39, in the first case, the output band-pass filter is tuned to the first harmonic of the input which means it attenuates higher harmonics. In the second case, the output band-pass filter passes the second harmonic and attenuates other frequency components, i.e., the first harmonic, the third harmonic and the higher ones at the output. For the ease of calculation, we define a large-signal transconductance (G_m) based on the harmonic number of the output. Our goal is to obtain an equation for the output signal in terms of G_m. For instance, one may write the output voltage for the first harmonic as

$$V_{o_1} = -G_{m_1} R_T V_i \tag{2.69}$$

We may write the amplitude of input signal as $V_i = xV_t$; so, we derive G_{m1} as

$$G_{m_1} = \frac{I_1}{V_i} = \frac{I_1}{xV_t} \tag{2.70}$$

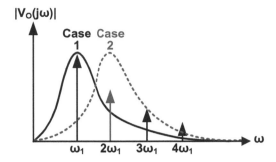

Figure 2.39: Typical output harmonics as well as load impedance variations for cases I and II.

With respect to the above definition, we can write the expression for G_{m1} as

$$G_{m_1} = \frac{I_{E_{bias}}}{V_i} \cdot \frac{2I_1(x)}{I_0(x)} = \frac{I_{E_{bias}}}{V_t} \cdot \frac{2I_1(x)}{xI_0(x)} = g_m \frac{2I_1(x)}{xI_0(x)} \qquad (2.71)$$

Equation 2.71 gives an explicit equation for the large-signal transconductance of the bipolar transistor. Thus, by using modified Bessel function table, we can easily calculate the G_m's. For computing any harmonic, we can define a conversion transconductance from the first harmonic to the nth harmonic as G_{m_n} where at the output we would have

$$G_{m_n} = \frac{I_n}{V_i} = \frac{I_n}{xV_t} = g_m \frac{2I_n(x)}{xI_0(x)} \quad \text{or} \quad \frac{V_{o_n}}{V_i} = -G_{m_n} Z_L (jn\omega_0) \qquad (2.72)$$

It is possible to generalize Equation 2.72 for the nth harmonic of the input frequency when the output band-pass filter is tuned at the mth harmonic of the input. In this case, the ratio of the output nth harmonic to the input phasor can be described as follows

$$\frac{V_{o_n}}{V_i} = -G_{m_n} \frac{R}{1 + jQ(\frac{n}{m} - \frac{m}{n})} = -g_m \frac{2I_n(x)}{xI_0(x)} \frac{R}{1 + jQ(\frac{n}{m} - \frac{m}{n})} \qquad (2.73)$$

Equation 2.73 describes a comprehensive equation to calculate the output voltage of a nonlinear transistor circuit for each of its harmonics with respect to the center frequency of the resonant circuit. Figure 2.40 illustrates the ratio of large-signal transconductance to its small-signal value as well as the conversion transconductances for the second and the third harmonics.

Table 2.4 shows the value of solid line in Figure 2.40.

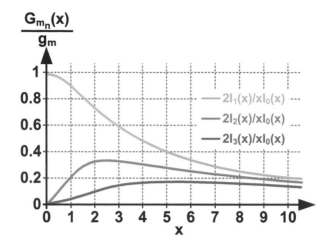

Figure 2.40: Ratio of large-signal transconductance normalized to small-signal transconductance as well as the conversion transconductances for the second and the third harmonics.

Table 2.4: Values of large-signal G_m normalized to small-signal g_m as well as the conversion transconductances for the second and the third harmonics.

x	0	1	2	3	4	5	6	7	8	9
$\dfrac{2I_1(x)}{xI_0(x)}$	1	0.893	0.698	0.540	0.432	0.357	0.304	0.264	0.234	0.209
$\dfrac{2I_2(x)}{xI_0(x)}$	0	0.214	0.302	0.307	0.284	0.257	0.232	0.210	0.192	0.176
$\dfrac{2I_3(x)}{xI_0(x)}$	0	0.035	0.093	0.131	0.148	0.152	0.149	0.144	0.138	0.131

Example 2.4 Consider Figure 2.33 where the input is $V_S = V_i \cos \omega_0 t$ with $\omega_0 = 2\pi\,(50\,\text{MHz})$. First, assume the band-pass filter is tuned to 50 MHz and then assume it is tuned to 150 MHz. Derive the relations for the output voltage at the first and the third harmonics.

Solution:
Using Equation 2.73, one may reach to Table 2.5.

Table 2.5: Normalized values of the first and the third harmonic voltages for the output tuned to either of the first or the third harmonics.

$H_1(x), n = 1$	$H_3(x), n = 3$
$\left\| \frac{V_{o_1}}{V_i} \right\| = g_{mQ} \frac{2I_1(x)}{xI_0(x)} R_L \cos\left(10^8 \pi t\right)$	$\left\| \frac{V_{o_3}}{V_i} \right\| = g_{mQ} \frac{2I_3(x)}{xI_0(x)} \left\| \frac{R_L}{1 + jQ\frac{8}{3}} \right\| \cos\left(3 \times 10^8 \pi t\right)$
$\left\| \frac{V_{o_1}}{V_i} \right\| = g_{mQ} \frac{2I_1(x)}{xI_0(x)} \left\| \frac{R_L}{1 - jQ\frac{8}{3}} \right\| \cos\left(10^8 \pi t\right)$	$\left\| \frac{V_{o_3}}{V_i} \right\| = g_{mQ} \frac{2I_3(x)}{xI_0(x)} R_L \cos\left(3 \times 10^8 \pi t\right)$

Apparently, in the first case, we have first harmonic at the output with a high gain and third harmonic at the output with a lower gain. In the second case, we have the first harmonic at the output with a low gain and the third harmonic at the output with a relatively higher gain. ∎

In RF communication circuits, mostly narrowband applications are of interest. However, investigation of harmonics and nonlinear behavior of circuits has a great influence on the performance of RF circuits. The aforementioned equations are for bipolar transistor; however, one may derive equations for MOS transistors as well. It is also possible to reject the undesired harmonics more efficiently with a typical matching band-pass network which is shown in Figure 2.41.

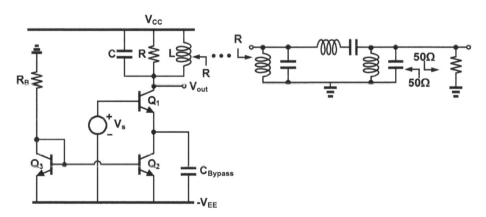

Figure 2.41: Typical matching network for filtering undesired harmonics.

In the next section, we focus on oscillators based on tapped capacitor and tapped inductor transformers. The inductive transformers are tunable with their number of turns and have a good isolation.

Example 2.5 An oscillator can be constructed using a tightly coupled RF transformer in the feedback circuit as well. Determine the complex Barkhausen oscillation condition in the common-base tuned circuit oscillator depicted in Figure 2.42. Note that C_B and C_E are considered as RF short circuit.

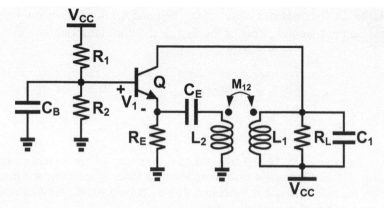

Figure 2.42: A common-base tuned circuit oscillator.

Solution:
The equivalent circuit for the above oscillator is depicted in Figure 2.43.

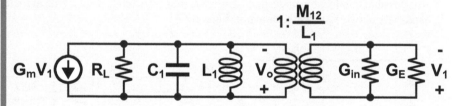

Figure 2.43: The equivalent circuit for the common-base tuned circuit oscillator.

Considering V_1 as the feedback voltage phasor, the output collector voltage of the oscillator is computed as

$$V_o = \frac{G_m V_1}{\frac{1}{R_L} + jC_1\omega - \frac{j}{L_1\omega} + \left(\frac{M_{12}}{L_1}\right)^2 (G_{in} + G_E)} \tag{2.74}$$

where G_{in} is the emitter's input conductance.
   Since

$$V_1 = \frac{M_{12}}{L_1} V_o \tag{2.75}$$

Given the fact that the transistor's emitter input conductance is $G_{in} = \frac{G_m}{\alpha}$, then the complex oscillation condition becomes

$$\frac{\frac{M_{12}}{L_1}G_m}{\frac{1}{R_L}+jC_1\omega-\frac{j}{L_1\omega}+\left(\frac{M_{12}}{L_1}\right)^2\left(\frac{G_m}{\alpha}+G_E\right)}=1 \tag{2.76}$$

Separating the real and the imaginary parts of Equation 2.76, one obtains two distinct equations

$$C_1\omega-\frac{1}{L_1\omega}=0 \tag{2.77}$$

$$G_m(x)=\frac{\frac{1}{R_L}+\left(\frac{M_{12}}{L_1}\right)^2\frac{1}{R_E}}{\frac{M_{12}}{L_1}\left(1-\frac{1}{\alpha}\frac{M_{12}}{L_1}\right)} \tag{2.78}$$

From the two above equations, the first one gives the oscillation frequency and the second one through $G_m(x)$ would determine the oscillation amplitude. ∎

2.11   Differential Bipolar Stage Large-Signal Transconductance

Figure 2.44 depicts a bipolar differential stage tuned amplifier.

Assuming exponential characteristic for either of the transistors, one can write

$$i_{c1}=\alpha I_{ES}e^{q(V_{BE0}+v_1)/kT}=\alpha I_{ES}e^{q\left(V_{BE0}+\frac{v}{2}\right)/kT} \tag{2.79}$$

$$i_{c2}=\alpha I_{ES}e^{q(V_{BE0}+v_2)/kT}=\alpha I_{ES}e^{q\left(V_{BE0}-\frac{v}{2}\right)/kT} \tag{2.80}$$

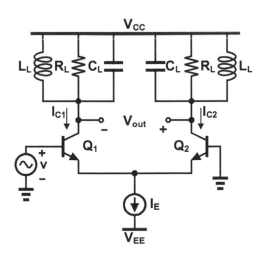

Figure 2.44: A differential pair tuned amplifier.

where V_{BE0} is the DC bias voltage of either of the transistors. Then

$$\frac{i_{c1}}{i_{c2}} = e^{qv/kT} \tag{2.81}$$

Given

$$i_{c1} + i_{c2} = \alpha I_E = I_C \tag{2.82}$$

One can compute either of the collector currents using the above two equations:

$$i_{c1} = \frac{I_C}{1 + e^{-z}} = \frac{I_C}{2}\left[1 + \tanh\left(\frac{z}{2}\right)\right] \tag{2.83}$$

$$i_{c2} = \frac{I_C}{1 + e^{+z}} = \frac{I_C}{2}\left[1 - \tanh\left(\frac{z}{2}\right)\right] \tag{2.84}$$

where

$$z = \frac{qv}{kT} = \frac{v}{V_t} \tag{2.85}$$

Assuming a large sinusoidal input voltage as

$$v = V_1 \cos(\omega t) \tag{2.86}$$

Either of the collector AC currents becomes

$$i_{c1,2} = \pm\frac{I_C}{2}\tanh\left(\frac{x}{2}\cos(\omega t)\right) \tag{2.87}$$

where

$$x = \frac{qV_1}{kT} = \frac{V_1}{V_t} \tag{2.88}$$

Now using the above equations, one can compute the harmonic components of the collector currents as

$$a_n(x) = \frac{1}{\pi}\int_{-\pi}^{\pi}\frac{1}{2}\tanh\left(\frac{x}{2}\cos(\theta)\right)\cos(n\theta)\,d\theta \tag{2.89}$$

Note that given the fact that the differential pair transfer characteristic has an odd symmetry, $a_n(x)$ functions would be zero for even values of n. Using the fundamental harmonic current of either of the collectors, one can calculate the large-signal transconductance of the differential bipolar stage:

$$G_m(x) = \frac{I_{C1}}{V_1} = -\frac{I_{C2}}{V_1} = \frac{qI_C}{kT}\frac{a_1(x)}{x} = g_m\frac{4a_1(x)}{x} \tag{2.90}$$

where

$$g_m = \frac{\partial i_{c1}}{\partial v} = -\frac{\partial i_{c2}}{\partial v} = \frac{qI_C}{4kT} = \frac{I_C}{4V_t} \tag{2.91}$$

Note that the DC and the fundamental harmonic output voltages at either of the collectors become

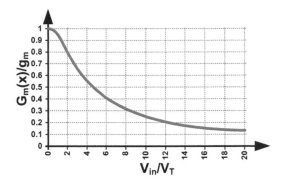

Figure 2.45: The evolution of the large-signal transconductance of a differential pair tuned amplifier as a function of the normalized input voltage amplitude.

$$V_{\text{out}}^- = V_{\text{CC}} - G_{\text{m}}(x) R_L V_1 \cos(\omega t) \tag{2.92}$$

$$V_{\text{out}}^+ = V_{\text{CC}} + G_{\text{m}}(x) R_L V_1 \cos(\omega t) \tag{2.93}$$

The differential fundamental harmonic voltage can be expressed as

$$V_{\text{out}} = V_{\text{out}}^+ - V_{\text{out}}^- = 2G_{\text{m}}(x) R_L V_1 \cos(\omega t) \tag{2.94}$$

Using this large-signal transconductance, one can compute the amplitude of oscillations in a differential pair oscillator.

2.12 Inductive and Capacitive Dividers (Impedance Transformers)

An ideal inductive transformer is depicted in Figure 2.46.

For the input impedance, we have

$$R_{\text{in}} = \frac{V_s}{i_s} = \frac{\frac{V_L}{m}}{m i_L} = \frac{1}{m^2} \frac{V_L}{i_L} = \frac{1}{m^2} R_L \tag{2.95}$$

For example, if the load resistance is $1\,\text{k}\Omega$, for $m = 5$, the input impedance will be $40\,\Omega$. Transformers play a crucial role in communication circuits for matching purposes. As stated earlier by tuning the resonant band-pass filter, one may attain a desired harmonic of the input signal at the output of a nonlinear circuit. The main role of transformers is to extract the desired signal from the resonant circuit without degrading its quality factor. However, it is instructive to know more about inductors before introducing their use in transformers. Discrete inductors have a good quality factor just for low-frequency applications. These inductors will fail at high frequencies due to their parasitic capacitances and resistances. Nowadays, RF engineers desire to integrate every thing on a single chip, thus on-chip inductors are of interest. However, because of their planar implementation, they will not have a good quality factor. Moreover, these elements are relatively huge and bulky in size and their fabrication occupies a huge area on the RF chip. It is possible to implement inductors at high frequencies (say at a few GHz) using microstrip or printed circuit transmission lines as well. Furthermore, realization of printed inductors is possible for monolithic microwave integrated circuits at frequencies well above 5 GHz. In the next section, we introduce circuits for impendence transformation and step-up and step-down voltage concepts.

2.12.1 Tapped Capacitive/Inductive Impedance Transformers

Figure 2.47 shows a tuned inductive transformer with a tap in the middle.

At it is depicted in Figure 2.47, a tap may separate the inductor in to two parts. Here, the winding ratio is $1 : m$ which divides the voltage by an m ratio. The inductive impedance transformer in Figure 2.47 can be modeled as Figure 2.48.

As depicted in Figure 2.48, by employing the impedance transformer, a resonant circuit is made which can have a high quality factor. Now consider an input source with $50\,\Omega$ impedance is applied at the input of Figure 2.48. This circuit is shown in Figure 2.49.

If we assume that R_P is a resistor modeling the loss of the resonant circuit, we can define an unloaded quality factor as

$$Q_{\text{unloaded}} = \frac{R_P}{L\omega} \tag{2.96}$$

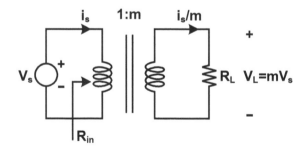

Figure 2.46: An ideal inductive transformer.

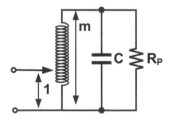

Figure 2.47: Inductive transformer with its loss and a capacitive load.

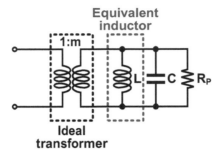

Figure 2.48: A model of the inductive transformer with its loss.

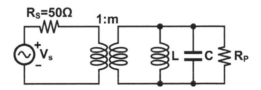

Figure 2.49: Applying signal source to the inductive transformer.

Then, the loaded quality factor becomes

$$Q_{\text{Loaded}} = \frac{R_P || (R_S \times m^2)}{L\omega} \tag{2.97}$$

Thus, as per Equation 2.96, the loading of the resonant circuit by the low impedance source degrades the quality factor of the resonant circuit. However, if the value of $m^2 R_S$ is much larger than R_P, the loaded quality factor would not be degraded as much.

> **Example 2.6** Is it possible to extract the signal of a resonant circuit by employing a buffer instead of impedance transformer?
> **Answer:**
> Buffer stages are not of interest in the RF receiver chains due to the very low available power of the sources; as such, we need the impedance matching at every stage (for the maximum power transfer purpose). Furthermore, bipolar buffer transistors are not of interest because of their finite base resistance. Similarly for MOS transistors, the gate–source capacitor introduces finite capacitance which may degrade the performance of the resonant circuit. Moreover, buffer design at high-frequency applications may introduce instability and unwanted oscillation in them. ∎

Capacitive impedance transformers act like the inductive ones except that they have capacitive reactances instead of inductive ones at their input. Note that the modeling and the behavior of capacitive/inductive coupling circuit are a straightforward procedure and we focus here on their basic behavior.

Figure 2.50 shows a capacitive impedance transformer.

At the resonance, one may obtain the input impedance of Figure 2.50 as

$$R_{\text{in}} = \frac{R}{m^2} \tag{2.98}$$

where

$$m = \frac{C_1 + C_2}{C_1} \tag{2.99}$$

Here, $m > 1$ means a step-up impedance transformation. Moreover, it is also possible to define an equivalent capacitance as

$$C_{\text{total}} = \frac{C_1 C_2}{C_1 + C_2} \tag{2.100}$$

Provided that for $R_{\text{in}}(C_1 + C_2)\omega \geq 10$, the above circuit can be modeled by a parallel RLC circuit along with a step-up transformer as depicted in Figure 2.51. Furthermore, the resonance condition becomes

$$L\omega - \frac{C_1 + C_2}{C_1 C_2 \omega} = 0 \tag{2.101}$$

The quality factor of this circuit will be

$$Q = \left(\frac{C_1 C_2}{C_1 + C_2} \right) \omega R_T \tag{2.102}$$

where R_T is the total equivalent parallel resistance of the tuned circuit. For large m, the relative values of the capacitors usually follow $\frac{C_2}{C_1} \gg 1$. For $C_2 \gg C_1$, Equation 2.102 can be approximately written as

$$Q \approx C_1 \omega R_T \tag{2.103}$$

For instance, if $C_1 = 10\,\text{pF}$ and $C_2 = 70\,\text{pF}$, then $m = 8$ and $C_{\text{total}} = 8.75\,\text{pF}$, and a load resistance of $1000\,\Omega$ will be transformed to $15.625\,\Omega$. The equivalent circuit of this step-up resonant impedance transformer is depicted in Figure 2.51 along with its dual counter part which consists of an inductive step-up impedance transformer.

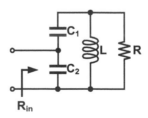

Figure 2.50: Capacitive impedance transformer with a parallel inductance.

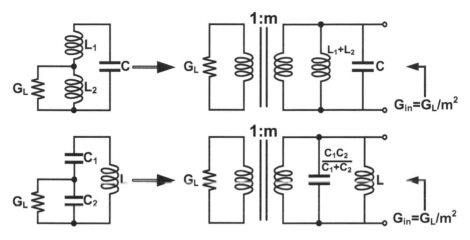

Figure 2.51:   Circuit models for inductive and capacitive impedance transformers.

Before using the models of Figure 2.51 for the step-up transformers, let's verify Equation 2.98 through 2.103 using the admittance or the impedance matrices of the corresponding circuits. Now, let's consider Figure 2.52.

One may obtain the admittance matrix of Figure 2.52 as

$$\begin{pmatrix} I_1 \\ I_2 \end{pmatrix} = j\omega \begin{pmatrix} C_1 + C_2 & -C_1 \\ -C_1 & C_1 \end{pmatrix} \begin{pmatrix} V_1 \\ V_2 \end{pmatrix} \tag{2.104}$$

Or in expanded form

$$I_1 = j\omega\,(C_1 + C_2)\,V_1 - j\omega C_1 V_2 \tag{2.105a}$$

$$I_2 = -j\omega C_1 V_1 + j\omega C_1 V_2 \tag{2.105b}$$

Now we derive the relation for the output admittance when the input is loaded with the conductance, G_L. The circuit is shown in Figure 2.53.

We can write at the input

$$I_1 = -G_L V_1 \tag{2.106}$$

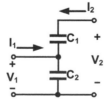

Figure 2.52: Step-up capacitive voltage transformer.

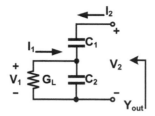

Figure 2.53: Step-up capacitive voltage transformer with loaded input.

Employing Equation 2.105a, we may obtain

$$I_1 = j\omega\left(C_1 + C_2\right)\frac{I_1}{-G_L} - j\omega C_1 V_2 \tag{2.107}$$

Thus, the ratio of I_1 to V_2 can easily be derived from Equation 2.107 as

$$\frac{I_1}{V_2} = \frac{-G_L V_1}{V_2} = \frac{-j\omega C_1}{1 + \frac{j\omega(C_1+C_2)}{G_L}} \tag{2.108}$$

Finally, we have

$$\frac{V_1}{V_2} = \frac{j\omega C_1}{G_L + j\omega\left(C_1 + C_2\right)} = A \tag{2.109}$$

Using Equation 2.105b, one obtains

$$I_2 = -j\omega C_1 A V_2 + j\omega C_1 V_2 \tag{2.110}$$

Thus, the output admittance will be

$$Y_{\text{out}} = \frac{I_2}{V_2} = -j\omega C_1 A + j\omega C_1 = -j\omega C_1\left(\frac{j\omega C_1}{G_L + j\omega\left(C_1 + C_2\right)}\right) + j\omega C_1 \tag{2.111}$$

We may rearrange Equation 2.111 as

$$Y_{\text{out}} = j\omega C_1\left(1 - \frac{j\omega C_1}{G_L + j\omega\left(C_1 + C_2\right)}\right) = j\omega C_1\left(\frac{G_L + j\omega C_2}{G_L + j\omega\left(C_1 + C_2\right)}\right) \tag{2.112}$$

Multiplying the numerator and the denominator of Equation 2.112 by the denominator's conjugate, one obtains

$$Y_{\text{out}} = j\omega C_1\left(\frac{G_L^2 + \omega^2 C_2\left(C_1 + C_2\right) - j\omega C_1 G_L}{G_L^2 + \omega^2(C_1 + C_2)^2}\right) \tag{2.113}$$

Now, considering a high quality factor for this circuit, it is imposed that the short-circuit quality factor should be larger than unity

$$\left(C_1 + C_2\right)\omega \gg G_L \tag{2.114}$$

Regarding Equation 2.109, then the voltage ratio would be approximated by

$$A = \frac{V_1}{V_2} \approx \frac{C_1}{C_1 + C_2} \tag{2.115}$$

Therefore, considering the fact that $(C_1 + C_2)^2 \omega^2 \gg G_L^2$, Equation 2.113 is reduced to Equation 2.116:

$$Y_{\text{out}} \approx \left(\frac{C_1}{C_1 + C_2}\right)^2 G_L + j\omega \frac{C_1 C_2}{C_1 + C_2} = \frac{G_L}{m^2} + j\omega C_{\text{total}} \tag{2.116}$$

where $m = \frac{C_1 + C_2}{C_1}$. Finally, the overall quality factor will be

$$Q = R_{\text{eq}} C_{\text{total}} \omega = \frac{1}{\left(\frac{C_1}{C_1 + C_2}\right)^2 G_L} \cdot \frac{C_1 C_2}{C_1 + C_2} \omega = \frac{(C_1 + C_2)\omega}{G_L} \cdot \frac{C_2}{C_1} \tag{2.117}$$

If $(C_1 + C_2)\omega > G_L$ and $C_2 > C_1$, and the quality factor is greater than 10, the equivalent circuit shown in Figure 2.51 will be valid. However, if the quality factor is less than 10, the precise relation for the calculation of the admittance (Equation 2.113) should be used.

The same method applies for inductive transformers. Consider Figure 2.54.
Impedance parameters of Figure 2.54 can be derived as

$$\begin{pmatrix} V_1 \\ V_2 \end{pmatrix} = j\omega \begin{pmatrix} L_2 & L_2 \\ L_2 & L_1 + L_2 \end{pmatrix} \begin{pmatrix} I_1 \\ I_2 \end{pmatrix} \tag{2.118}$$

Or in expanded form

$$V_1 = j\omega L_2 I_1 + j\omega L_2 I_2 \tag{2.119a}$$
$$V_2 = j\omega L_2 I_1 + j\omega (L_1 + L_2) I_2 \tag{2.119b}$$

The loaded circuit for inductive transformer is shown in Figure 2.55.
At the input, we have

$$I_1 = -G_L V_1 \tag{2.120}$$

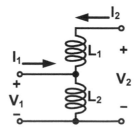

Figure 2.54: Step-up inductive voltage transformer.

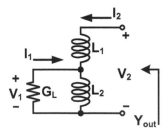

Figure 2.55: Step-up inductive voltage transformer with loaded input.

Using Equation 2.118, we can write

$$V_1 = j\omega L_2 \left(-G_L V_1 + I_2\right) \tag{2.121}$$

Thus, we will have

$$\frac{V_1}{I_2} = \frac{j\omega L_2}{1 + j\omega L_2 G_L} = B \tag{2.122}$$

Again, using Equation 2.119b gives

$$V_2 = j\omega L_1 I_2 + B I_2 \tag{2.123}$$

Finally, the output admittance will be

$$Y_{\text{out}} = \frac{I_2}{V_2} = \frac{1}{j\omega L_1 + \frac{j\omega L_2}{1 + j\omega L_2 G_L}} = \frac{1 + j\omega L_2 G_L}{j\omega L_2 + j\omega L_1 - \omega^2 L_1 L_2 G_L} \tag{2.124}$$

Given the fact that, in the tapped inductor transformer, the short-circuit quality factor should be greater than unity, it is deduced that

$$\frac{L_1 + L_2}{L_1 L_2 \omega} > G_L \tag{2.125}$$

After some manipulations, Equation 2.124 will be reduced to Equation 2.126

$$Y_{\text{out}} \approx \frac{1}{j\left(L_1 + L_2\right)\omega} + \left(\frac{L_2}{L_1 + L_2}\right)^2 G_L \tag{2.126}$$

As stated earlier in capacitive step-up transformer, with the condition stated in relation 2.125, if the quality factor of the inductive transformer is greater than 10 and $L_1 > L_2$, the above approximation will be valid, and the equivalent circuit shown in Figure 2.51 can be used. Otherwise, the precise relation for the admittance (Equation 2.124) should be employed.

Example 2.7 Calculate the output impedance of the circuit depicted in Figure 2.56 at 1 GHz for two cases

(a) $R = 50\,\Omega$ and $C_1 = C_2/9 = 3.18\,\text{pF}$.

(b) $R = 50\,\Omega$ and $C_1 = C_2/9 = 0.318\,\text{pF}$.

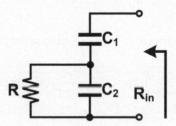

Figure 2.56: Calculation of the output impedance of the loaded capacitive transformer.

Solution:

(a) Here the short-circuit quality factor becomes

$$\frac{(C_1 + C_2)\,\omega}{G_L} = 10 \tag{2.127}$$

Equation 2.113 gives an exact value of the output admittance as

$$Y_{\text{out}} = j\omega C_1 \left(\frac{G_L^2 + \omega^2 C_2 (C_1 + C_2) - j\omega C_1 G_L}{G_L^2 + \omega^2 (C_1 + C_2)^2} \right) = 0.0002 + j0.0180 \tag{2.128}$$

$$\text{or} \quad Z_{\text{out}} = \frac{1}{Y_{\text{out}}} = 0.6172 - j55.55$$

If we use the approximation stated in Equation 2.116, we reach to

$$Y_{\text{out}} = \left(\frac{C_1}{C_1 + C_2} \right)^2 G_L + j\omega \frac{C_1 C_2}{C_1 + C_2} = 0.0002 + j0.01797 \tag{2.129}$$

$$\text{or} \quad Z_{\text{out}} = \frac{1}{Y_{\text{out}}} = 0.6193 - j55.64$$

As it is obvious from Equations 2.128 and 2.129, the results are quite close to each other.

(b) Here the short circuit quality factor becomes

$$\frac{(C_1 + C_2)\,\omega}{G_L} = 1 \tag{2.130}$$

Equation 2.113 gives the exact value of the output admittance as

$$Y_{out} = j\omega C_1 \left(\frac{G_L^2 + \omega^2 C_2 (C_1 + C_2) - j\omega C_1 G_L}{G_L^2 + \omega^2 (C_1 + C_2)^2} \right) = 0.0001 + j0.0018 \quad (2.131)$$

$$\text{or} \quad Z_{out} = \frac{1}{Y_{out}} = 30.77 - j553.8$$

If we use the approximation stated in Equation 2.116, we reach to

$$Y_{out} = \left(\frac{C_1}{C_1 + C_2} \right)^2 G_L + j\omega \frac{C_1 C_2}{C_1 + C_2} = 0.0002 + j0.0018 \quad (2.132)$$

$$\text{or} \quad Z_{out} = \frac{1}{Y_{out}} = 60.975 - j548.78$$

As it is obvious from Equations 2.131 and 2.132, the results don't match completely, that is, the approximation is not valid for a low quality factor circuit. ∎

2.13   Analysis of Large-signal Loop Gain of an Oscillator

Now that we have got acquainted to the large-signal transconductance of a nonlinear device, we are able to analyze the oscillator behavior more precisely. Consider Figure 2.57 where parasitic capacitances are ignored for the sake of simplicity.

The AC model of Figure 2.57 is shown in Figure 2.58.

Here the resistance R_P represents the RF losses of the inductor and the capacitors. Using the equivalent circuit of the capacitive impedance transformer of Figure 2.51, we can reduce Figure 2.58 to the equivalent circuit of Figure 2.59.

The output collector voltage of the oscillator can be calculated as

$$V_{out} = \frac{G_m V_i}{j\omega C_{eq} + \frac{1}{j\omega L_P} + \frac{1}{R_P} + n^2 G_{in}} \quad (2.133)$$

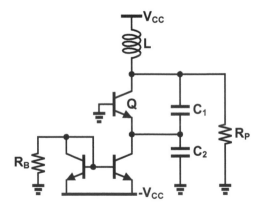

Figure 2.57: A Colpitts oscillator core circuit.

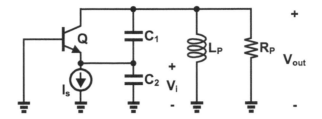

Figure 2.58: An AC model of the oscillator core of Figure 2.57.

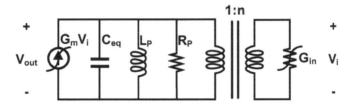

Figure 2.59: Equivalent circuit of the Colpitts oscillator.

Here, we have

$$n = \frac{C_1}{C_1 + C_2}, \quad C_{eq} = \frac{C_1 C_2}{C_1 + C_2} \tag{2.134}$$

The input emitter voltage V_i is (note that here $m < 1$)

$$V_i = n V_{out} \tag{2.135}$$

Now, replacing V_{out} in Equation 2.133, we arrive at an expression for the closed-loop gain of the oscillator:

$$A_{CL}(j\omega) = \frac{n G_m}{j\omega C_{eq} + \frac{1}{j\omega L_P} + \frac{1}{R_P} + n^2 G_{in}} = 1 \tag{2.136}$$

Equation 2.136 describes the oscillation condition of the oscillator (Barkhausen's oscillation condition). Indeed, the large-signal input at the emitter is amplified by the large-signal gain of the oscillator and the output voltage at the collector is divided by the capacitive division ratio of the transformer m, and is fed back to the emitter. The whole loop gain in the large-signal regime and at the oscillation frequency should become equal to unity ($1\angle0$). Note that in the small-signal regime, the oscillation begins with the small-signal noise voltage at the emitter, which is amplified by the small-signal gain of the transistor (note that normally in every oscillator the small-signal gain is much larger than the large-signal gain). As the feedback signal grows, the small signal-gain is gradually compressed to its large-signal value. Furthermore, the large-signal input conductance of the emitter can be approximated by the large-signal transconductance divided by α. That is

$$G_m = \frac{G_{in}}{\alpha} \tag{2.137}$$

As it is seen in Equation 2.136, the right-hand side of the equation is purely real, so we can separate the real and the imaginary parts of Equation 2.136 and obtain the following pair of equations:

$$\omega C_{eq} - \frac{1}{\omega L_P} = 0 \tag{2.138}$$

$$G_m = \frac{\frac{1}{R_P}}{n\left(1 - \frac{n}{\alpha}\right)} \tag{2.139}$$

Equation 2.138 gives the oscillation frequency and by Equation 2.139, we can obtain the oscillation amplitude through the large-signal transconductance. Here our focus was on the first harmonic, because other harmonics are attenuated by the high-Q tuned resonant circuit to some extent.

Example 2.8 Consider Figure 2.60, where the transistor has a current gain $\alpha = 0.99$ and a parasitic collector–base capacitance of 0.2 pF and a parasitic base–emitter capacitance of 5 pF. Given the 1.5 mA emitter current source, compute the oscillation amplitude and the oscillation frequency in this circuit.

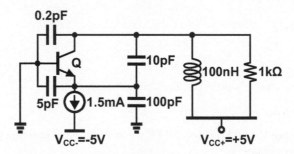

Figure 2.60: A Colpitts oscillator with parasitic capacitances.

Solution:
An important point in this example is the absorption of parasitic capacitances in the resonant circuit. First, we calculate the capacitive transformer ratio n as

$$n = \frac{10}{10 + 100 + 5} = 0.087 \tag{2.140}$$

From Equation 2.139, we obtain

$$G_{m_1} = \frac{\frac{1}{1000}}{0.087\left(1 - \frac{0.087}{0.99}\right)} = 12.6\,\text{m}\mho \tag{2.141}$$

We can easily calculate the small-signal g_m as

$$g_{mQ} = 0.99 \times \frac{1.5\,mA}{26\,mV} = 57.1\,m\mho \tag{2.142}$$

Thus, using Equations 2.141 and 2.142, we will have

$$\frac{G_{m1}}{g_{mQ}} = 0.220 \tag{2.143}$$

Using Equation 2.143 and Table. 2.4, x can be obtained as

$$x = 8.5\,V_t = 8.5\,(26\,mV) = 221\,mV \tag{2.144}$$

The voltage obtained from Equation 2.144 is that of the base–emitter junction. The collector voltage is higher by the ratio of $1/m$, thus

$$V_C = \frac{x}{n} = \frac{0.221}{0.087} = 2.54\,V \tag{2.145}$$

By assuming the supply voltage equal to 5 V, the output voltage will be

$$V_{C_{total}} = 5 + 2.54\cos\left(2\pi f_0 t + \varphi\right) \tag{2.146}$$

Note that the parasitic collector–base capacitance will be added to the equivalent capacitance of the capacitive divider. So, the total capacitance would become

$$C_{total} = \frac{C_1\,(C_2 + C_{BE})}{C_1 + C_2 + C_{BE}} + C_{BC} = 9.33\,pF \tag{2.147}$$

The oscillation frequency for Equation 2.146 will be

$$f_0 = \frac{1}{2\pi} \frac{1}{\sqrt{100\,nH \times 9.33\,pF}} = 164.77\,MHz \tag{2.148}$$

Here we verify the required condition for the capacitor tapped transformer model that is

$$\frac{(C_1 + C_2)\,\omega}{G_L} = 114 \tag{2.149}$$

which is much larger than unity and therefore, the tapped transformer equivalent circuit is valid here. ∎

2.13.1 Increasing the Quality Factor and the Frequency Stability with a Crystal

Consider Figure 2.61.

As it is depicted in Figure 2.61, for increasing the quality factor, a crystal is placed in series within the feedback loop. Normally, the parallel RLC circuit's resonance frequency is chosen the same as that of the crystal. However, if the frequency of

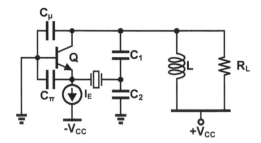

Figure 2.61: A Colpitts oscillator with series crystal.

resonant circuit is slightly different from the crystal series resonance, the oscillation frequency will change to satisfy the Barkhausen's oscillation condition. Any phase change in the loop should be compensated by the crystal. However, due to sharp phase characteristic of the crystal, this will result in a very small frequency change. It is possible to show the series crystal equivalent model as presented in Figure 2.62.

Now, we are going to investigate the resonant circuit detuning more precisely. This will result in Q degradation. Figure 2.63 shows the crystal impedance behavior about its series resonance.

Similarly, the impedance of the parallel resonant circuit is shown in Figure 2.64.

In this case, assume that the resonant frequency of parallel RLC is higher than the crystal resonant frequency. The circuit will oscillate near the crystal resonant frequency (f_S) due to its higher quality factor. The tank circuit introduces a finite phase change as $\Delta\Phi$ in the loop gain. Thus, the crystal phase characteristics should compensate this phase by introducing $-\Delta\Phi$ in the loop gain to maintain the Barkhausen's oscillation condition. While the phase characteristics of the crystal is sharp, this phase change does not alter the oscillation frequency significantly. It is noteworthy that the amplitude of oscillation might be altered a little bit as well. The interested reader is referred to section 2.20 for further details.

In another topology, the crystal might be used as an inductive reactance within the resonant circuit of a Colpitts oscillator as depicted in Figure 2.65.

Here, with a slight shift of the oscillation frequency with respect to the crystals' resonant frequency, the Barkhausen oscillation condition can be satisfied. Assuming the oscillation frequency near to f_s, and further neglecting the effect of C_P, the crystal impedance can be represented by

$$Z_X = R + jX \tag{2.150}$$

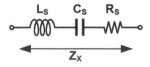

Figure 2.62: An equivalent model for series crystal.

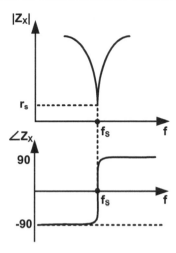

Figure 2.63: Impedance behavior of the crystal about its series resonance.

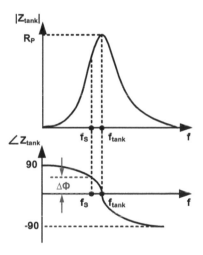

Figure 2.64: Variations of the impedance of the parallel resonant circuit about the resonance frequency.

where

$$R = r_s \tag{2.151a}$$

$$X = 2Q_0 r_s \frac{\Delta f}{f_s} \tag{2.151b}$$

Here r_s is the crystal's series resistance and $Q_0 = \frac{L_s \omega_s}{r_s}$ is the crystal's unloaded quality factor. For a series resonant circuit, one can write

$$Z_X = r_s + jL_s\omega - j\frac{1}{C_\omega} \tag{2.152}$$

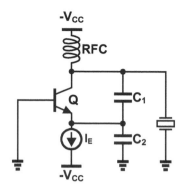

Figure 2.65: Colpitts-like crystal oscillator; the crystal operates in an inductive mode.

Or

$$Z_X = r_s \left[1 + jQ_0 \left(\frac{\omega}{\omega_s} - \frac{\omega_s}{\omega} \right) \right] = r_s \left[1 + j2Q_0 \frac{\Delta\omega}{\omega_s} \right] \tag{2.153}$$

Here, $\omega = \omega_s + \Delta\omega$. For the resonance condition, we should have

$$X = \frac{C_1 + C_2}{2\pi f_s C_1 C_2} \tag{2.154}$$

or

$$\frac{C_1 + C_2}{2\pi f_s C_1 C_2} = 2Q_0 r_s \frac{\Delta f}{f_s} \tag{2.155}$$

The point is that we have put f_s instead of f_0 in the left-hand side of Equation 2.155. The reason being the fact that Δf is extremely small compared to f_s. By resolving Equation 2.155, one simply obtains Δf and the oscillation frequency is determined as

$$f_0 = f_s + \Delta f \tag{2.156}$$

The oscillation amplitude could be obtained from the following

$$G_m(x) = \frac{mG_X}{\left(1 - \frac{1}{m\alpha}\right)} \tag{2.157}$$

where

$$m = \frac{C_1 + C_2}{C_1} \tag{2.158}$$

and the crystal conductance is

$$G_X = \frac{R}{R^2 + X^2} \tag{2.159}$$

With the above-mentioned procedure, given the crystal parameters (r_s, Q_0, and f_s), one can easily determine the oscillation frequency and the oscillation amplitude. The same procedure can be used for a Hartley-like crystal oscillator.

2.13.2 Oscillator Harmonics Calculation

It is also instructive to calculate other harmonics in a bipolar transistor oscillator circuit. The oscillator voltage harmonics are proportional to the harmonic currents as well as the load impedance at the harmonic frequencies. As such, the ratio of the second harmonic to the first harmonic can be written as Equation 2.160:

$$\frac{V_2}{V_1} = \frac{I_2(x)}{I_1(x)} \cdot \frac{Z_L(2\omega_0)}{Z_L(\omega_0)} \approx \frac{I_2(x)}{I_1(x)} \frac{2}{3jQ} \tag{2.160}$$

Equation 2.160 describes the amplitude ratio as well as the phase difference. A more general form of Equation 2.160 for the kth harmonic will be

$$\frac{V_k}{V_1} = \frac{I_k(x)}{I_1(x)} \cdot \frac{Z_L(jk\omega_0)}{Z_L(j\omega_0)} \approx \frac{I_k(x)}{I_1(x)} \frac{k}{jQ(k^2-1)} \tag{2.161}$$

It is possible to calculate the ratio of each harmonic to the main harmonic by Equation 2.161. In Equation 2.161, the load is impedance and is the one which is seen at the collector of the transistor. Given the fact that harmonic currents are smaller than the fundamental current and Q is large, it is obvious from Equation 2.161 that the harmonic voltages are quite smaller than the fundamental voltage.

In order to extract the oscillator signal, we should not load the tank circuit directly because its quality factor will be degraded. It is possible to use either an impedance transformer or extract the output from the emitter (where there is a low output impedance). This point is illustrated in Figure 2.66.

As it is stated earlier, the output impedance of Figure 2.66 will decrease by p^2 and could reach to 50Ω. An appropriate value of p ($p < 1$) will not degrade the quality factor of tank circuit as much.

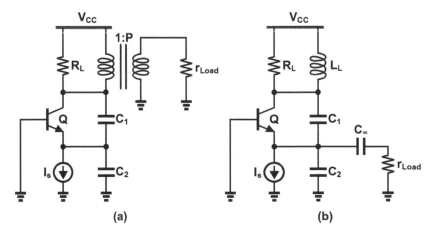

Figure 2.66: Extracting the output signal of an oscillator without degrading its quality factor.

2.14 Colpitts Oscillator with Emitter Degeneration

Consider Figure 2.67 where the equivalent bias circuit of the transistor is shown at the input.

In this configuration, the operating point of the transistor will be modified by the input large-signal by a certain coefficient. That is the input DC current and consequently the emitter DC current will increase due to the large-signal imposition. It can be shown that the overall DC current in the presence of the large-signal input voltage will take the following form

$$I_{E_0} = I_{EQ} \left(1 + \frac{\ln\left(I_0\left(x\right)\right)}{\frac{V_\lambda}{V_t}} \right) \tag{2.162}$$

where $I_0(x)$ is the zeroth-order modified Bessel function of the first kind, I_{EQ} is the operating point DC current of the emitter (in the absence of the large-signal), and V_λ is defined as

$$V_\lambda = V_{\text{dropped}_{\text{base}}} + V_{\text{dropped}_{\text{emitter}}} = R_B I_{BQ} + R_E I_{EQ} \approx R_E I_{EQ} \tag{2.163}$$

Actually, V_λ is the DC voltage drop across R_E. Consequently, the large-signal transconductance of the transistor will be modified by the same coefficient as in Equation 2.164:

$$G_m = \frac{\alpha I_{EQ}}{V_t} \left(1 + \frac{\ln(I_0(x))}{\frac{V_\lambda}{V_t}} \right) \frac{2I_1(x)}{xI_0(x)} = g_{mQ} \left(1 + \frac{\ln(I_0(x))}{\frac{V_\lambda}{V_t}} \right) \frac{2I_1(x)}{xI_0(x)} \tag{2.164}$$

As we increase R_E, the voltage drop across it for a constant current will increase (V_λ increases and the coefficient approaches unity), and as a result, it acts approximately as a current source. The evolution of the transconductance of a bipolar transistor stage biased with an emitter resistor instead of a current source is depicted in Figure 2.68.

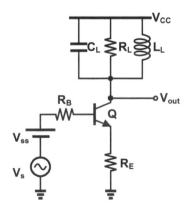

Figure 2.67: Common-emitter amplifier with emitter degeneration.

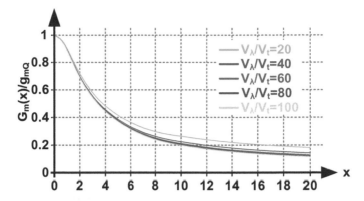

Figure 2.68: The evolution of the large-signal transconductance of a bipolar transistor biased with an emitter resistor instead of a current source (normalized DC voltage drop across the emitter resistor as the parameter).

2.15 MOS Stage Large-Signal Transconductance

In a similar derivation, it is possible to compute the large-signal transconductance of a MOS transistor having its $I-V$ characteristics. This point is shown in Figure 2.69.

A MOS tuned amplifier stage is depicted in Figure 2.70. We assume a large signal is applied to the gate input. The large capacitors C_G and C_S are considered as AC short-circuit at the RF carrier frequency. The output RLC circuit is considered to have a high quality factor and is tuned to the carrier frequency.

Assuming above the threshold bias voltage and a square law characteristics in the saturation region, one can compute the drain source current from the square law characteristics:

$$I_{DS} = k(V_{GS} - V_{TH})^2 \tag{2.165}$$

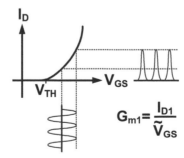

Figure 2.69: Typical $I-V$ characteristic of MOS transistor ($\tilde{V}_{GS}$ is the gate–source's AC voltage phasor).

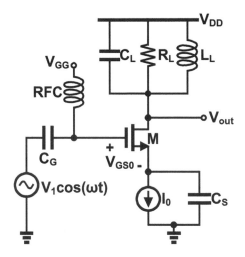

Figure 2.70: Constant current MOS stage tuned amplifier for computation of the large-signal transconductance.

The drain–source DC current at the bias voltage is

$$I_{DS0} = k(V_{GS0} - V_{TH})^2 \tag{2.166}$$

The small-signal transconductance can be computed as

$$g_m = 2k(V_{GS0} - V_{TH}) \tag{2.167}$$

The total drain–source current under the large-signal excitation can be computed as

$$i_{DS} = k(V_{GS0} - V_{TH} + V_1 \cos(\omega_0 t))^2 \tag{2.168}$$

The DC, the fundamental, and the second harmonic currents are obtained as

$$i_{DS} = k\left[(V_{GS0} - V_{TH})^2 + \frac{V_1^2}{2}\right] + 2k(V_{GS0} - V_{TH})V_1 \cos(\omega_0 t) + k\frac{V_1^2}{2}\cos(2\omega_0 t) \tag{2.169}$$

Now, neglecting the second harmonic, one can express the output drain–source current as

$$i_{DS} \approx I_0 \left[1 + \frac{2(V_{GS0} - V_{TH})V_1 \cos(\omega_0 t)}{\left[(V_{GS0} - V_{TH})^2 + \frac{V_1^2}{2}\right]}\right] \tag{2.170}$$

where I_0 is the current source's bias current. Now, the large-signal transconductance is defined as

$$G_m = \frac{I_1}{V_1} = I_0 \left[\frac{2(V_{GS0} - V_{TH})}{\left[(V_{GS0} - V_{TH})^2 + \frac{V_1^2}{2}\right]}\right] \tag{2.171}$$

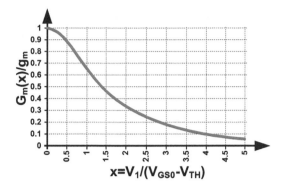

Figure 2.71: Normalized transconductance variations of a MOS tuned amplifier stage as a function of the normalized input voltage.

The normalized large-signal transconductance becomes

$$\frac{G_m}{g_m} = \frac{I_0}{k\left[(V_{GS0} - V_{TH})^2 + \frac{v_1^2}{2}\right]} \qquad (2.172)$$

Or in another form

$$\frac{G_m}{g_m} = \frac{I_0}{k(V_{GS0} - V_{TH})^2\left[1 + \frac{v_1^2}{2(V_{GS0}-V_{TH})^2}\right]} \qquad (2.173)$$

Given the fact that at the operating point, one can write

$$I_0 \approx k(V_{GS0} - V_{TH})^2 \qquad (2.174)$$

The normalized large-signal transconductance is simplified to the following

$$\frac{G_m}{g_m} = \frac{1}{\left[1 + \frac{v_1^2}{2(V_{GS0}-V_{TH})^2}\right]} \qquad (2.175)$$

Let $x = \frac{v_1}{V_{GS0}-V_{TH}}$,

$$\frac{G_m(x)}{g_m} = \frac{1}{1 + \frac{x^2}{2}} \qquad (2.176)$$

The normalized MOS stage transconductance is depicted in Figure 2.71.

This large-signal transconductance can be employed in the MOS oscillator circuit design/analysis for computation of the amplitude of oscillation.

2.16 Differential MOS Stage Large-Signal Transconductance

A differential MOS tuned amplifier is depicted in Figure 2.72.

Considering a square law transfer characteristic of the MOS transistors and assuming that the differential voltage is equally divided between the pair of transistors (at least for a limited range of differential voltage), one can write

$$I_1 = k\left(V_{GS0} - V_{TH} + \frac{v}{2}\right)^2 \quad \text{for} \quad \left|\frac{v}{2}\right| < |V_{GS0} - V_{TH}| \tag{2.177}$$

$$I_2 = k\left(V_{GS0} - V_{TH} - \frac{v}{2}\right)^2 \quad \text{for} \quad \left|\frac{v}{2}\right| < |V_{GS0} - V_{TH}| \tag{2.178}$$

Then

$$\frac{I_1}{I_2} = \frac{\left(V_{GS0} - V_{TH} + \frac{v}{2}\right)^2}{\left(V_{GS0} - V_{TH} - \frac{v}{2}\right)^2} \tag{2.179}$$

Given

$$I_1 + I_2 = I_0 \tag{2.180}$$

The difference in current becomes

$$I_1 - I_2 = 2k\left(V_{GS0} - V_{TH}\right)v \tag{2.181}$$

The normalized difference current can be calculated as

$$\frac{\Delta I}{I_0} = \frac{\left(V_{GS0} - V_{TH}\right)v}{\left(V_{GS0} - V_{TH}\right)^2 + \frac{v^2}{4}} \tag{2.182}$$

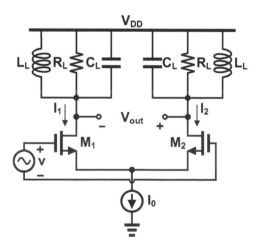

Figure 2.72: A differential MOS pair tuned amplifier.

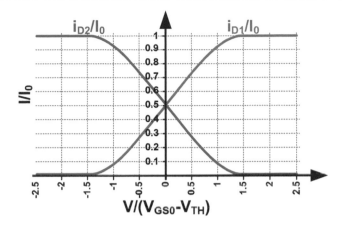

Figure 2.73: Variations of the differential pair drain currents as a function of the normalized differential voltage in a MOS differential pair.

Either of the drain currents can be expressed as

$$I_1 = \frac{I_0}{2} \left[1 + \frac{\frac{v}{V_{GS0} - V_{TH}}}{1 + \frac{v^2}{4(V_{GS0} - V_{TH})^2}} \right] \tag{2.183}$$

$$I_2 = \frac{I_0}{2} \left[1 - \frac{\frac{v}{V_{GS0} - V_{TH}}}{1 + \frac{v^2}{4(V_{GS0} - V_{TH})^2}} \right] \tag{2.184}$$

The normalized difference current can be expressed as

$$\Delta i_{DD} = I_0 \frac{\frac{v}{(V_{GS0} - V_{TH})}}{1 + \frac{v^2}{4(V_{GS0} - V_{TH})^2}} \tag{2.185}$$

This model describes approximately the nonlinear behavior of a differential MOS transistor pair. The variations of a real differential pair drain currents as a function of the differential input voltage are depicted in Figure 2.74. Although the mathematical expressions are quite different, it is noticeable that these currents' variations are quite similar to those of the bipolar differential pair.

Note that Equations 2.183 through 2.185 are valid for $\left| \frac{v}{V_{GS0} - V_{TH}} \right| \leq 2$. If $\frac{v}{V_{GS0} - V_{TH}} > 2$, then $I_1 = I_0$ and $I_2 = 0$, and if $\frac{v}{V_{GS0} - V_{TH}} < -2$, then $I_1 = 0$ and $I_2 = I_0$. As such, a nonlinear transfer characteristic has been specified for a MOS differential pair for the whole span of possible input voltages.

The differential pair small-signal transconductance becomes

$$g_{md} = \frac{I_0}{V_{GS0} - V_{TH}} \tag{2.186}$$

For large-signal AC drive, we can extract the large-signal transconductance of the differential MOS stage:

$$\Delta i_{DD} = I_0 \frac{\frac{V_1 \cos(\omega t)}{(V_{GS0} - V_{TH})}}{1 + \frac{V_1^2 \cos^2(\omega t)}{4(V_{GS0} - V_{TH})^2}} \tag{2.187}$$

Defining the normalized AC input voltage as

$$x = \frac{V_1}{V_{GS0} - V_{TH}} \tag{2.188}$$

The harmonic components of the large-signal output current can be computed as

$$b_n(x) = \frac{1}{\pi} \int_{-\pi}^{\pi} \frac{x \cos\theta}{1 + \frac{x^2}{4}\cos^2\theta} \cos n\theta \, d\theta \tag{2.189}$$

Care should be taken that these computations are valid for $x \leq 2$. Note that given the fact that the differential MOS pair transfer characteristic has an odd symmetry, $b_n(x)$ functions would be zero for even values of n. The fundamental harmonic current becomes

$$I_1 = I_0 b_1(x) \tag{2.190}$$

Now the large-signal differential transconductance can be calculated as

$$G_{md}(x) = \frac{I_1}{V_1} = \frac{I_0 b_1(x)}{(V_{GS0} - V_{TH})x} = g_{md} \frac{b_1(x)}{x} \tag{2.191}$$

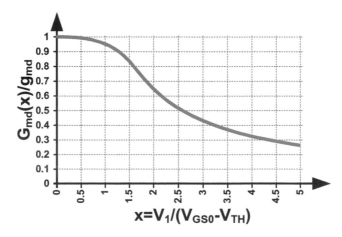

Figure 2.74: Normalized transconductance variations of a MOS differential tuned amplifier stage as a function of the normalized input voltage.

or

$$\frac{G_{md}(x)}{g_{md}} = \frac{b_1(x)}{x} \tag{2.192}$$

For the case where the large input signal does not satisfy the condition $x \leq 2$, the transistor pair drains would switch between zero and I_0. As such, the first harmonic currents would have a value as

$$I_1 = \frac{4}{\pi} I_0 \tag{2.193}$$

The large-signal transconductance becomes

$$G_{md}(x) = \frac{I_1}{V_1} \approx \frac{4}{\pi} \frac{I_0}{V_1} = \frac{4}{\pi} \frac{\frac{I_0}{V_{GS0}-V_{TH}}}{\frac{V_1}{V_{GS0}-V_{TH}}} = \frac{4}{\pi x} g_{md} \tag{2.194}$$

or

$$\frac{G_{md}(x)}{g_{md}} \approx \frac{4}{\pi x} \tag{2.195}$$

This is valid for $x \geq 2\sqrt{2}$.

As such, the overall normalized differential MOS stage transconductance is depicted in Figure 2.74.

This large-signal transconductance can be employed in the differential MOS oscillator circuit design/analysis for computation of the amplitude of oscillation.

2.17 An Oscillator With a Hypothetical Model

Consider Figure 2.75 where a hypothetical active element is shown within a Colpitts oscillator circuit. The $I - V$ characteristics of the hypothetical element, about the bias point, are given in Equation 2.196

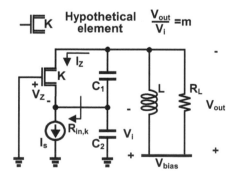

Figure 2.75: Oscillator with a hypothetical amplifier.

$$I_Z = B_0 + B_1 V_Z + B_2 V_Z^2 + B_3 V_Z^3 \tag{2.196}$$

In Equation 2.196, V_Z is considered as a sinusoidal signal as $V_Z = V_i \cos \omega t$, so we have

$$I_Z = B_0 + B_1 V_i \cos(\omega_0 t) + \frac{B_2}{2} V_i^2 (1 + \cos(2\omega_0 t)) + B_3 V_i^3 \cos^3(\omega_0 t) \tag{2.197}$$

We are looking for a gain at the fundamental frequency in Equation 2.197. Thus, one may obtain the first harmonic large-signal transconductance as

$$G_{m_1} = \frac{B_1 V_i + \frac{3}{4} B_3 V_i^3}{V_i} = B_1 + \frac{3}{4} B_3 V_i^2 \tag{2.198}$$

Therefore, the large-signal loop gain will be

$$A_{Ls} = G_{m_1} \left(R_L || m^2 R_{in,k} \right) \frac{1}{m} = \left(B_1 + \frac{3}{4} B_3 V_i^2 \right) \left(R_L || m^2 R_{in,k} \right) \frac{1}{m} \tag{2.199}$$

Here, $m = \frac{C_1 + C_2}{C_1}$. Or the oscillation condition becomes explicitly as

$$\frac{m R_L R_{in}}{R_L + m^2 R_{in}} \left(B_1 + \frac{3}{4} B_3 V_i^2 \right) = 1 \tag{2.200}$$

and

$$\frac{1}{L \omega_0} - \frac{C_1 C_2 \omega_0}{C_1 + C_2} = 0 \tag{2.201}$$

2.18   A MOS Oscillator with Differential Gain Stage

Consider Figure 2.76 where a differential MOS oscillator is depicted with a transformer-type feedback.

The $I - V$ characteristics of the circuit are also shown in Figure 2.76. As it was described in section 2.17, the large-signal transconductance G_m of this stage can be obtained from the $I - V$ characteristics of the differential pair. The oscillation condition for this oscillator becomes as described in Equation 2.202:

$$A_{Ls}(j\omega) = \frac{m G_m}{jC\omega - \frac{j}{L\omega} + \frac{1}{R} + m^2 Y_{in}} = 1\angle 0 \tag{2.202}$$

where Y_{in} is the input admittance of the differential pair. The above equation describes the closed-loop gain of the oscillator. Considering the fact that the input admittance of

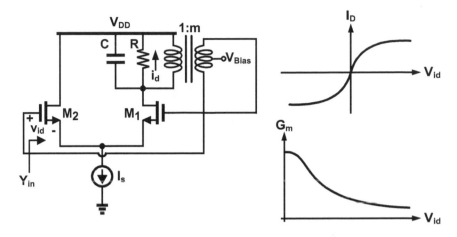

Figure 2.76: Differential amplifier for the oscillator.

the MOS differential pair is mainly capacitive, at the resonant frequency, regarding the
closed loop gain, one can write

$$C\omega_0 - \frac{1}{L\omega_0} + m^2 Y_{in}(j\omega_0) = 0 \qquad (2.203)$$

$$A_{Ls}(j\omega_0) \approx mG_m R = 1 \qquad (2.204)$$

Having the large-signal transconductance of the differential MOS pair as a function of
the input fundamental harmonic voltage amplitude, the oscillation amplitude can be
computed from Equation 2.204.

2.19 Voltage-Controlled Oscillators

Voltage-controlled oscillators are widely used in RF communication circuits. VCO
is a block where we can change the oscillation frequency by a certain input voltage.
Historically, this was done with mechanically variable capacitances where a sample of
it is shown in Figure 2.77.

However, nowadays instead of VCOs frequency synthesizers are widely used. In
integrated circuits, we employ variable capacitances which are tunable with voltage.
Figure 2.78 shows a typical VCO where the transistors M_3 and M_4 are used as voltage-
controlled capacitors (the control voltage being V_C). These varactors would be in
parallel with L_1 and L_2, respectively, AC wise (C_1 is relatively large), and as such they
would directly affect the oscillation frequency along with C_3 and C_2. The transistors
M_1 and M_2 provide the required loop gain in this oscillator.

2.19.1 Different Types of Varactors and their Bias

Voltage variable capacitors can be implemented via reverse-biased diodes as well. The
area of the diode is sufficient to achieve a certain reverse capacitance. Figure 2.79
shows a typical $C - V$ characteristic of a diode.

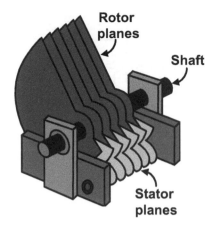

Figure 2.77: Mechanically variable capacitor.

The capacitor is biased by a negative voltage where the reverse DC current of the diode is negligible. The negative voltage should be less than the breakdown voltage of the diode junction. Moreover, it is possible to switch between capacitors to change the frequency coarsely. Figure 2.80 shows the implementation of a variable capacitance in an oscillator.

At low frequency, the bias current of the variable capacitors passes thorough the inductor (note that the inductor is short-circuit at low frequencies). The oscillation frequency of the circuit is determined by the resonance of the total capacitance of

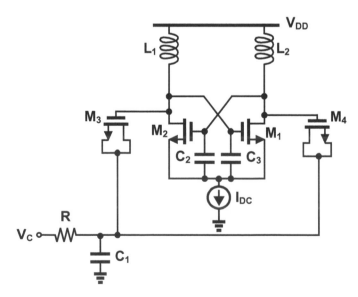

Figure 2.78: Cross-coupled VCO with voltage variable MOS capacitors.

C_1 and C_2 plus the capacitances of the varactor with the inductor, L. Furthermore, it is possible to modulate the frequency of oscillation by a time-varying voltage at the cathode of the varactors. The bias resistance of the varactors (R_{bias}) is assumed to be sufficiently large in order to avoid the loading of the tuned circuit or else an RFC should be added in the bias circuit. At high frequencies, the varactor capacitances are in series.

As Figure 2.79 suggests, the variable capacitor is nonlinear which may result in signal distortion (generation of RF harmonics). In Figure 2.80, the oscillation voltage is divided between the two varactors, resulting in better linearity. Indeed, a varactor pair allows for double AC voltage swing, and thus, a lower distortion can be achieved by a varactor pair at the output (with respect to a single varactor oscillator). In fact, the bias voltage across these varactors changes their values and these changes will result in resonant frequency variations. Three different kinds of VCOs with their varactor implementation are shown in Figure 2.81.

In Figure 2.81(a), the tuning voltage can select any value greater than zero and variable capacitance will experience the substrate noise. Tuning voltage in Figure 2.81(b) can have only values smaller than supply voltage for staying in reverse bias. It also experiences the supply voltage noise and its ripples. Finally, Figure 2.81(c) will have the same tuning voltage as Figure 2.81(b). In this case, from the supply point of view, the two varactors are in parallel; however, from the RF point of view, those are in series, therefore, the supply noise voltage is canceled out in the RF circuit. Furthermore, the nonlinear behavior of the VCO is ameliorated in this case.

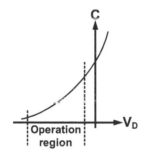

Figure 2.79: Capacitance variations of a diode as a function of its voltage.

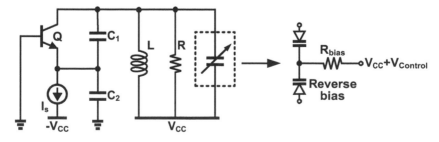

Figure 2.80: A Colpitts voltage-controlled oscillator using a pair of varactors.

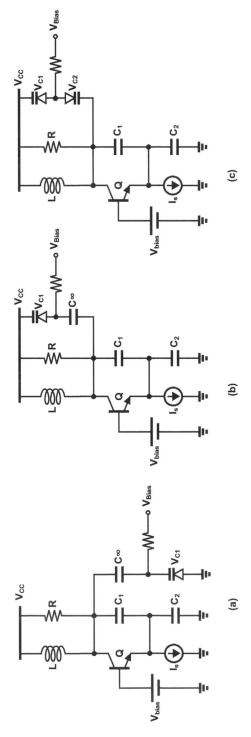

Figure 2.81: Voltage-controlled oscillators with different kinds of varactor implementations.

Example 2.9 Consider the given differential MOS pair in the circuit of Figure 2.82.

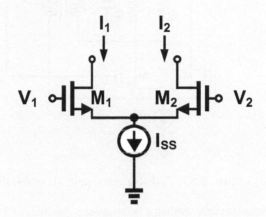

Figure 2.82: Differential MOS pair circuit.

(a) Considering the quadratic $I - V$ characteristic of MOS device in the following equation

$$I = k(V_{gs} - V_{th})^2, \quad V_d = V_1 - V_2, \quad k = \frac{1}{2}\mu_n C_{ox}\frac{w}{L} \tag{2.205}$$

prove that

$$I_2 = \frac{I_{SS}}{2}\left(1 - V_d\sqrt{\frac{2k}{I_{SS}}}\sqrt{1 - \frac{k}{2I_{SS}}V_d^2}\right) \tag{2.206}$$

(b) Now, suppose that the input is a differential signal with $V_i = V_m\cos(\omega t)$. Find the expansion of Equation 2.206, and then find the large-signal transconductance. (Large-signal transconductance is the ratio of the fundamental harmonic current to the input voltage amplitude.)

(c) Now, suppose that with the circuit of part (a), we have implemented the oscillator of Figure 2.83.

With the assumption of ideal transformer and with the derived equation for the large-signal transconductance, find the loop gain and then the oscillation criteria. Finally, with $R = 3\,\text{k}\Omega, k = 1\,\text{m}\frac{A}{V^2}$, and $I_{SS} = 1\,\text{mA}$, find the coupling coefficient n for the oscillation amplitude of 600 mV.

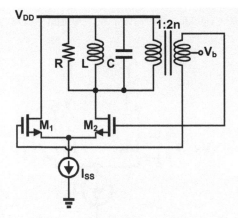

Figure 2.83: A differential pair MOS oscillator.

(d) Suppose that this oscillator is designed to oscillate at 1 GHz with the values of $L = 5\,\text{nH}$, and $C = 5\,\text{pF}$. To change the oscillator, to the VCO of Figure 2.84, we need to add the varactors in parallel to the capacitors. The characteristic of the varactor is shown in Figure 2.85.

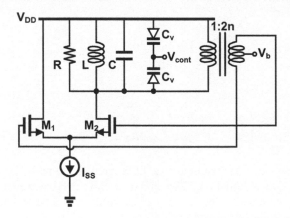

Figure 2.84: A VCO based on a differential pair.

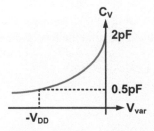

Figure 2.85: Varactor characteristics.

Find the frequency range of oscillation.
(e) Using the derived equation of part (a) for the amplitude of third harmonic and considering the frequency response of the resonant circuit, find the ratio of third harmonic to the fundamental harmonic.

Solution
(a) By writing KCL at the source of the transistor of Figure 2.82, we have

$$I_{SS} = I_1 + I_2 \tag{2.207a}$$

$$I_{1,2} = k(V_{GS_{1,2}} - V_{TH})^2 \tag{2.207b}$$

and using KVL, we reach to

$$-V_1 + V_{GS_1} - V_{GS_2} + V_2 = 0 \tag{2.208}$$

Equation 2.208 can be modified as

$$V_1 - V_2 = V_d = (V_{GS_1} - V_{TH}) - (V_{GS_2} - V_{TH}) = V_{GS_1} - V_{GS_2} = \sqrt{\frac{I_1}{k}} - \sqrt{\frac{I_2}{k}} \tag{2.209}$$

Substituting the value for I_2 from Equation 2.207a into Equation 2.209, we obtain the following equation

$$I_2^2 - I_2 I_{SS} + \frac{(I_{SS} - kV_d^2)^2}{4} = 0 \tag{2.210}$$

The roots of this equation will have the following forms

$$I_2 = \frac{I_{SS}}{2}\left(1 - V_d\sqrt{\frac{2k}{I_{SS}}}\sqrt{1 - \frac{k}{2I_{SS}}V_d^2}\right) \tag{2.211}$$

$$I_1 = \frac{I_{SS}}{2}\left(1 + V_d\sqrt{\frac{2k}{I_{SS}}}\sqrt{1 - \frac{k}{2I_{SS}}V_d^2}\right) \tag{2.212}$$

Note that the above two equations are valid for $V_d < \sqrt{\frac{2I_{SS}}{k}}$.

(b) We can use the approximation of Equation 2.213 for a relatively large differential signal (however, satisfying $V_d < \frac{1}{2}\sqrt{\frac{2I_{SS}}{k}}$) as

$$\sqrt{1 - \frac{k}{2I_{SS}}V_d^2} \simeq 1 - \frac{k}{4I_{SS}}V_d^2 \tag{2.213}$$

Therefore, the drain current of M_2 for relatively large signals will be

$$I_2 \approx \frac{I_{SS}}{2}\left(1 - V_d\sqrt{\frac{2k}{I_{SS}}} + \sqrt{\frac{k^3}{8I_{SS}^3}}V_d^3\right) \tag{2.214}$$

For sinusoidal input (i.e., $V_d = V_m \cos(\omega t)$), I_2 becomes

$$I_2 = \frac{I_{SS}}{2} - \frac{I_{SS}}{2}\sqrt{\frac{2k}{I_{SS}}}V_m\left(1 - \frac{3k}{16I_{SS}}V_m^2\right)\cos(\omega t) \tag{2.215}$$

$$+ \frac{I_{SS}}{32}\sqrt{\frac{2k}{I_{SS}}}\frac{k}{I_{SS}}V_m^3 \cos(3\omega t)$$

Thus, the large-signal transconductance will be

$$G_m = -\frac{\frac{I_{SS}}{2}\sqrt{\frac{2k}{I_{SS}}}V_m\left(1 - \frac{3k}{16I_{SS}}V_m^2\right)}{V_m}$$

$$= -\frac{I_{SS}}{2}\sqrt{\frac{2k}{I_{SS}}}\left(1 - \frac{3k}{16I_{SS}}V_m^2\right) \tag{2.216}$$

As Equation 2.216 suggests, when the oscillation amplitude increases, the large-signal transconductance decreases which results in stable oscillation.

(c) As we stated earlier, for the oscillation condition, the resonant circuit impedance will become real at the oscillation frequency. Now, bearing this in mind, we write the loop gain. The operation of the oscillator is as follows, the differential pair converts the input voltage to the output current at the opposite drain, and then this current flows through the resonant circuit and generates the output voltage. Finally, the transformer returns a part of the output voltage to the input with the same phase (positive feedback). Therefore, one may write the loop gain (assuming that the input of the MOS stage does'nt load the output transformer), H_L, as

$$H_L = 2nG_mR = 1\angle 0° \tag{2.217}$$

By substituting the 600 mV amplitude in Equation 2.216, we compute $G_m = 0.66$ mA/V. By substituting the large-signal transconductance in Equation 2.217, we obtain $n = 0.25$. For instance, if the number of turns at the drain of the transistor is 8, then the number of turns at the gate must be about 2.

(d) Half the value of the varactor adds up with the constant capacitor (why?), and therefore we are able to obtain the frequency range as

$$f_{max} = \frac{1}{2\pi}\frac{1}{\sqrt{L\left(C + \frac{C_{V,min}}{2}\right)}} = 982.3\,\text{MHz} \tag{2.218}$$

$$f_{min} = \frac{1}{2\pi} \frac{1}{\sqrt{L\left(C + \frac{C_{V,max}}{2}\right)}} = 918.8\,\text{MHz} \tag{2.219}$$

(e) Using Equation 2.215, we can compute the third harmonic from Equation 2.215. Then, the ratio of the third harmonic to the first one becomes

$$\frac{|H_3|}{|H_1|} = \frac{kV_m^2}{16I_{SS} - 3kV_m^2} \frac{3}{8Q} \approx 9.5 \times 10^{-5} \quad \text{or} \quad -80.4\,\text{dBc} \tag{2.220}$$

∎

Example 2.10 Consider the oscillator circuit depicted in Figure 2.86 which is a VCO.

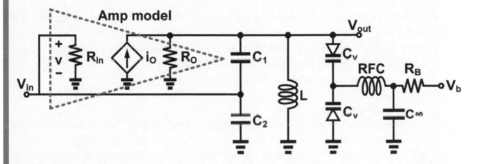

Figure 2.86: A voltage-controlled oscillator with a power series nonlinear amplifier characteristics.

Moreover, the varactor's characteristics are depicted in Figure 2.87.

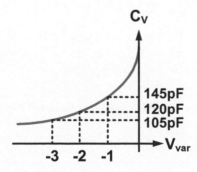

Figure 2.87: Varactor characteristics.

The circuit parameters are as follows:

$$L = 10\,\text{nH}, C_1 = 150\,\text{pF}, C_2 = 1350\,\text{pF}, R_B = 20\,\text{k}\Omega,$$
$$R_{\text{in}} = 10\,\text{k}\Omega, R_o = 5\,\text{k}\Omega \tag{2.221}$$

$$Q_L = 50, L_{\text{RFC}} = 10\,\mu, a = 0.2\frac{A}{V}, c = -5.8\frac{A}{V^3}, e = 1\frac{A}{V^5} \tag{2.222}$$

and the nonlinear characteristics are

$$i_o = av + cv^3 + ev^5 \tag{2.223}$$

Note that

$$\cos^3\theta = \frac{3\cos\theta + \cos 3\theta}{4}, \cos^5\theta = \frac{10\cos\theta + 5\cos 3\theta + \cos 5\theta}{16} \tag{2.224}$$

(a) Investigate the effect of finite resistance of RFC if its equivalent parallel resistor is 2.5 kΩ.
(b) Find the amplitude and the frequency of oscillations for $V_b = 1$ V.
(c) Find the range of oscillation frequency for $1 < V_b < 3$.
(d) Find the amplitude of the fifth harmonic in case (a).

Solution:
(a) The RFC resistance is added in parallel to the output. To calculate its effect, we should first determine the total output resistance at the operating frequency. The effective loading resistance of the RFC which would appear at the output of the oscillator would be

$$R'_{\text{P}_{\text{RFC}}} = n^2 R_{\text{P}_{\text{RFC}}} = 10\,\text{k}\Omega \tag{2.225}$$

where $n = 2$ because of the existence of the double varactors. The effect of the RFC parallel resistance would be to reduce the output resistance and the overall gain as described in part (b).

(b) The total capacitance at the output (at $V_b = 1$ V) is computed as

$$C_T = \frac{C_1 C_2}{C_1 + C_2} + \frac{C_V}{2} = 207.5\,\text{pF} \tag{2.226}$$

By calculating the resonant frequency at $V_b = 1$ V

$$f = \frac{1}{2\pi\sqrt{LC_T}} = 110.487\,\text{MHz} \tag{2.227}$$

The equivalent parallel resistance of the inductor L becomes

$$R_{\text{PL}} = Q_L L \omega_0 = 347\,\Omega \tag{2.228}$$

The closed loop-gain at the oscillation frequency will be

$$Gain_{Loop} = G_{m_1} \left(R_o \parallel R_{P_L} \parallel R'_{P_{RFC}} \right) \frac{1}{m} = 1 \tag{2.229}$$

Then

$$G_{m_1} = \frac{m}{R_T} = 31.8 \frac{mA}{V} \tag{2.230}$$

where

$$R_T = R_o \parallel R_{P_L} \parallel R'_{P_{RFC}} = 314.3 \tag{2.231}$$

To calculate the large-signal transconductance out of the nonlinear characteristics, one can write

$$V_{in} = V_1 \cos(\omega_0 t) \tag{2.232}$$
$$i_o = aV_1 \cos(\omega_0 t) + cV_1^3 \cos^3(\omega_0 t) + eV_1^5 \cos^5(\omega_0 t) \tag{2.233}$$

By expanding Equation 2.232 and considering only the main harmonic component of the current, we obtain

$$i_{o_1} = V_1 \cos(\omega_0 t) \left(a + \frac{3c}{4} V_1^2 + \frac{5e}{8} V_1^4 \right) \tag{2.234}$$

Finally, the large-signal transconductance will be

$$G_{m_1} = a + \frac{3c}{4} V_1^2 + \frac{5e}{8} V_1^4 = 0.0318 \tag{2.235}$$

Resolving Equation 2.235, one obtains two possible solutions:

$$V_1 = 0.197\,V \tag{2.236a}$$

or

$$V_1 = 2.63\,V \tag{2.236b}$$

Only one of these solutions is acceptable and that is the one for which $\frac{\partial G_{m_1}}{\partial V_1} < 0$. Actually, the loop gain should decrease with the amplitude at a stable point (why?), and here only the first solution has such a characteristic.

(c) For the frequency range of the VCO, we can compute the upper bound and the lower bound frequency of the oscillation from Equation 2.227 for $C_{Vmin} = 105\,pF$ and $C_{Vmax} = 145\,pF$, respectively. Therefore, the frequency range will be from 116.230 MHz to 110.487 MHz .

(d) To calculate the amplitude of the fifth harmonic, we have

$$\left|\frac{V_5}{V_1}\right| = \frac{G_{m_5}}{G_{m_1}} \cdot \left|\frac{Z_L(5\omega_0)}{Z_L(\omega_0)}\right| \tag{2.237}$$

The total output quality factor is

$$Q_T = R_T C_T \omega_0 = 45.27 \tag{2.238}$$

The ratio of the fifth harmonic to the first harmonic becomes

$$\left|\frac{V_5}{V_1}\right| = \frac{\frac{V_1^4}{16}e}{a + \frac{3}{4}cV_1^2 + \frac{5}{8}eV_1^4}\left|\frac{1}{1 + jQ_T\left(5 - \frac{1}{5}\right)}\right| \approx \frac{1}{73110} \tag{2.239}$$

Finally, the amplitude of the fifth harmonic will be

$$V_5 = 0.197 \times \frac{1}{73110} = 2.7\,\mu V \tag{2.240}$$

∎

2.20 Special Topic: Nonlinear Device Fed by Sinusoidal Large-Signal Current

Till now in most of the oscillators which we have studied, the nonlinear element has been considered as a nonlinear transconductance fed by a large-signal sinusoidal voltage and the loop gain has been computed in terms of the fundamental voltage. In some cases, one can consider a nonlinear element as a resistance or a transresistance fed by a large-signal sinusoidal current. In this case, a large-signal resistance or transresistance can be defined as the ratio of the output fundamental voltage to the amplitude of the input sinusoidal current. Furthermore, it is possible to compute the loop gain in terms of the fundamental current instead of the fundamental voltage. The oscillator depicted in Figure 2.88 illustrates this concept.

If we consider the crystal model as a high-Q series resonant circuit, it is obvious that it has a low impedance at the resonant frequency and has a high impedance at other frequencies (harmonic frequencies). As such, one could say that only the fundamental current passes through the crystal (it is approximately considered as open circuit for the harmonics). Therefore, the expression for the emitter current would be

$$i_E = I_{EQ} + I_{E1}\cos(\omega t) \tag{2.241}$$

As the transistor's input nonlinear characteristics are

$$i_E = I_{ES}e^{qV_{BE}/kT} \tag{2.242}$$

Equating Equations 2.241 and 2.242, one obtains a relation between the base–emitter voltage and the emitter current.

$$v_{BE} = \frac{kT}{q} \ln \frac{I_{EQ}}{I_{ES}} + \frac{kT}{q} \ln \left[1 + \frac{I_{E1}}{I_{EQ}} \cos(\omega t) \right] \qquad (2.243)$$

Or in another form

$$v_{BE} = V_{BEQ} + \frac{kT}{q} \ln[1 + y\cos(\omega t)] \qquad (2.244)$$

where

$$y = \frac{I_{E1}}{I_{EQ}} \qquad (2.245)$$

The fundamental voltage component at the base–emitter can be expressed as [1]

$$V_{BE1} = \frac{kT}{q} \left[\frac{2}{y} \left(1 - \sqrt{1 - y^2} \right) \right] \qquad (2.246)$$

As such, the large-signal input resistance seen from the emitter can be obtained

$$R_{in}(y) = \frac{V_{BE1}}{I_{E1}} = \frac{V_{BE1}}{I_{EQ}} \frac{I_{EQ}}{I_{E1}} = \frac{kT}{qI_{EQ}} \times \frac{2\left(1 - \sqrt{1 - y^2}\right)}{y^2} \qquad (2.247)$$

Contrary to the large-signal transconductance which is compressed with the input voltage amplitude, the large-signal resistance expands with input current amplitude. For this reason, we draw the inverse normalized large-signal resistance. The inverse normalized emitter input resistance can be expressed as

$$\frac{r_{in}}{R_{in}(y)} = \frac{y^2}{2\left(1 - \sqrt{1 - y^2}\right)} \qquad (2.248)$$

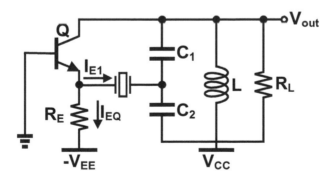

Figure 2.88: A crystal Colpitts oscillator (Butler oscillator) where the input current at the emitter is approximately sinusoidal.

where r_{in} is the small signal input resistance seen through the emitter:

$$r_{in} = \frac{kT}{qI_{EQ}} = \frac{V_t}{I_{EQ}} \tag{2.249}$$

The variations of the inverse normalized emitter resistance as a function of the normalized input current are depicted in Figure 2.89. Note that $0 < y < 1$ in essence.

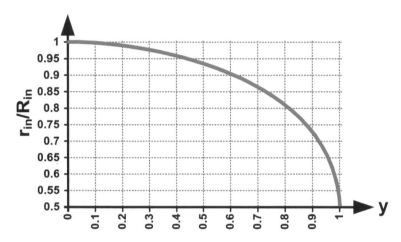

Figure 2.89: Variations of the normalized inverse large-signal input resistance as a function of the input current amplitude.

The application of the above concept is illustrated in Example 2.11.

Example 2.11 Consider the crystal Colpitts oscillator of Figure 2.90.

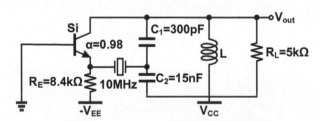

Figure 2.90: Crystal Colpitts oscillator.

In this circuit, if the resonant frequency of the tank circuit and the crystal resonant frequency are not exactly the same, a slight change in frequency and the amplitude will occur. Find the amplitude and the frequency of the oscillations for two cases: (i) determine the value of inductance L for the oscillations at $10\,\text{MHz}$, (ii) if the resonance frequency of the tank circuit is reduced by an amount of $10\,\text{kHz}$, for example, due to temperature or process variations, determine the new frequency and amplitude of oscillations. The crystal parameters are $f_s = 10\,\text{MHz}$, $r_s = 35\,\Omega$, and $Q_0 = 49000$ and the supply voltage is $V_{CC} = V_{EE} = 5\,\text{V}$. Assume $V_{BEQ} = 0.7\,\text{V}$.

Solution:
First, we assume that both resonant frequencies are the same, and therefore we attain

$$n = \frac{C_1}{C_1 + C_2} = \frac{1}{51} \tag{2.250}$$

and the equivalent capacitor will be

$$C_{eq} = \frac{C_1 C_2}{C_1 + C_2} - 294\,\text{pF} \tag{2.251}$$

Thus, the overall quality factor at $10\,\text{MHz}$ will be

$$Q = R_L C_{eq} \omega_0 \approx 92.4 \tag{2.252}$$

and for the value of inductor, one may write

$$L = \frac{1}{C_{eq} \omega_s^2} = 861\,\text{nH} \tag{2.253}$$

and

$$I_{EQ} = \frac{V_{EE} - V_{BEQ}}{R_{EE}} - 0.512\,\text{mA} \tag{2.254}$$

and

$$r_{in} = \frac{kT}{q I_{EQ}} = 50.7\,\Omega \tag{2.255}$$

In Figure 2.91, the current loop gain at the resonant frequency can be written as

$$\frac{\alpha I_{E1} R_L n}{r_s + R_{in}} = I_{E1} \tag{2.256}$$

Note that for $n \ll 1$, the secondary of the transformer in Figure 2.91 doesn't load the primary. The oscillation condition is

$$\frac{\alpha R_L n}{r_s + R_{in}} = 1 \tag{2.257}$$

Therefore

$$R_{in} = n \alpha R_L - r_s = 61.1\,\Omega \tag{2.258}$$

So the normalized inverse large-signal resistance becomes

$$\frac{r_{in}}{R_{in}(y)} = 0.829 \tag{2.259}$$

Using Figure 2.89, one obtains $y = 0.75$. Therefore, for the oscillation amplitude, we attain

$$V_{osc} = \alpha I_{E_1} R_L = \alpha y I_{E_Q} R_L = 1.89\,\text{V} \tag{2.260}$$

Finally, without considering the phase change, output voltage will be

$$V_{out} = V_{CC} + 1.89 \cos(\omega_s t) \tag{2.261}$$

Now, suppose that due to process variation, the resonant frequency of the tank is 10 kHz lower than the series resonance frequency of the crystal, i.e., $f_o = f_{crystal} - 10$ kHz. The tank circuit introduces a phase-change which must be compensated by the crystal. Since the rate of change of the reactance of the crystal is extremely high near the resonant frequency, a slight change in the frequency will compensate the aforementioned phase shift. To calculate the oscillation frequency alongside the oscillation amplitude, consider Figure 2.91.

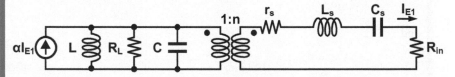

Figure 2.91: Crystal Colpitts oscillator equivalent circuit seen through the emitter.

The oscillation criteria mandate that the overall current loop gain to be unity with zero phase, therefore we obtain (neglecting the loading of the secondary impedance on the primary of the transformer, given the fact that $n \ll 1$)

$$\frac{\alpha I_{E_1} R_L n}{1 + jQ_t\left(\frac{\omega}{\omega_0} - \frac{\omega_0}{\omega}\right)} \times \frac{1}{r_s + R_{in} + jX} = I_{E_1} \tag{2.262}$$

Or

$$\frac{\alpha R_L n}{1 + jQ_t\left(\frac{\omega}{\omega_0} - \frac{\omega_0}{\omega}\right)} \times \frac{1}{r_s + R_{in} + jX} = 1\angle 0 \tag{2.263}$$

where in Equation 2.262, X is the crystal reactance and R_{in} is the emitter input resistance. Therefore, from Equation 2.263, we have

$$X \approx 2Q_L \left(r_s + R_{in}\right) \frac{\omega - \omega_s}{\omega_s} = 2Q_0 r_s \frac{\omega - \omega_s}{\omega_s} \tag{2.264}$$

where Q_L is the loaded quality factor and Q_0 is the unloaded quality factor of the crystal. The magnitude of Equation 2.263 can be written as

$$\frac{(\alpha R_L n)^2}{1 + \left(Q_t \left(\frac{\omega}{\omega_0} - \frac{\omega_0}{\omega}\right)\right)^2} \times \frac{1}{\left(r_s + R_{in}\right)^2 + X^2} = 1 \tag{2.265}$$

For the phase criteria, one may write

$$\tan^{-1} \left(Q_t \left(\frac{\omega}{\omega_0} - \frac{\omega_0}{\omega}\right)\right) = -\tan^{-1} \left(\frac{X}{r_s + R_{in}}\right) \tag{2.266}$$

Equation 2.265 gives us

$$Q_t \left(\frac{\omega}{\omega_0} - \frac{\omega_0}{\omega}\right) - \frac{X}{r_s + R_{in}} \tag{2.267}$$

Substituting Equation 2.267 into Equation 2.265, we then obtain

$$r_s + R_{in} = \frac{\alpha R_L n}{1 + \left(Q_t \left(\frac{\omega}{\omega_0} - \frac{\omega_0}{\omega}\right)\right)^2} \tag{2.268}$$

It is possible to replace ω in Equation 2.268 with the series resonant frequency of the crystal, i.e., $\omega \approx \omega_s$ (why?). Thus, for input resistance, we have

$$R_{in} = \frac{\alpha R_L n}{1 + \left(Q_t \left(\frac{\omega_s}{\omega_0} - \frac{\omega_0}{\omega_s}\right)\right)^2} - r_s \approx 57.9 \,\Omega \tag{2.269}$$

and correspondingly, for the reactance X, one may write

$$X = -\left(r_s + R_{in}\right) Q_t \left(\frac{\omega}{\omega_0} - \frac{\omega_0}{\omega}\right) = -17.17 \,\Omega \tag{2.270}$$

With respect to Equation 2.264, we obtain

$$\Delta f = X \frac{f_s}{2 Q_0 r_s} \approx -63\,\text{Hz} \tag{2.271}$$

and finally, the oscillation frequency will be

$$f_{\text{osc}} = 10\,\text{MHz} - 63\,\text{Hz} = 9.999937\,\text{MHz} \tag{2.272}$$

Now for computing the new oscillation amplitude, we should consider the new value for R_{in}. The inverse normalized large-signal resistance becomes

$$\frac{r_{\text{in}}}{R_{\text{in}}(y)} = 0.875 \tag{2.273}$$

Using Figure 2.89, one obtains $y = 0.66$. The output tuned circuit voltage can be calculated as

$$|v_t| = \alpha I_{E1} |Z_L| = \alpha y I_{EQ} \frac{R_L}{\sqrt{1 + Q_t^2 \left(\frac{\omega}{\omega_0} - \frac{\omega_0}{\omega} \right)^2}} = 1.69\,\text{V} \tag{2.274}$$

■

2.21 Datasheet of a Voltage-Controlled Oscillator

Model name: ZX95-2536C+

- Maximum Ratings

Operating Temperature	-55°C to 85°C
Storage Temperature	-55°C to 100°C
Absolute Max. Supply Voltage (VCC)	5.6 V
Absolute Max. Tuning Voltage (Vtune)	7.0 V
All specifications	$50\,\Omega$ system

- Electrical Specifications

		Min.	2315
Frequency (MHz)		Max.	2536
Power output (dBm)		Typ.	+6
Typical phase noise (dBc/Hz) SSB at offset frequencies, kHz		1	-75
		10	-105
		100	-128
		1000	-148
Tuning voltage range (V)		Min.	0.5
		Max.	5

Sensitivity (MHz/V)	Typ.	57–77
Port cap (pF)	Typ.	36
Modulation bandwidth, 3 dB (MHz)	Typ.	70
Nonharmonic spurious (dBc)	Typ.	−90
Harmonics (dBc)	Typ.	−18
	Max.	−10
DC operating power	VCC (V)	5
	Max. current (mA)	45

- Performance Data

V tune	Tune Sens (MHz/V)	Frequency (MHz)			Output Power (dBm)	Harmonics (dBc)		
		−55°C	+25°C	+85°C	+25°C	F2	F3	F4
0.0	81.90	2267.6	2257.4	2249.2	5.14	−21.7	−19.0	−36.6
0.5	74.61	2306.7	2297.3	2289.5	5.23	−30.5	−20.4	−35.5
1.0	73.76	2344.0	2334.4	2326.4	5.32	−32.0	−22.3	−36.4
1.5	74.01	2381.6	2371.3	2362.6	5.43	−25.6	−22.5	−39.9
2.0	74.15	2419.7	2408.5	2398.9	5.58	−22.2	−23.5	−44.0
2.5	71.91	2456.9	2445.3	2435.4	5.69	−20.0	−23.9	−43.5
3.0	68.45	2492.6	2481.0	2471.1	5.80	−18.5	−25.3	−44.9
3.5	61.36	2525.2	2514.5	2504.9	5.91	−17.3	−27.7	−46.3
4.0	53.56	2554.4	2544.2	2535.4	6.01	−16.3	−30.1	−48.7
4.5	45.62	2579.9	2570.1	2561.7	6.10	−15.6	−33.0	−49.1
5.0	36.26	2601.0	2591.8	2583.8	6.17	−15.2	−35.9	−51.1

- Curves

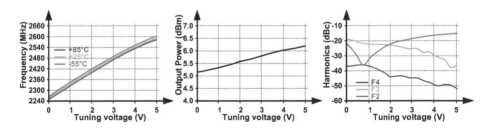

Figure 2.92: Oscillation frequency, output power, and harmonic levels of the ZX95 VCO as a function of the tuning voltage.

Model name: POS-100+

- Maximum Ratings

Operating Temperature	$-55°C$ to $85°C$
Storage Temperature	$-55°C$ to $100°C$
Absolute Max. Supply Voltage (VCC)	16 V
Absolute Max. Tuning Voltage (Vtune)	18 V
All specifications	50 Ω system

- Electrical Specifications

	Min.	50
Frequency (MHz)	Max.	100
Power output (dBm)	Typ.	+8.3
	1	−83
Typical phase noise (dBc/Hz)	10	−107
SSB at offset frequencies, kHz	100	−130
	1000	−150
	Min.	1
Tuning voltage range (V)	Max.	16
Sensitivity (MHz/V)	Typ.	4.2-4.8
Modulation bandwidth, 3 dB (MHz)	Typ.	0.1
	Typ.	−23
Harmonics (dBc)	Max.	−18
	VCC (V)	12
DC operating power	Max. current (mA)	20

- Performance Data

V tune	Tune Sens (MHz/V)	Frequency (MHz)			Output Power (dBm)	Harmonics (dBc)		
		$-55°C$	$+25°C$	$+85°C$	$+25°C$	F2	F3	F4
1.00	3.80	45.55	44.40	43.93	9.42	−40.40	−38.20	−48.70
2.00	4.20	49.41	48.60	48.31	9.40	−46.40	−40.80	−46.60
3.00	4.50	53.98	53.15	52.83	9.32	−58.40	−40.90	−44.50
4.00	4.10	58.08	57.27	56.90	9.22	−50.30	−39.80	−43.40
5.00	4.00	62.10	61.31	60.88	9.12	−45.60	−38.60	−42.50
6.00	4.10	66.20	65.43	64.96	8.99	−43.40	−37.40	−41.50
7.00	4.20	70.43	69.62	69.13	8.86	−42.60	−36.20	−40.50
8.00	4.30	74.80	73.93	73.43	8.67	−43.10	−35.00	−39.60
9.00	4.40	79.26	78.33	77.81	8.46	−44.70	−34.10	−38.70
10.00	4.40	83.78	82.77	82.25	8.22	−48.50	−33.10	−38.30
11.00	4.50	88.26	87.23	86.68	8.00	−53.80	−32.50	−38.10
12.00	4.50	92.73	91.69	91.10	7.77	−54.90	−31.80	−37.70
13.00	4.40	97.13	96.06	95.46	7.52	−51.10	−31.50	−37.40
14.00	4.40	101.53	100.45	99.81	7.31	−48.00	−30.90	−37.30
15.00	4.40	105.98	104.84	104.18	7.13	−46.20	−30.60	−37.30
16.00	4.40	110.43	109.22	108.55	6.93	−44.60	−30.20	−37.30

- Curves

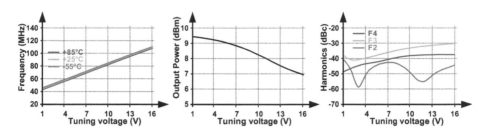

Figure 2.93: Oscillation frequency, output power, and harmonic levels of the POS-100 VCO as a function of the tuning voltage.

2.22 Conclusion

In this chapter, we have studied the basic operation of sinusoidal oscillators. Oscillators generally operate by means of amplification of circuit noise in a relatively high-gain frequency selective closed-loop circuit. Here, the noise as an initial signal contributes to the build up of the oscillator sinusoidal signal and the loop gain of the oscillator is consequently compressed (reduced) by the generated large signal. For a stable oscillation, it is necessary to satisfy Barkhausen's criteria. That is to say, to achieve a unity closed-loop gain with 2π or zero phase. In general, an active element in addition to a frequency selective (resonating) circuit is needed in an oscillator. It is noteworthy that the oscillators generally operate in large-signal regime. So it is important to have a nonlinear model for the device in order to compute adequately the amplitude and the frequency of the oscillation.

In this chapter, different oscillator topologies including CE, CB, CC, (or CS, CG, and CD for MOS transistors) as well as Colpitts-like or Hartely-like oscillators were studied. The study of oscillator circuit is essentially divided into two parts: In the first part, the resonant dividing circuits were studied where RLC resonant circuits are used with either capacitive or inductive dividers. In the second part, the nonlinear behavior of the active elements used in the oscillator circuits should be studied. Here we presented large-signal models for the bipolar transistors, differential bipolar pairs, MOS transistors, and the MOS differential pairs where large-signal transconductances were computed for either of the active elements. Using variable capacitors or varactors in the oscillator circuits permits the frequency tuning of them. As such, voltage-controlled oscillators (VCO) were presented. VCOs are one of the main building blocks of the phase-locked loops which will be discussed in the next chapter. The main figures of merits of an oscillator are its frequency stability, its spectral purity, low harmonic level, and its low phase noise.

2.23 References and Further Reading

1. K.K. Clarke, D.T. Hess, *Communication Circuits, Analysis and Design*, United States: Krieger Publishing Company, 1994.

2. F. Farzaneh, *RF Communication Circuits* (in Persian), Tehran: Sharif University Press, 2005.

3. H.L. Krauss, C.W.Bostian, F.H. Raab, *Solid State Radio Engineering*, New York, NY: J. Wiley & Sons, Inc., 1980.

4. B. Razavi, *Design of Analog CMOS Integrated Circuits*, Boston, MA: McGraw-Hill, 2001.

5. B. Razavi, *RF Microelectronics*, second edition, Castleton, NY: Prentice-Hall, 2011.

6. J.R. Smith, *Modern Communication Circuits*, second edition, New York, NY: McGraw Hill, 1997.

7. J. Everard, *Fundamentals of RF Circuit Design with Low Noise Oscillators*, United Kingdom: J. Wiley & Sons, Inc., 2000.

8. R. Chi-Hsi Li, *RF Circuit Design*, Hoboken, NJ: J. Wiley & Sons, Inc., 2009.

9. R. Dehghani, *Design of CMOS Operational Amplifiers*, Norwood, MA: Artech House, 2013.

10. A. Hajimiri, T.H. Lee, *The Design of Low Noise Oscillators*, Boston, MA: Kluwer Academic, 1999.

11. D.H. Wolaver, *Phase-Locked Loop Circuit Design*, United Kingdom: Prentice Hall, 1991.

12. M. Tohidian, A. Fotowat-Ahmady, Mahmoud Kamarei, "A simplified method for phase noise calculation," *IEEE Custom Integrated Circuits Conference (CICC 2009)*, San Jose, CA, September 2009.

13. H. Teymoori, A. Fotowat-Ahmady, A. Nabavi, "A new low phase noise LC-tank CMOS cascode Cross-coupled oscillator," *IEEE Iranian Conference Electrical Engineering (ICEE 2010)*, Isfahan University of Technology, Isfahan, May 2010.

14. Mini-circuits, RF designers handbook, VCO datasheets (http://www.mini circuits.com/).

2.24 Problems

Problem 2.1 Consider the resonant circuit depicted in Figure 2.94 which is normally used in Clapp oscillators.

1. We know that the oscillation will occur where the impedance of resonant circuit is pure real which corresponds to zero phase shift. In the given resonant circuit, find at which frequency the impedance will be pure real?

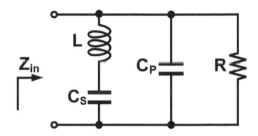

Figure 2.94: Clapp-type resonant circuit.

2. With the results of part 1, find the oscillation frequency of Figure 2.95 for $L_1 = 50\,\text{nH}$ and $C_0 = C_1 = C_2 = 3\,\text{nF}$.

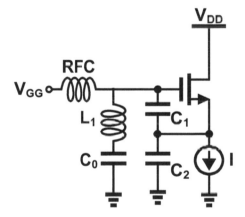

Figure 2.95: Common-drain Clapp oscillator.

3. In the Colpitts-like oscillator, we can change the oscillation frequency by varying capacitors C_1 or C_2; however, this will change the loop gain. Explain what is the advantage of Clapp oscillator with respect to its counterpart.

Problem 2.2 Consider the Colpitts oscillator depicted in Figure 2.96 with the given values of parasitic capacitances, namely, base–collector $C_\mu = 15\,\text{pF}$, collector–substrate $C_{\text{CS}} = 15\,\text{pF}$, base–emitter $C_\pi = 30\,\text{pF}$, $I_C = 3\,\text{mA}$, and C_B is RF short.

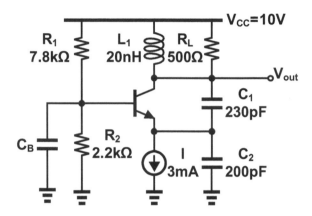

Figure 2.96: Bipolar Colpitts oscillator.

1. Find the oscillation frequency and the fundamental harmonic amplitude.
2. Find the second and third harmonics' amplitudes.
3. If we substitute the current source with the resistor, find the value of the resistor in such a way that the emitter current is 3 mA. Moreover, recalculate parts 1 and 2. In all parts, assume that the transistor is silicon-type with $V_{\text{BE,Q}} = 0.7$.

Problem 2.3 In the given Colpitts oscillator of Figure 2.97, assume that the MOS transistor is ideal with $V_{\text{th}} = 0.7v$ and $\mu_n C_{\text{ox}} = 0.134\frac{mA}{v^2}$, $\frac{W}{L} = 100$. The circuit parameters are $L_1 = 2\mu H, C_1 = C_2 = 10nF, C_g = 200nF$ and the transistor is biased in the square law region. Note that in the MOS transistor, $K = \frac{1}{2}\mu_n C_{\text{ox}}\frac{W}{L}$. Find the oscillation frequency and amplitude in this oscillator.

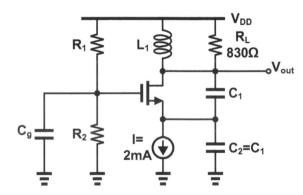

Figure 2.97: MOS common-gate Colpitts oscillator.

Problem 2.4 For the given Colpitts oscillator depicted in Figure 2.98 which is supposed to oscillate at $100\,\text{MHz}$? Consider the transistor's Early voltage is $40\,\text{V}$ and $\beta = 100$.

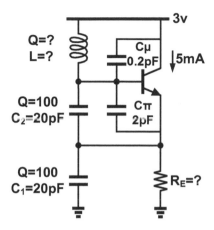

Figure 2.98: Bipolar common-collector Colpitts oscillator with corresponding C_μ and C_π parasitic capacitances.

1. Find the required emitter resistance, R_E.
2. Find the proper value of the inductor, L.
3. Determine the oscillation condition, and then find the minimum quality factor for the inductor to sustain the oscillations.

Problem 2.5 Find the equivalent capacitor in such a way that the circuit depicted in Figure 2.99 oscillates at $50\,\text{MHz}$. Find the ratio of the capacitors in order to have an oscillation amplitude of $200\,\text{mV}$ and as such determine the values of either of the capacitors.

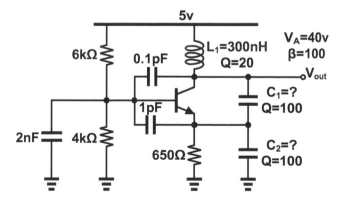

Figure 2.99: Bipolar common-base Colpitts oscillator.

Problem 2.6 Figure 2.100 depicts the block diagram of a hypothetical oscillator.

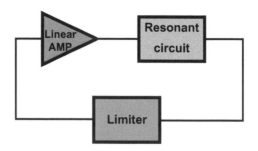

Figure 2.100: A hypothetical oscillator block diagram.

Each block in Figure 2.100 can be modeled by an ideal element (for the amplifier, you may put a voltage-dependent current source with g_m transconductance and finite output resistance R, and for the resonant circuit, an LC resonator with infinite quality factor). Suppose that the limiter characteristics follow $V_o = \tanh(bV_{in})$ where V_{in} and V_{out} are the input and output voltages of the limiter, respectively. Moreover, assume we have a $g_m = 4\,\text{m}\mho, R = 500\,\Omega, C = 5\,\text{pF}, L = 5\,\text{nH}$, and $|b| = 10\,\text{V}^{-1}$.

1. Draw an equivalent circuit diagram for the oscillator, and determine the oscillation frequency.
2. Determine the effective gain of the limiter.
3. Determine the sign of the parameter b in the characteristic of the limiter for positive feedback.

Problem 2.7 Common topologies of the MOS oscillators are shown in Figure 2.101.

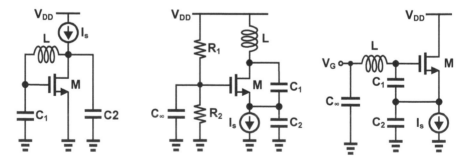

Figure 2.101: Different types of the MOS Colpitts oscillators, common-source, common-gate, and common-drain.

All transistors have the transconductance and their parasitic capacitances are $C_{gs} = 231\,\text{fF}, C_{gd} = 94\,\text{fF}, C_{sb} = 24\,\text{fF}$, and $C_{db} = 19\,\text{fF}$. Moreover, for other parameters, we have $L = 1.5\,\text{nH}, C_1 = 20\,\text{pF}$, and $C_2 = 5\,\text{pF}$. With the given values, find the oscillation frequency and compare them in the three topologies.

Problem 2.8 In the circuit depicted in Figure 2.102, assume that the input large signal of the stage is $V_i = V_m \cos(\omega_0 t)$, and the $I - V$ characteristics of the active device follow $I = \frac{1}{2}K(V_{gs} - V_{th})^2$ for $V_{gs} \geq V_{th}$, and $I = 0$ for $V_{gs} \leq V_{th}$. Assume $Q_L = 50$, the output circuit is tuned to ω_0 and $V_b = V_{th}$. Find the conduction angle in the output current and then find the first to the fifth output current harmonics and the first to the fifth output voltage harmonics.

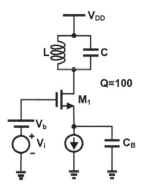

Figure 2.102: MOS tuned amplifier driven by a large signal.

Problem 2.9 The crystal oscillator depicted in Figure 2.103 is named after its designer as Driscoll oscillator. In this oscillator contrary to other types of oscillators, at the oscillation condition, transistor Q_1 will not be driven to the nonlinear regime and the diodes D_1 and D_2 will be driven to the nonlinear region and as such will limit the signal level. Using the exponential $I - V$ characteristics of diodes and using the Bessel function expansion, find the loop gain, and find the amplitude of the oscillation (an important feature of the circuit in Figure 2.103 is the separation of the resonant circuit from the limiter which results in better phase noise of the oscillator). Assume that V_{b1} and V_{b2} are adequate positive voltages to maintain Q_1 and Q_2 in their active region. Furthermore, the phase shifter block has a voltage gain of unity and it doesn't load the output tuned circuit.

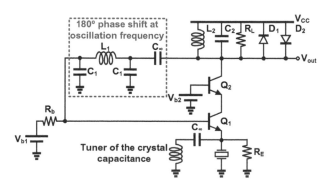

Figure 2.103: Driscoll oscillator.

Problem 2.10 Assuming a square law characteristics for the MOS transistors as in Equation 2.275, one can derive the drain current of the MOS differential stage as Equation 2.276. Using the polynomial expansion of Equation 2.276, find the large-signal transconductance of the stage and the large-signal loop gain to deduce the amplitude of oscillation.

$$I_D = \frac{1}{2} \mu_n C_{ox} \frac{W}{L} \left(V_{gs} - V_{th} \right)^2 \tag{2.275}$$

$$I_{D_1} = \frac{I_{SS}}{2} - V_{g2} \frac{\mu_n C_{ox}}{4} \left(\frac{W}{L} \right) \sqrt{\frac{4 I_{SS}}{\mu_n C_{ox} \frac{W}{L}} - V_{g2}^2} \tag{2.276}$$

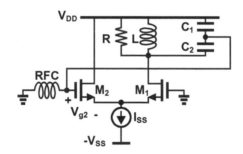

Figure 2.104: MOS differential oscillator.

Problem 2.11 Consider the Colpitts oscillator depicted in Figure 2.105 and assume that $V_{control} = 6$ V, bipolar transistor's $\beta = 100, L = 2\,\mu H, C_1 = 55$ pF, and $C_2 = 550$ pF. Moreover, Figure 2.106 depicts the variable capacitance characteristics. Assume that the MOS transistor is off.

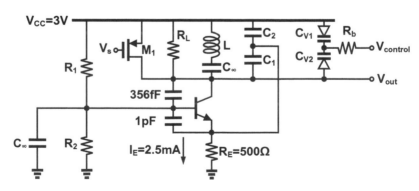

Figure 2.105: A BiCMOS VCO.

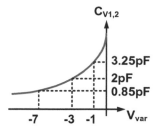

Figure 2.106: Variable capacitance characteristics.

1. Find the oscillation frequency.
2. If the control voltage varies between 4 V and 10 V, find the range of the oscillation frequency.
3. Find the load resistance R_L to have an oscillation amplitude of 500 mV at the collector.
4. If the transistor M_1 is actuated by a feedback network depicted in Figure 2.107, and goes into the triode region, find the ratio of the third harmonic to the first harmonic in this case. Note that if $I_{sd} = \frac{1}{2}K(V_{gs} - V_{th})^2$, then in the triode region, we would have $g_{ds} = K(V_{gs} - V_{th})$. Assume $K = 500 \frac{\mu A}{V^2}$ and $V_{th} = -1$ V.

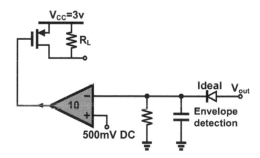

Figure 2.107: Feedback amplitude control loop.

Problem 2.12 In a nonlinear amplifier depicted in Figure 2.108, the input–output relation follows $V_{out} = 5V_{in} - 0.65V_{in}^2 - 0.3V_{in}^3$. This amplifier is placed in a feedback loop and the combination results in a stable oscillation.

1. Find the oscillation frequency.
2. Find the -3 dB bandwidth of the resonant circuit.
3. Find the small-signal and the large-signal loop gain.
4. Find the oscillation amplitude alongside the second harmonic amplitude in dBc.

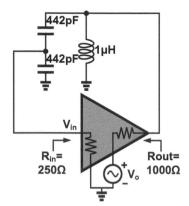

Figure 2.108: An oscillator with a nonlinear amplifier in feedback.

Problem 2.13 In an oscillator depicted in Figure 2.109 where the nonlinear transfer characteristics of the device are given by Equation 2.278, find the oscillation frequency as well as the amplitude of the fundamental and the second harmonic.

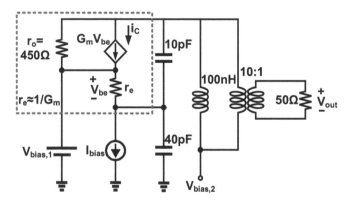

Figure 2.109: Colpitts oscillator.

$$v_1 = v_{be} \tag{2.277}$$

$$i_C = \frac{1}{100} + \frac{v_1}{50} + \frac{v_1{}^2}{300} - \frac{v_1{}^3}{400} \quad A \tag{2.278}$$

Problem 2.14 **Design problem.** In the reference Driscoll oscillator depicted in Figure 2.110,

1. How the values of C_1, C_2, L_4, and L_5 are determined, in such a way that the circuit oscillates at the third series resonance of the crystal (f_S). Describe the corresponding relations (write the oscillation condition).

2. The oscillation amplitude at the emitter of Q_1 is much smaller than the thermal voltage V_T, and at the collector of Q_2, it is larger than a few V_T's, and the third overtone of the crystal is the dominant impedance. The amplitude of oscillations at the collector of Q_2 is determined by the Schottky diodes impedances. The $I - V$ characteristic of diodes follows $I_D = I_S e^{\frac{V_D}{V_t}}$. Find the harmonic content of the current by the Bessel function expansion of the output characteristics and determine an expression for the loading impedance of the diodes.

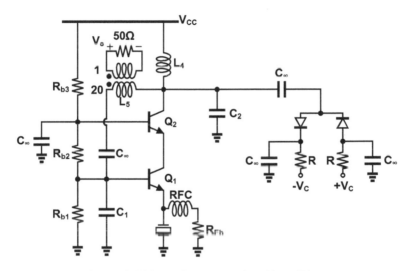

Figure 2.110: Reference Driscoll oscillator.

Problem 2.15 For the oscillator depicted in Figure 2.111, using the large-signal model of the transistor, draw the equivalent circuit and write the complex relation describing Barkhausen's oscillation criteria. Assume that the crystal's admittance is represented by a complex value Y_x. Note: do not use the equivalent transformer model for the capacitive divider in this case.

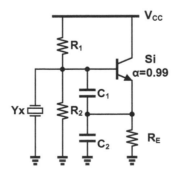

Figure 2.111: Common-collector crystal oscillator.

Problem 2.16 Writing the oscillation condition, find the frequency and the amplitude of oscillations in the circuit of Figure 2.112 at the collector of the transistor Q_1. Furthermore, find the third harmonic amplitude at the same node.

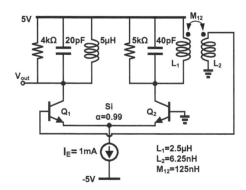

Figure 2.112: Differential pair tuned circuit oscillator.

Problem 2.17 The amplifier model in Figure 2.113 is representative of an operational amplifier where R_i is quite large and μ is adequately large and r_o is quite small. Find the complex oscillation condition in Figure 2.113 as a function of the circuit parameters. Assume that the crystal impedance is represented by $R + jX$. Furthermore, explain how does the signal amplitude is limited in this topology.

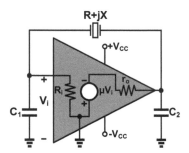

Figure 2.113: Pierce-like oscillator using an ideal amplifier.

Problem 2.18 In the Pierce crystal oscillator depicted in Figure 2.114, the crystal is inductive and will resonate with the input and the output capacitors resulting in a sinusoidal oscillation. Assume the nonlinear element has the given $I - V$ characteristics. First find the large-signal effective G_m as a function of AC voltage amplitude and draw it to the scale. Then, write the complex oscillation condition in Figure 2.114 as a function of the circuit parameters. What is the required G_m for an oscillation amplitude of 1 V?

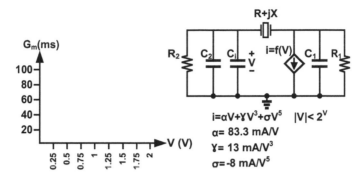

Figure 2.114: Pierce crystal oscillator with a nonlinear transconductance.

Part 2

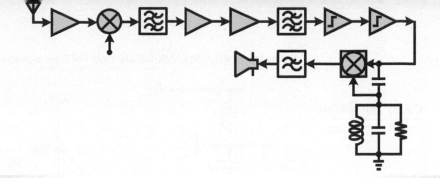

3. PLL, FM Modulation, and FM Demodulation

As discussed in Chapter 2, voltage-controlled oscillators are widely used in many applications such as phase-locked loops (PLLs), frequency modulation (FM), and demodulation. Baseband signal transmission cannot be realized without modulation. For instance, voice signals need a carrier to be transmitted through the transmission medium. Moreover, each standard specifies a carrier frequency and a bandwidth for the specified service.

3.1 Frequency Modulation

It is possible to change the frequency of an oscillator by varying its tuned circuit varactor voltage. As an example, in a Colpitts-like oscillator circuit, as depicted in Figure 3.1, the total capacitance of the tuned circuit is varied by the varactor control voltage.

As it is obvious from Figure 3.1, we can easily change the oscillation frequency by the control voltage of the varactor which changes the resonant frequency of the circuit. Tuning voltage passes through a low-pass circuit and by a specific time constant changes the output frequency. Voice signal adds through a high-pass filter (C_1 and R_2) to its common-mode level and as a result, the baseband signal modulates the oscillator frequency. Modulation index is based on the amplitude of the baseband signal which also modulates the oscillator nonlinearly. The modulation index is determined by the derivative of $C_v(v)$ and the amplitude of the voice signal, as shown in Figure 3.1. It is possible to obtain the instantaneous frequency of the oscillator in Figure 3.1 as

$$f = \frac{1}{2\pi\sqrt{L\left(\frac{C_1 C_2}{C_1 + C_2} + C_V\right)}} \tag{3.1}$$

The instantaneous value of the varactor capacitance can be described as

$$C_v(V_B + V_m g(t)) = C_v(V_B) + \frac{\partial C_v}{\partial v} V_m g(t) = C_v(V_B) + C(t) \tag{3.2}$$

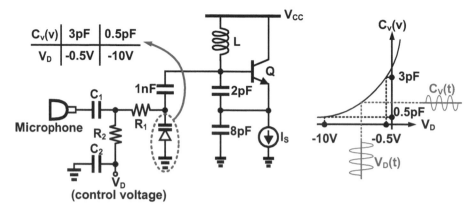

Figure 3.1: A typical voltage-controlled oscillator (VCO) with voice input for frequency modulation.

where V_m is the maximum value of the baseband voltage, V_B is the varactor's bias voltage, and $g(t)$ is a random normalized function (varying between $+1$ and -1) proportional to the input information. The function $g(t)$ may be a continuously valued analog signal for FM or a two discrete level valued digital signal for FSK modulation. The instantaneous frequency can be computed as

$$f = \frac{1}{2\pi\sqrt{L\left(\frac{C_1 C_2}{C_1+C_2}+C_v(V_B)+C(t)\right)}} \approx \frac{\left(1-\frac{1}{2}\frac{C(t)}{\frac{C_1 C_2}{C_1+C_2}+C_v(V_B)}\right)}{2\pi\sqrt{L\left(\frac{C_1 C_2}{C_1+C_2}+C_v(V_B)\right)}} = f_0 + \Delta f(t) \tag{3.3}$$

Here, the carrier frequency is

$$f_0 = \frac{1}{2\pi\sqrt{L\left(\frac{C_1 C_2}{C_1+C_2}+C_v(V_B)\right)}} \tag{3.4}$$

and the frequency deviation becomes

$$\Delta f(t) = \frac{\left(-\frac{1}{2}\frac{C(t)}{\frac{C_1 C_2}{C_1+C_2}+C_v(V_B)}\right)}{2\pi\sqrt{L\left(\frac{C_1 C_2}{C_1+C_2}+C_v(V_B)\right)}} \tag{3.5}$$

Here the carrier frequency and the frequency deviation are clearly described as a function of circuit parameter values. Now with the definition of frequency modulation, we can write the FM signal as

$$v_{FM}(t) = A \sin \left(\overbrace{\omega_0 t + \int \Delta\omega_m(t)dt}^{\Theta} \right) \tag{3.6}$$

We can obtain the instantaneous frequency of Equation 3.6 by the derivation of the argument of the sinusoidal signal:

$$\omega(t) = \frac{d\Theta}{dt} \tag{3.7}$$

where in Equation 3.7, Θ is called the total phase. Moreover, we know that

$$\omega(t) = 2\pi f = \omega_0 + \Delta\omega_m(t) \tag{3.8}$$

where in Equation 3.8 ω_0 is the carrier radian frequency and $\Delta\omega_m(t)$ is a function of the baseband or the radian frequency deviation. Comparing Equations 3.7 and 3.8, we reach to

$$\Theta = \int (\omega_0 + \Delta\omega_m(t))dt = \omega_0 t + \int \Delta\omega_m(t)dt \tag{3.9}$$

Thus far, we have calculated the signal phase for Equation 3.6 and also introduced the frequency modulation. As Equation 3.6 suggests, we have defined the frequency-modulated signal by a constant amplitude sinusoid. The information is embedded in $\Delta\omega_m(t)$ which changes the VCO frequency proportionally. It should be noted that the bigger the amplitude of the input baseband signal the more will be the frequency deviation. Thus, in many applications, we employ a limiter in the baseband circuit to limit the bandwidth occupancy. Therefore, the information is merely in a specific bandwidth. As an example, we can inspect the specifications of the commercial FM radio. The standard obligates the bandwidth of this radio to be limited to 200 kHz. The Carson bandwidth equation predicts that

$$BW_{FM} \cong 2(f_{dev} + f_{BB}) \tag{3.10}$$

where in Equation 3.10, f_{dev} is the frequency deviation and f_{BB} is the maximum baseband signal frequency. As an example, for a frequency deviation of 75 kHz, and a maximum baseband frequency of 20 kHz, the Carson bandwidth becomes

$$BW_{FM} \cong 2(75 + 20) = 190 \, \text{kHz} \tag{3.11}$$

which is within the specified bandwidth of 200 kHz. As another example, the time variation of an FM signal is illustrated in Figure 3.2. Here for the sake of illustration, the carrier frequency is chosen as 500 kHz and the frequency deviation is chosen as 75 kHz.

As depicted in Figure 3.2, when the amplitude of the baseband signal is high, the frequency increases and when this amplitude is low, the frequency decreases.

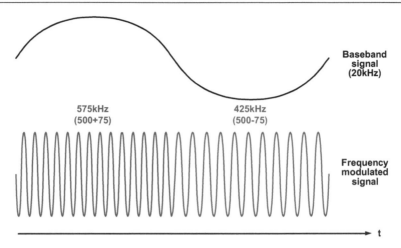

Figure 3.2: Typical baseband and the corresponding FM signal variations with time.

3.2 Frequency Demodulation

We can easily detect frequency-modulated signals by employing a time derivative and then amplitude detection:

$$\frac{dv_{\mathrm{FM}}}{dt} = A\left(\omega_0 + \Delta\omega_{\mathrm{m}}\left(t\right)\right)\cos\left(\omega_0 t + \int_0^t \Delta\omega_{\mathrm{m}}\left(t\right)dt\right) \tag{3.12}$$

The above signal can be detected using an envelope detector. As such, it is possible to detect an FM signal through a differentiator followed by an AM detector. We now continue with the other concepts for frequency demodulation.

3.2.1 Phase Detector

One of the methods to detect FM signals is through the use of a phase detector circuit. Thus, if we find a circuit which gives the phase difference between two inputs, we can demodulate the FM signal. This procedure can be done by an XOR (Exclusive-OR). Consider Figure 3.3.

If two signals in Figure 3.3 completely overlap, the output will be zero. If the two signals have a slight phase shift, the output will be nonzero (proportional to the DC component of the output pulse train). Furthermore, the maximum value of the phase detector output occurs when the signals have 180° phase difference. The main drawback in Figure 3.3 is the logic level which is not appropriate for phase detection (the DC component of the output is proportional to the absolute value of the phase difference). Thus, we introduce another circuit for this purpose which is the Gilbert cell.

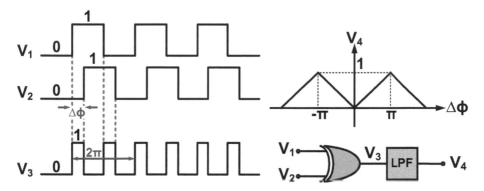

Figure 3.3: XOR (phase detection) characteristics driven by two signals with a common frequency and a fixed phase shift.

3.2.2 Gilbert Cell as a Phase Detector

Figure 3.4 shows the Gilbert cell circuit. We can assert that the Gilbert cell is composed of three differential pairs. The lower differential transistors are called the lower tree and the upper pairs are called the upper tree. The Gilbert cell is capable of being a phase detector. The advantage of this circuit is that it doesn't need a certain logic level to function as a phase detector. In bipolar transistors, the required voltage for correct behavior of phase detector is at least $4V_t$ where V_t is the thermal voltage. In the lower tree of the Gilbert cell shown in Figure 3.4(a), one can write

$$I_{C1,C2} = \frac{I_0}{2}\left(1 \pm \tanh\left(\frac{qv_1}{2kT}\right)\right)$$ (3.13)

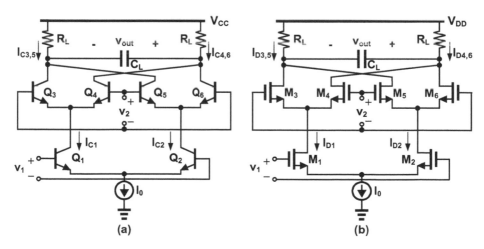

Figure 3.4: Gilbert cell (analog multiplier), (a) by bipolar transistor pairs and (b) by MOS transistor pairs, used as a phase detector.

The differential current in the upper tree becomes

$$\Delta I = I_{C3,5} - I_{C4,6} = I_0 \tanh\left(\frac{qv_1}{2\,\mathrm{kT}}\right) \tanh\left(\frac{qv_2}{2\,\mathrm{kT}}\right) = I_0 \tanh\left(\frac{v_1}{2V_t}\right) \tanh\left(\frac{v_2}{2V_t}\right) \quad (3.14)$$

Given the large load capacitance, C_L, the output voltage will be proportional to the low-pass component of $R_L \Delta I$. Note that the Gilbert cell produces a differential current proportional to the analog multiplication of the input voltages. The point here is that the hyperbolic tangent function saturates to ± 1 once its argument (or the input voltage) is large, either positive or negative. As such, the output current will exhibit a bipolar XOR (Exclusive-OR) function of the two large inputs (an Exclusive-OR with ± 1 logic levels). This way the low-pass component of the output will be proportional to the phase difference of the inputs, provided that the inputs are large signal, that is larger than $4V_t$ (Figure 3.5).

In an analytical manner, one can describe the all-pass components of the output (if the capacitance, C_L did not exist) as

$$V_{\mathrm{out}} = I_0 R_L \tanh\left(\frac{V_1 \cos\left(\omega_0 t\right)}{2V_t}\right) \tanh\left(\frac{V_2 \cos\left(\omega_0 t + \phi\right)}{2V_t}\right) \quad (3.15)$$

For large signal inputs, that is, $\frac{V_1}{V_t} \gg 1$ and $\frac{V_2}{V_t} \gg 1$, the hyperbolic tangent of sinusoidal signals turn into square-wave signals of the same frequency and phase. That is

$$V_{\mathrm{out}} = I_0 R_L S\left(\omega_0 t\right) S\left(\omega_0 t + \phi\right) \quad (3.16)$$

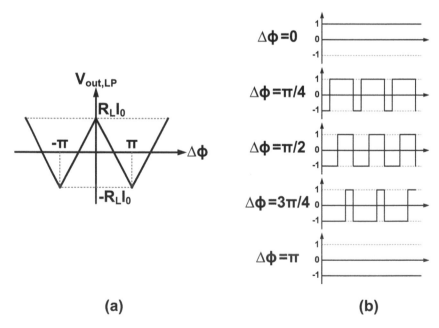

(a)   **(b)**

Figure 3.5: Gilbert cell function as a phase detector. (a) The phase detector output characteristics. (b) Signal waveforms for phase detector operation.

where $S(\omega_0 t)$ is a bipolar square wave of a unity amplitude. The Fourier expansion of the square waves gives

$$V_{\text{out}} = \frac{4}{\pi}I_0 R_L \left[\cos(\omega_0 t) - \frac{1}{3}\cos(3\omega_0 t) + \frac{1}{5}\cos(5\omega_0 t) - \cdots \right]$$
$$\times \frac{4}{\pi} \left[\cos(\omega_0 t + \phi) - \frac{1}{3}\cos(3\omega_0 t + 3\phi) + \frac{1}{5}\cos(5\omega_0 t + 5\phi) - \cdots \right] \quad (3.17)$$

The low-pass component of the output becomes

$$V_{\text{out,LP}} = \frac{8}{\pi^2}I_0 R_L \left[\cos(\phi) + \frac{1}{9}\cos(3\phi) + \frac{1}{25}\cos(5\phi) + \cdots \right] \quad (3.18)$$

With a coefficient of $I_0 R_L$, this is evidently a Fourier expansion of a triangular function of ϕ whose value is unity at 0 radians and its value is 0 at $\pm\pi/2$ radians (see Figure 3.5(a) and compare it with Figure 3.3).

With MOS transistors (with a typical 700 mV threshold voltage), this can be roughly done by about 200 mV–300 mV bias above the threshold voltage. The capacitor in Figure 3.4(b) realizes a low-pass response to suppress undesired higher frequency components.

Considering a square law characteristics for the MOS transistors

$$I_D = K(v_{\text{GS}} - V_{\text{TH}})^2 \quad \text{for} \quad v_{\text{GS}} > V_{\text{TH}} \quad (3.19)$$

Here we assume that $v_{\text{GS}_1} = V_{\text{GS0}_1} + v_1/2$ and $v_{\text{GS}_2} = V_{\text{GS0}_1} - v_1/2$. The ratio of the currents in the lower tree transistors, with the above assumption, becomes (with $\frac{v_1}{2} < V_{\text{GS0}_1} - V_{\text{TH}}$ to remain in the square law region)

$$\frac{I_1}{I_2} = \frac{\left(V_{\text{GS0}_1} + \frac{v_1}{2} - V_{\text{TH}}\right)^2}{\left(V_{\text{GS0}_1} - \frac{v_1}{2} - V_{\text{TH}}\right)^2} = \frac{\left(V_{\text{eff}} + \frac{v_1}{2}\right)^2}{\left(V_{\text{eff}} - \frac{v_1}{2}\right)^2} \quad (3.20)$$

where $V_{\text{eff}} = V_{\text{GS0}_1} - V_{\text{TH}}$, and

$$I_1 + I_2 = I_0 \quad (3.21)$$

and the current in the either drains of the lower tree transistors can be described as

$$I_1 = \frac{I_0}{2}\left[1 + \frac{\frac{v_1}{V_{\text{GS0}_1} - V_{\text{TH}}}}{1 + \frac{1}{4}\left(\frac{v_1}{V_{\text{GS0}_1} - V_{\text{TH}}}\right)^2} \right] = \frac{I_0}{2}\left[1 + \frac{\frac{v_1}{V_{\text{eff}}}}{1 + \frac{1}{4}\left(\frac{v_1}{V_{\text{eff}}}\right)^2} \right] \quad (3.22)$$

$$I_2 = \frac{I_0}{2}\left[1 - \frac{\frac{v_1}{V_{\text{GS0}_1} - V_{\text{TH}}}}{1 + \frac{1}{4}\left(\frac{v_1}{V_{\text{GS0}_1} - V_{\text{TH}}}\right)^2} \right] = \frac{I_0}{2}\left[1 - \frac{\frac{v_1}{V_{\text{eff}}}}{1 + \frac{1}{4}\left(\frac{v_1}{V_{\text{eff}}}\right)^2} \right] \quad (3.23)$$

These equations are valid for $v_1 \leq 2\,(V_{GS0_1} - V_{TH})$.

In a similar manner, the differential current of the upper tree can be described as a function of two differential voltages, v_1 and v_2 (provided $\frac{v_1}{2} < V_{GS0_1} - V_{TH}$ and $\frac{v_2}{2} < V_{GS0_2} - V_{TH}$)

$$\Delta I = I_{D3,5} - I_{D4,6} = I_0 \left[\frac{\left(\frac{v_1}{V_{GS0_1} - V_{TH}} \right) \left(\frac{v_2}{V_{GS0_2} - V_{TH}} \right)}{\left[1 + \frac{1}{4} \left(\frac{v_1}{V_{GS0_1} - V_{TH}} \right)^2 \right] \left[1 + \frac{1}{4} \left(\frac{v_2}{V_{GS0_2} - V_{TH}} \right)^2 \right]} \right] \quad (3.24)$$

where, V_{GS0_1} is the DC bias voltage of the lower tree transistors and V_{GS0_2} is the DC bias voltage of the upper tree transistors. Here again, the output voltage will be proportional to the low-pass component of $R_L \Delta I$. Note that the MOS Gilbert cell produces a differential current proportional to the analog multiplication of the input voltages approximately. This function is again a saturating function of the input voltages and tends approximately to ± 1 once its argument (input voltage) is large. As such, again the output current will exhibit a bipolar XOR (Exclusive-OR) function of the two large inputs (an Exclusive-OR with ± 1 logic levels). This way the low-pass component of the output will be proportional to the phase difference of the inputs, provided that the inputs are large signal, that is, larger than $2V_{\text{eff}}$. The same analytical procedure as described in Equations 3.16 through 3.18 holds for MOS phase detector as well and consequently, a triangular output characteristics is produced here again. The phase detector characteristics, that is, the output voltage versus the phase difference of the two input signals, are shown in Figure 3.5(b).

In general, the Gilbert cell will be a phase detector when both inputs are driven to the large signal regime. If the lower tree is driven by a small signal and the upper tree experiences hard switching, the circuit changes to a mixer operation (which we will study in Chapter 4). However, if both trees are driven by small-signal inputs, we will have an analog multiplier. Now we desire to design a phase detector with a Gilbert cell. We know that the data lie in the phase or the frequency of the carrier.

3.2.3   Quadrature Phase (FM) Detector

Figure 3.6 shows a receiver with quadrature FM detection. The main objective in this circuit is to detect a frequency-modulated signal through a phase shift of 90° degrees of the signal.

In Figure 3.6, the 104 MHz input signal is first amplified and then downconverted to the IF frequency of 10.7 MHz. Then the IF signal again is amplified and using an external filter is fed to a limiter. The limiter keeps the phase information and removes any variations on the amplitude of the signal. Finally, the limited signal goes through a quadrature detector for frequency demodulation. The point is that the capacitive impedance of the series capacitor C_S is much larger than the impedance of the resonator at its center frequency. As such, we would observe a phase shift of approximately 90° between V_2 and V_1 at the center frequency. Note that, for this purpose, we should have $R_P C_S \omega_1 \ll 1$. For instance, consider a signal with the limited amplitude of $1V$ at 10.7 MHz with a frequency deviation of ± 75 KHz which has experienced a phase shift of 90° (V_2 is in quadrature with V_1) and the two signals are fed to the phase

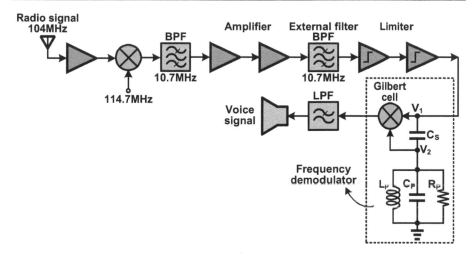

Figure 3.6: Signal receiver for FM.

detector. Here, a Gilbert cell can be used as the phase detector. Assuming a high input impedance for both inputs of the phase detector, we can easily show that

$$\frac{V_2}{V_1} = \frac{C_S L_P s^2}{\frac{s^2}{\omega_1^2} + \frac{s}{Q\omega_1} + 1} \tag{3.25}$$

where

$$\omega_1 = \frac{1}{\sqrt{L_P (C_P + C_S)}} \tag{3.26a}$$

$$Q = R_P (C_P + C_S) \omega_1 \tag{3.26b}$$

Here Q is the detector's quality factor and ω_1 is the center frequency of the detector. In the frequency domain, one can write

$$\frac{V_2}{V_1} = \frac{jQ\frac{C_S}{C_S + C_P}\frac{\omega}{\omega_1}}{1 + jQ\frac{\omega_1}{\omega}\left(\frac{\omega^2}{\omega_1^2} - 1\right)} \approx \frac{jQ\frac{C_S}{C_S + C_P}}{1 + j2Q\frac{\Delta\omega}{\omega_1}} \tag{3.27}$$

where $\Delta\omega = \omega - \omega_1$, and it is considered that $\omega \approx \omega_1$ in Equation 3.27. Figure 3.7 shows the frequency response of the a quadrature detector.

As such, the phase difference between the voltages in the quadrature phase detector becomes

$$\Delta\phi = \angle V_2 - \angle V_1 = \frac{\pi}{2} - \tan^{-1}\left(2Q\frac{\Delta\omega}{\omega_1}\right) \tag{3.28}$$

Assume the input FM signal has the following form

$$V_1 = A\cos\left(\omega_1 t + \Delta\omega \int_0^t f(\tau)d\tau\right) \tag{3.29}$$

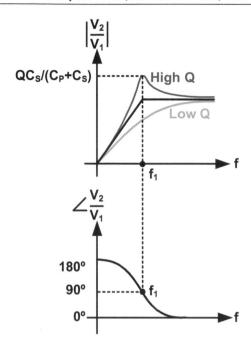

Figure 3.7: Frequency response of the phase detector.

Then, the instantaneous frequency of this signal would be

$$\omega = \omega_1 + \Delta\omega f(t) \tag{3.30}$$

where $f(t)$ is the normalized baseband signal, that is, $-1 \leq f(t) \leq +1$. The approximate formula for Equation 3.28 can be written as

$$\Delta\phi = \tan^{-1}\left(\frac{-\frac{\omega_1}{(\omega_1 + \Delta\omega f(t))Q}}{1 - \left(\frac{\omega_1}{\omega_1 + \Delta\omega f(t)}\right)^2}\right) \approx \frac{\pi}{2} - \frac{2Q}{\omega_1}(\Delta\omega)f(t) \tag{3.31}$$

As it is seen in Equation 3.31, the phase difference between the two voltages V_1 and V_2 is proportional to the instantaneous frequency deviation $\Delta\omega f(t)$ (while the two voltages are at quadrature at the resonant frequency). Or in other words, as Figure 3.7 suggests, the phase difference between two inputs of phase detector at f_1 is $90°$. Due to the frequency deviation of the frequency-modulated signal, the output voltage varies with frequency deviation and consequently with the slope of the phase characteristics of the quadrature tank. Figure 3.8 depicts the phase characteristics of the quadrature tank for different quality factors.

As it is obvious from Figure 3.8, for a higher quality factor, we will attain a higher sensitivity for a specific frequency deviation and the characteristic of Figure 3.8 will be sharper. We can approximate the phase variations in Figure 3.8 near the center frequency linearly as depicted in Figure 3.9.

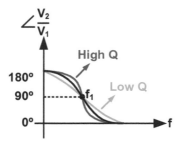

Figure 3.8: Phase characteristics of the quadrature tank by different quality factors.

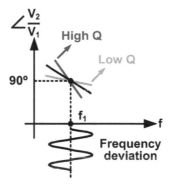

Figure 3.9: A linear approximation for phase characteristic of the quadrature tank.

This linear approximation is valid for a small frequency range. In our example, the IF frequency is 10.7 MHz which is normally used in FM receivers. Moreover, as stated earlier, the frequency deviation is ±75 KHz. The mentioned frequency deviation is the maximum value; however, its instantaneous value depends on the input baseband signal. We can write the phase difference in Figure 3.9 with Equation 3.31 as

$$\Delta\phi = \frac{\pi}{2} - \frac{2Q}{\omega_1}(\Delta\omega)f(t) \tag{3.32}$$

where in Equation 3.32, $\Delta\omega$ is the frequency deviation, $f(t)$ is the voice signal, Q is the quality factor, and ω_1 is the center frequency. The quality factor is the following

$$Q = (C_S + C_P)\,\omega_1 R_P \tag{3.33}$$

In this circuit, the frequency deviation is translated to $\Delta\phi$ and the phase detector translates the phase difference to a voltage proportional to the baseband. Now, let's analyze quantitatively the output from the multiplication occurring in the phase detector. Consider the FM input applied to port1 of the phase detector as

$$V_1 = A\cos\left(\omega_1 t + \Delta\omega \int_0^t f(\tau)d\tau\right) \tag{3.34}$$

One may obtain V_2 at ω_1 as

$$V_2 = \frac{C_S}{C_S + C_P} QA \cos\left(\omega_1 t + \Delta\omega \int_0^t f(\tau)d\tau + \frac{\pi}{2} - \frac{2Q}{\omega_1}(\Delta\omega)f(t)\right) \qquad (3.35)$$

It should be noted that higher quality factor results in more nonlinearity and distortion in the detection process. For a moment if we assume the phase detector as a multiplier and with the inputs of Equations 3.34 and 3.35, then the output would have a form like

$$V_1 V_2 = \frac{QA^2}{2} \frac{C_S}{C_S + C_P}\left[\cos\left(\frac{\pi}{2} - \frac{2Q}{\omega_1}(\Delta\omega)f(t)\right)\right.$$
$$\left. - \cos\left(2\omega_1 t + 2\Delta\omega \int_0^t f(\tau)d\tau + \frac{\pi}{2} - \frac{2Q}{\omega_1}(\Delta\omega)f(t)\right)\right] \qquad (3.36)$$

But considering the amplitude A is large enough to make a hard switching for the upper tree of the Gilbert cell, and the amplitude of V_2 that is $QA\frac{C_S}{C_S+C_P}$ is still large enough to make a hard switching of the lower tree, then the output of the phase detector would be proportional to $R_L I_0$. As such, the low-pass component of the output of the phase detector will have the following form

$$v_{\text{out}_{\text{LPF}}} = R_L I_0 \sin\left(\frac{2Q}{\omega_1}(\Delta\omega)f(t)\right) \approx R_L I_0\left(\frac{2Q}{\omega_1}(\Delta\omega)f(t)\right) \qquad (3.37)$$

The linear approximation is valid for $2Q\Delta\omega/\omega_1 < \pi/4$. As Equation 3.36 suggests, higher quality factor of the resonant circuit results in larger amplitude of V_2. However, linear approximation of the frequency response will be violated for large values of the quality factor and distortion in the baseband data will emerge at the output due to response nonlinearity. To mitigate this issue, one may decrease the quality factor. There is another frequency demodulation scheme which is discussed in the PLL section.

Example 3.1 In the given FM detector circuit
(a) Find the transfer function of the frequency demodulator. The carrier signal frequency is 455 kHz and the Sallen–Key filter has two poles at 455 kHz. Moreover, the Gilbert cell is a multiplier circuit which experiences complete switching for its transistors.
(b) Suppose the modulation frequency of 1 kHz and the frequency deviation of 8 kHz, find the amplitude of the second and the third harmonics of 1 kHz at the output. $R_1 = R_2 = 1\,k\Omega$, $C_1 = C_2 = 350\,pF$ and you may use the given Taylor's expansion.

$$\tan^{-1}(1+x) \approx \left(\frac{\pi}{4} + \frac{1}{2}x - \frac{1}{4}x^2 + \frac{1}{12}x^3\right) \quad \text{for } x < 1 \qquad (3.38)$$

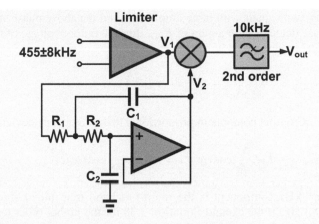

Figure 3.10: Frequency demodulator by quadrature phase detector using a Sallen–key filter.

Solution:
(a) Given $R_1 = R_2 = R$, and $C_1 = C_2 = C$, the transfer function of the Sallen–Key filter can be written as

$$\frac{v_2}{v_1} = \frac{1}{(1 + jRC\omega)^2} \tag{3.39}$$

Then

$$\angle \frac{v_2}{v_1} = -2\tan^{-1}(RC\omega) = -2\tan^{-1}\left(\frac{\omega}{\omega_0}\right) = -2\tan^{-1}\left(\frac{f}{f_0}\right) \tag{3.40}$$

We also know that $f = f_0 + \Delta f$, therefore Equation 3.39 shrinks to

$$\angle \frac{v_2}{v_1} = -2\tan^{-1}\left(1 + \frac{\Delta f}{f_0}\right) \tag{3.41}$$

Note that

$$\left|\frac{V_2}{V_1}\right| \approx \frac{1}{2} \quad \text{for} \quad \frac{\Delta f}{f_0} \ll 1 \tag{3.42}$$

If we expand Equation 3.41, we then reach to

$$\angle \frac{v_2}{v_1} = -2\left(\frac{\pi}{4} + \frac{\Delta f}{2f_0} - \cdots\right) = -\frac{\pi}{2} - \frac{\Delta f}{f_0} \tag{3.43}$$

Now two signals with large amplitudes and the above phase shift are applied to a phase detector with a gain of K_{PD}, thus the output voltage of this block for a sinusoidal modulation will be

$$v_{OUT} = -K_{PD}\left(\frac{\pi}{2} + \frac{\Delta f}{f_0}\cos(\omega_m t)\right) \tag{3.44}$$

For the AC output component proportional to the frequency deviation, we have

$$v_{out} = -K_{PD}\frac{\Delta f}{f_0}\cos(\omega_m t) = -K_{PD}\frac{8}{455}\cos(\omega_m t) \tag{3.45}$$

(b) The 1 kHz component is the main baseband transmitted signal, i.e., $\omega_m = 2\pi(1000)$ Hz. If we expand Equation 3.38 for the higher order terms (nonlinear terms) as well for v_{OUT}, then we reach to

$$v_{OUT} = -K_{PD}\left(\frac{\pi}{2} + \frac{\Delta f}{f_0}\cos(\omega_m t) - \frac{1}{2}\left(\frac{\Delta f}{f_0}\right)^2\cos^2(\omega_m t)\right.$$
$$\left. + \frac{1}{6}\left(\frac{\Delta f}{f_0}\right)^3\cos^3(\omega_m t)\right) \tag{3.46}$$

The all-pass filter translates this frequency deviation to a specific phase shift, and consequently to the corresponding voltage at the output of the phase detector. Moreover, we know that the nonlinear characteristic of phase transfer function results in harmonic generation of the baseband signal. As such, the output voltage will be

$$v_{OUT} = -k_{PD}\left[\left(\frac{\pi}{2} - \frac{1}{4}\left(\frac{\Delta f}{f_0}\right)^2\right) + \left(1 + \frac{1}{8}\left(\frac{\Delta f}{f_0}\right)^2\right)\left(\frac{\Delta f}{f_0}\right)\right. \tag{3.47}$$
$$\left.\cos(\omega_m t) - \frac{1}{4}\left(\frac{\Delta f}{f_0}\right)^2\cos(2\omega_m t) + \frac{1}{24}\left(\frac{\Delta f}{f_0}\right)^3\cos(3\omega_m t) + \cdots\right]$$

Therefore, the amplitudes of the second and the third harmonics are

$$H_2 = \frac{K_{PD}}{4}\left(\frac{8}{455}\right)^2 \tag{3.48a}$$

$$H_3 = \frac{K_{PD}}{24}\left(\frac{8}{455}\right)^3 \tag{3.48b}$$

Example 3.2 Consider the FM detector depicted in Figure 3.6. The FM signal carrier is at 455 kHz with 2 V amplitude. If the frequency deviation is 8 kHz, and with the assumption of an ideal multiplier with a load resistance of $R_L = 1\,k\Omega$ and the total bias current of $I_0 = 1\,mA$, and assuming $C_P = 1\,nF$, $C_S = 10\,pF$, and $Q = 5$, obtain the detected output signal amplitude.

Solution:
With the given parameters, we can write

$$V_1 = 2\sin\left(\omega_0 t + \Delta\omega \int_0^t f(\tau)d\tau\right) \quad \text{Volts} \tag{3.49}$$

For V_2, we have

$$V_2 = Q\frac{C_S}{C_S + C_P} \times 2\sin\left(\omega_0 t + \Delta\omega \int_0^t f(\tau)d\tau + \frac{\pi}{2} - \frac{2Q}{\omega_0}(\Delta\omega)f(t)\right) \quad \text{Volts} \tag{3.50}$$

$$V_2 = 0.099\sin\left(\omega_0 t + \Delta\omega \int_0^t f(\tau)d\tau + \frac{\pi}{2} - \frac{2Q}{\omega_0}(\Delta\omega)f(t)\right) \quad \text{Volts} \tag{3.51}$$

Considering that the Gilbert cell multiplier is driven to its saturation level by both input signals, the low-pass component of the output becomes

$$v_{\text{out}_{LPF}} = R_L I_0 \sin\left(\frac{2Q}{\omega_1}(\Delta\omega)f(t)\right) \tag{3.52}$$

Now noting that the sinusoidal argument is less than 1 rad, we then reach to

$$v_{\text{out}_{LPF}} \approx R_L I_0 \left(\frac{2Q}{\omega_0}(\Delta\omega)f(t)\right) = 0.176 f(t) \quad \text{Volts} \tag{3.53}$$

where by substituting the parameters, the output voltage amplitude is 176 mV. ∎

3.3 Basics of PLLs and their Application as an FM Demodulator

In this section, we introduce the PLL as a frequency demodulator. First of all, let's survey some important characteristics of PLLs. Consider Figure 3.11.

The input in Figure 3.11 consists of an FM sinusoidal signal whose frequency alternates in a time interval ΔT. For the sake of simplicity, assume that the VCO is operating at its free-running frequency, say, 10.7 MHz. The VCO signal is multiplied by the input frequency-modulated signal by an analog multiplier. If we assume an ideal multiplication, we will reach to the second harmonic (21.4 MHz $\pm$ 70 kHz) and the low-pass component whose frequency of variations is proportional to $1/\Delta T$. This low-pass component appears at the output of the low-pass filter. The feedback loop tends to

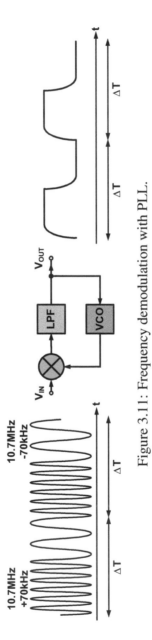

Figure 3.11: Frequency demodulation with PLL.

change the VCO frequency toward the instantaneous frequency of $10.7\,\text{MHz} \pm 70\,\text{kHz}$. In other words, the VCO in this loop tries to follow the input frequency. This behavior is called phase locking and we call this loop as PLL. While the input frequency varies in the PLL, the loop tries to generate an error voltage to correct the VCO frequency. This error voltage is proportional to the baseband modulating signal and by this virtue, the PLL output will be the FM detected signal. Note that if the variations are fast (that is, faster than the loop bandwidth), the loop would not be capable of following the input frequency and the loop will not operate properly. Indeed, the PLL is a low-pass system.

> **Example 3.3** A student asks whether the PLL is the same as frequency-locked loop (FLL), i.e., at the steady state, the frequencies will be the same as the phases are the same. Is he/she right?
>
> **Answer:**
> Yes, in a sense that once the loop is locked the reference and the VCO output frequencies would be the same but with a constant phase shift existing between them. But if the loop is not locked the VCO will act as a free-running oscillator. ∎

Now, consider Figure 3.12.

Assume in Figure 3.12, the multiplier is a Gilbert cell phase detector, then we may obtain the phase detector gain as

$$K_{\text{PD}} = \frac{V_{O_{\text{PD}}}}{\phi_{\text{in}} - \phi_{\text{out}}} \tag{3.54}$$

where in Equation 3.36, K_{PD} is the phase detector gain. We may obtain the transfer function of the low-pass filter (for a single-pole low-pass filter) easily as

$$\frac{V_{\text{out}}}{V_{\text{in}_{\text{LPF}}}} = \frac{1}{1 + \frac{s}{\omega_{\text{LPF}}}} \tag{3.55}$$

We can also write the output signal of the oscillator as

$$V_{\text{osc}} = A \sin \Theta(t) = A \sin \left(\omega_{\text{fr}} t + K_{\text{VCO}} \int V_{\text{out}}(t) dt \right) \tag{3.56}$$

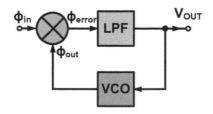

Figure 3.12: PLL schematic diagram.

The oscillator in its static mode oscillates at the free-running frequency, ω_{fr}. Here K_{VCO} is the VCO gain in rad/s per $Volts$ or equivalently $\frac{Hz}{V}$. Depending on the VCO gain and the input signal, the output frequency changes. We can write the expression for the instantaneous frequency of the oscillator as

$$\omega_{osc} = \omega_{fr} + K_{VCO}V_{out}(t) \tag{3.57}$$

We can reach to time-dependent frequency by taking the derivative of the total phase which results in

$$\frac{d}{dt}\Theta(t) = \omega_{fr} + K_{VCO}V_{out}(t) = \omega_{osc}(t) \tag{3.58}$$

Taking the Laplace transform of both sides of Equation 3.58 gives

$$s\phi_{out}(s) = K_{VCO}V_{out}(s) = \omega_{osc}(s) \tag{3.59}$$

Finally, the output phase in s-domain

$$\phi_{out}(s) = \frac{K_{VCO}V_{out}(s)}{s} \tag{3.60}$$

As Equation 3.60 suggests, VCO acts as an integrator in the PLL.

3.3.1 The Transfer Function of the First-Order PLL

Figure 3.12 can be modeled as Figure 3.13.

The phase detector output is the result of multiplication of two square-wave signals which the low-pass filter extracts its average value. As depicted in Figure 3.13, the feedback is of unity gain. Note, signals in this loop are both considered as voltage and phase. However, our transfer function of interest is the output phase as a function of the input phase. One may obtain the open loop gain as

$$a(s) = \left(\frac{1}{1+\frac{s}{\omega_{LPF}}}\right) K_{PD}\frac{K_{VCO}}{s} \tag{3.61}$$

Then, the closed-loop gain (as a negative feedback loop) which is the transfer function of interest can be written as

$$\frac{\phi_o}{\phi_i} = \frac{a(s)}{1+fa(s)} \tag{3.62}$$

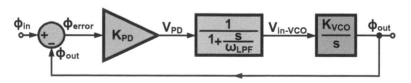

Figure 3.13: PLL model.

Substituting Equation 3.61 into Equation 3.62 and setting the feedback gain $f = 1$ give

$$\frac{\phi_o}{\phi_i} = \frac{\left(\dfrac{1}{1+\frac{s}{\omega_{LPF}}}\right)\dfrac{K_{PD}K_{VCO}}{s}}{1+\left(\dfrac{1}{1+\frac{s}{\omega_{LPF}}}\right)\dfrac{K_{PD}K_{VCO}}{s}} \tag{3.63}$$

We can write Equation 3.63 as

$$\frac{\phi_o}{\phi_i} = \frac{1}{\dfrac{s^2}{\omega_{LPF}K_{VCO}K_{PD}} + \dfrac{s}{K_{VCO}K_{PD}} + 1} \tag{3.64}$$

Equation 3.64 suggests that the transfer function gain at low frequencies is unity which means that the loop follows the input phase at the output for low-frequency variation. However, for a high-frequency input, the gain of the loop will be decreased. Thus, the transfer function has a low-pass behavior. We can rewrite the transfer function as

$$\frac{\phi_o}{\phi_i} = \frac{1}{\left(\dfrac{s}{\omega_n}\right)^2 + \dfrac{s}{Q\omega_n} + 1} \tag{3.65}$$

where parameters in Equation 3.65 can be derived as

$$\omega_n = \sqrt{\omega_{LPF}K_{VCO}K_{PD}} \tag{3.66a}$$

$$Q = \sqrt{\frac{K_{VCO}K_{PD}}{\omega_{LPF}}} \tag{3.66b}$$

The parameter Q in Equation 3.65 has an important effect. If Q is equal to $1/2$, the poles of the loop coincide, if it is greater than $1/2$, we will have complex conjugate poles, and if Q is lower than $1/2$, the loop consists of real poles. It is instructive to know that at ω_n, the loop exhibits an overshoot which is illustrated in Figure 3.14.

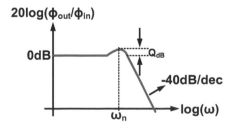

Figure 3.14: Overshoot in the frequency response of the PLL near the natural frequency.

Example 3.4 To implement the phase detector, we use the given Gilbert cell.
(a) Find the load resistance and the load capacitance to have a phase-detector gain of $\frac{1}{\pi}\frac{\text{Volts}}{\text{Radian}}$.
(b) With the phase detector characteristics depicted in Fig. 3.16, we implement a PLL as in Figure 3.17. Suppose $R_f = 100\,\Omega$, find the value of C_f and the transfer function of the loop.
(c) If the input frequency suddenly changes from $100\,\text{MHz}$ to $100.1\,\text{MHz}$, draw the control voltage as a function of time.
Assume $K_{VCO} = 500\,\text{kHz/V}$, and $Q = 1/(2\zeta) = 0.5$.

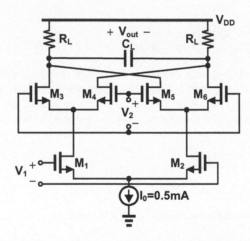

Figure 3.15: The Gilbert cell used as the phase detector.

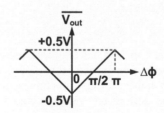

Figure 3.16: The desired transfer function of the phase detector.

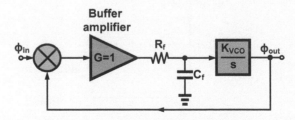

Figure 3.17: Simple PLL (Type I).

Solution:

(a) When a current completely flows to one side, we have $R_L I_0 = 0.5\,\text{V}$, which gives the load resistance of $1\,\text{k}\Omega$, and consequently (Figure 3.16) the gain of the phase detector will be $1/\pi$.

As the output of the phase detector should be low pass

$$\frac{1}{2\pi R_L C_L} \ll 100\,\text{MHz} \tag{3.67}$$

Let

$$\frac{1}{4\pi R_L C_L} = 5\,\text{MHz} \tag{3.68}$$

Then, $C_L = 16\,\text{pF}$.

(b) We have the expression for Q as

$$Q = \sqrt{\frac{K_{PD} K_{VCO}}{\omega_{LPF}}} \Rightarrow 0.5 = \sqrt{\frac{\frac{1}{\pi}(2\pi \times 0.5\,\text{MHz})}{\omega_{LPF}}} \tag{3.69}$$

$$\omega_{LPF} = 2\pi \times 6.25 \times 10^5 \,\frac{\text{rad}}{\text{sec}} \tag{3.70}$$

For the value of the capacitor, we have

$$C_f = \frac{Q^2}{2\pi K_{PD} K_{VCO} R_f} = 400\,\text{pF} \tag{3.71}$$

The transfer function of the PLL can be expressed as

$$\frac{\phi_o}{\phi_i} = \frac{1}{\left(\frac{s}{\omega_n}\right)^2 + \frac{s}{Q\omega_n} + 1} \tag{3.72}$$

The natural frequency of the loop can be calculated as

$$\omega_n = \sqrt{K_{PD} K_{VCO} \omega_{LPF}} = \sqrt{\frac{1}{\pi} \times 2\pi \times 0.5\,\text{MHz} \times 2\pi \times 6.25 \times 10^5 \tfrac{\text{rad}}{\text{sec}}}$$

$$= 2M\,\tfrac{\text{rad}}{\text{sec}} \tag{3.73}$$

and thus the output phase relation will be

$$\frac{\phi_o}{\phi_i} = \frac{\left(2 \times 10^6\right)^2}{\left(s + 2 \times 10^6 \tfrac{\text{rad}}{\text{sec}}\right)^2} \tag{3.74}$$

As Equation 3.74 suggests, the system is critically damped here, and therefore, it will have the fastest response without overshoot.

(c) The relation between the output frequency and the input frequency can be written as

$$\frac{f_o(s)}{f_i(s)} = \frac{s\phi_o(s)}{s\phi_i(s)} \tag{3.75}$$

Thus, the output frequency varies with double pole as $1/(s+2\,\text{Mrad/sec})^2$, and we will reach to Figure 3.18.

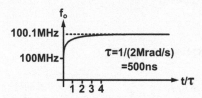

Figure 3.18: Time response of the PLL Frequency.

Moreover, one may write the output phase as

$$\phi_o = \frac{K_{\text{VCO}}}{s} v_{\text{in,control}} \Rightarrow s\phi_o = K_{\text{VCO}} v_{\text{in,control}} \tag{3.76}$$

and for the frequency, we have

$$f(s) = K_{\text{VCO}} v_{\text{in,control}}(s) \Rightarrow f_1(t) = K_{\text{VCO}} v_{\text{in,control}_1}(t) \tag{3.77}$$

Finally, Figure 3.19 depicts how the control voltage varies with time and reaches to its final value.

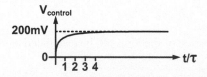

Figure 3.19: The control voltage of the PLL as a function of time.

Example 3.5 In the FM detector circuit depicted in Figure 3.20, the output of the limiter has an amplitude of 200 mV. With a maximum frequency variation rate (modulation rate) of 5 MHz, the frequency deviation is 7 MHz. The IF carrier frequency is at 140 MHz.

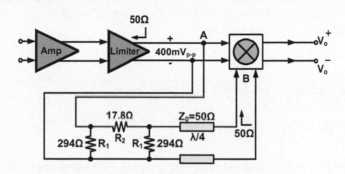

Figure 3.20: Quadrature FM demodulator.

(a) Determine the input signal amplitude at the point B.
(b) Secondly, given the multiplier circuit, find the amplitude of the detected signal.
(c) If the transmission line had a phase shift of 70° instead of 90° at 140 MHz, what would be the DC value across the 7.6 pF capacitor (at the output of circuit depicted in Fig. 3.21).

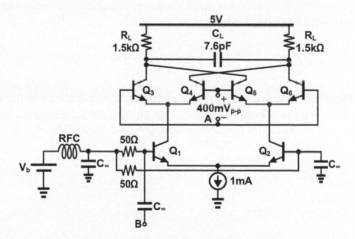

Figure 3.21: Gilbert cell phase detector.

Solution:
(a) In this part, the input signal is attenuated through a π-section resistive attenuator. As the attenuator is matched at the input and the output, the output voltage will become

$$V_B = \frac{R_1 \parallel Z_0}{R_2 + R_1 \parallel Z_0} V_A = 0.706 V_A = 141 \, mV \tag{3.78}$$

(b) Here, given the instantaneous FM signal frequency, the quarter-wave transmission line acts as a 90° phase shifter in the following manner.

The instantaneous frequency is

$$\omega_i = \omega_0 + \Delta\omega f(t) \tag{3.79}$$

The instantaneous phase shift would be

$$\Delta\phi = \frac{\pi}{2}\left(1 + \frac{\Delta\omega}{\omega_0} f(t)\right) \tag{3.80}$$

where $f(t)$ is the baseband modulating signal, with unity amplitude. Now, given the low-pass output circuit of the multiplier and the fact that V_A and V_B are quite larger than V_t, the Gilbert Cell acts as an ideal phase detector (recall section 3.2.2), so its output would be proportional to the phase difference of the in-phase and the quadrature signals. Considering the cut-off frequency of the output RC circuit as

$$f_{\text{cut-off}} = \frac{1}{4\pi R_L C_L} \approx 7\,\text{MHz} \tag{3.81}$$

Therefore, given the fact that the modulating signal is band limited to 5 MHz, the output would have the following form

$$V_{\text{out}} = I_0 R_L \left(\frac{\pi}{2}\frac{\Delta\omega}{\omega_0} f(t)\right) = 117 f(t) \, \text{mV} \tag{3.82}$$

(c) The phase detector works such that it gives a zero DC output for a $\pi/2$ phase shift between the two input signals. Therefore, if the transmission line has a 70° phase shift at the center frequency, the DC output would become

$$V_{\text{DC,out}} = I_0 R_L \left(1 - \frac{\Delta\phi}{\frac{\pi}{2}}\right) = I_0 R_L \left(1 - \frac{\frac{7\pi}{18}}{\frac{\pi}{2}}\right) = \frac{2}{9} I_0 R_L \approx 333\,\text{mV} \tag{3.83}$$

■

Some applications mandate high-speed PLLs; however, others may use slow loops. It is possible to control the loop speed by proper choice of ω_n. Moreover, one may change the bandwidth of the low-pass filter to control the loop bandwidth. Equation 3.66 suggests that lowering the low-pass filter bandwidth results in increase in Q which may be undesirable and also may make the loop unstable with any additional parasitic pole. It can be stated that the flat gain is mostly obtained up to ω_n frequency. If one increases the bandwidth of the low-pass filter in order to achieve a fast loop, the bandwidth will not be extended because of the fact that the poles move farther from each other.

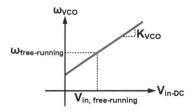

Figure 3.22: VCO characteristics.

Thus, it seems that we should increase the gain of the phase detector or the VCO gain to maintain the quality factor constant. This loop is called Type-1 loop, because its open-loop gain has a pole at the origin (note that the order of the transfer function of the PLL is always equal to the order of the transfer function of low-pass filter plus one). Now, consider a tone with 10 MHz frequency is applied at the input of a Type-I PLL. Moreover, the initial phase difference between the input and the output is 90°. If this input is applied to a Gilbert cell, the output voltage will be zero. If the oscillator is at its free-running frequency, the loop will be stable. Note that, if the input frequency changes to 11 MHz, an input voltage must be applied to the VCO to shift its frequency to 11 MHz. Thus, the phase difference between the input and the output will diverge from 90°, it may be, say, 85°. Thus, a PLL is not inherently capable of locking to any frequency. This phenomenon occurs due to the limited locking and capturing range in PLLs that is due to transfer function of the phase detector. The consequence of this phenomenon is that a PLL may not be locked.

Figure 3.22 shows the transfer function of the VCO. It is imperative that the designer must take into account the voltage range of the phase detector output and the VCO transfer function to allow the loop to lock.

Example 3.6 Given the initial conditions of the PLL transfer function, how is that the input and the output frequencies will be equal in steady state?

$$\frac{\phi_o}{\phi_i} = \frac{\frac{2\pi}{s} f_o}{\frac{2\pi}{s} f_i} = \frac{f_o}{f_i} = \frac{1}{\left(\frac{s}{\omega_n}\right)^2 + \frac{s}{Q\omega_n} + 1} \tag{3.84}$$

Solution:
Bearing in mind that the initial condition must be considered in Laplace transform, since we describe here the equation about the free-running frequency of the VCO ($V_{in-VCO} = 0$), the phase initial condition is not important here. This point is shown in the time domain as follows

$$\varphi_o(t) - \varphi_i(t) = cte. \tag{3.85}$$

Taking the derivative of both sides of Equation 3.85, we then reach to

$$\frac{d}{dt}\varphi_o(t) - \frac{d}{dt}\varphi_i(t) = 0 \Rightarrow f_o(t) = f_i(t) \tag{3.86}$$

Thus, the frequencies will be equal.   ∎

3.4   Further PLL Applications

In Figure 3.23, different inputs are applied to the PLL and the output is shown.

As depicted in Figure 3.23, the input signal bears different conditions. First, noise is added to the input signal, then the signal continues as usual, then the signal disappears (goes to zero), and finally a large signal, with a same frequency, is applied to the input. As the transfer function of the PLL has a low-pass characteristic, it passes the low-frequency component of the noisy signal; however, the overall noise is averaged out and the loop continues its normal behavior. Then, once the input signal has vanished, one of the phase detector inputs goes to zero. As the phase detector output will be equal to the product of the inputs, and if one of the inputs is zero, the output of the phase detector will be zero, the oscillator should tend to its free-running frequency by zero control voltage. However, the loop will maintain its current frequency if it has a sufficiently slow response (the time duration of the signal cut-off is much shorter than the loop time constant). Nonetheless, for high-speed loops, it may result in frequency change and movement to the free-running frequency of the oscillator. For the large-signal input (Figure 3.23), as the phase detector is principally insensitive to the input amplitude its output will remain unchanged, and consequently the PLL output will be unchanged.

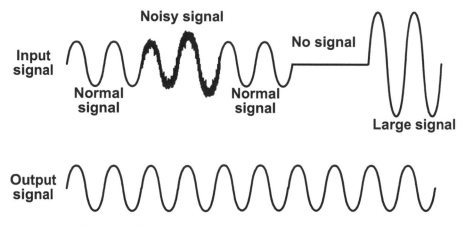

Figure 3.23: PLL response for different input signals.

Example 3.7 Does the transfer function relating the output phase to the input phase of the PLL infer unconditional stability, because of the fact that the output phase reaches to $-180°$ at positive infinite frequency?

Solution:
This is the simplified transfer function of the system with two poles; however, due to nonidealities, the order of the system might be increased. Furthermore, the phase margin is defined for an open loop, and we write it for the open loop to predict the closed-loop behavior. Moreover, since the transient response of the loop is of great importance, we need to take care of the phase margin.   ∎

Example 3.8 Consider the given type I PLL in Figure 3.24 with the Gilbert cell as the phase detector and with the following parameters.

$$VCO_{\text{Free-running frequency}} = 100\,\text{MHz}, K_{\text{VCO}} = 200\,\text{MHz/V} \tag{3.87}$$

$$\omega_{\text{LPF}} = 2\pi \times 1\,\text{MHz}, \quad K_{\text{PD}} = 2\,\text{V/rad} \tag{3.88}$$

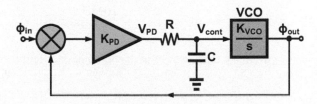

Figure 3.24: Type I PLL.

(a) Find the closed-loop transfer function.
(b) Find the loop phase margin.
(c) If the loop locks at 100 MHz, what is the phase difference between ϕ_i and ϕ_o?
(d) If the loop locks at 110 MHz, what is the phase difference between ϕ_i and ϕ_o?

Solution:
(a) For the transfer function of the PLL, we can write

$$H(s) = \frac{\varphi_{\text{out}}}{\varphi_{\text{in}}} = \frac{K_{\text{PD}}\frac{1}{RCs+1}\frac{K_{\text{VCO}}}{s}}{1 + K_{\text{PD}}\frac{1}{RCs+1}\frac{K_{\text{VCO}}}{s}} = \frac{K_{\text{PD}}K_{\text{VCO}}}{RCs^2 + s + K_{\text{PD}}K_{\text{VCO}}} \tag{3.89}$$

(b) To calculate the phase margin, we should find the point where the open-loop gain reaches unity. Then at that point, we compute the phase. Therefore

$$|H_{\text{OL}}(j\omega)| = 1 \Rightarrow \left| K_{\text{PD}}\frac{1}{RCj\omega+1}\frac{K_{\text{VCO}}}{j\omega} \right| = 1 \tag{3.90}$$

Then, the unity gain frequency will be

$$\left| \frac{K_{\text{PD}}K_{\text{VCO}}}{\omega\sqrt{1+R^2C^2\omega^2}} \right| = 1 \tag{3.91}$$

$$\Rightarrow \omega^2 = \frac{-1+\sqrt{1+4R^2C^2K_{\text{PD}}^2K_{\text{VCO}}^2}}{2R^2C^2} = 1.58 \times 10^{16} \tag{3.92}$$

$$\omega = 125.69 \times 10^6 \frac{\text{rad}}{\text{sec}} \Rightarrow f = 20\,\text{MHz}$$

Finally, the phase at this frequency will be $\varphi = -\frac{\pi}{2} - \tan^{-1}(RC\omega) = -177.1°$ and the resulting phase margin is $180 - 177.1 = 2.9°$.

(c) Since the given frequency is equal to the free-running frequency of the oscillator, the phase difference will be 90° and the control voltage will be zero.

$$\varphi_i - \varphi_o = 90°$$
(3.93)

(d) For this case, we have

$$\Delta V \times K_{VCO} = \Delta f \Rightarrow \Delta V = \frac{\Delta f}{K_{VCO}} = \frac{10\,\text{MHz}}{200\frac{\text{MHz}}{V}} = 50\,\text{mV}$$
(3.94)

and the phase difference with respect to the previous case will be

$$\Delta\varphi = \frac{V_{PD}}{K_{PD}} = \frac{50 \times 10^{-3}}{2} = 0.025\,\text{rad} = 1.43°$$
(3.95)

$$\varphi_i - \varphi_o = 90 + \Delta\varphi$$
(3.96)

Thus, the obtained phase difference will be added to 90° ($\varphi = 91.43°$). ∎

Example 3.9 In the previous example, using the ADS simulation tool, compute the following. The reference signal at first has a frequency of f_1 and then it experiences a frequency step and goes to a frequency of f_2,
(a) Draw the control voltage (V_{cont}), V_{PD}, V_{in}, V_{out}, and f_{out}.
(b) Suppose $f_1 = 100\,\text{MHz}$, $f_2 = 110\,\text{MHz}$, and $K_{VCO} = 200$ MHz/V. Find the final value of V_{cont} with respect to its initial value.

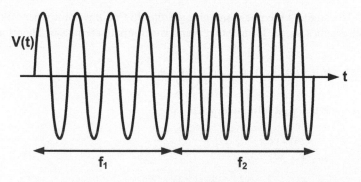

Figure 3.25: Time variations depicting a frequency step.

(c) If the input signal with the frequency of f_1, where f_1 is not the free-running frequency of the oscillator, vanishes, describe qualitatively what happens in the PLL.

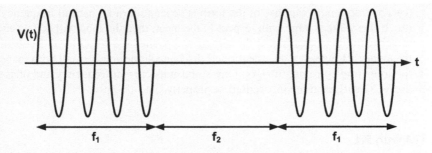

Figure 3.26: The signal vanishes in short step.

Solution:
(a) Figure 3.27 depicts the wanted signals.
(b) We can write

$$\Delta V \times K_{VCO} = \Delta f \Rightarrow \Delta V = \frac{\Delta f}{K_{VCO}} = \frac{10\,\text{MHz}}{200\,\frac{\text{MHz}}{\text{V}}} = 50\,\text{mV} \tag{3.97}$$

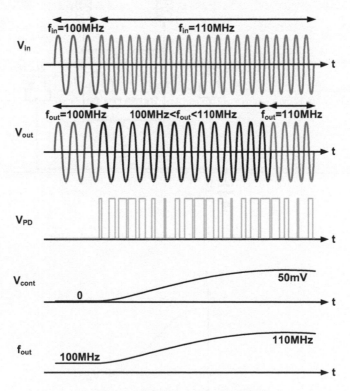

Figure 3.27: Desired signals.

(c) The frequency response of the loop is dependent of its natural frequency, i.e., ω_n. If this value is large with respect to the input, then the loop will be fast enough to sense the disappearing of the signal and pushes the VCO to its free-running frequency. However, if the mentioned disconnection time is small with respect to the loop time constant, the loop may stand at its current frequency and phase and the VCO will continue its oscillation properly.   ∎

3.4.1 FM with PLL

Consider the oscillator in Figure 3.28. In Figure 3.28, the MOS transistors M_3 and M_4 are used as varactors where their drain sources are short circuited. Figure 3.29 depicts the characteristics of the varactors (here C_1 has a large capacitance which is considered as a short circuit at oscillation frequency). As Figure 3.28 suggests, the varactors are in parallel with the inductors and make the resonant circuit. The transistors M_1 and M_2 realize the positive feedback, or otherwise, make a negative resistance across the

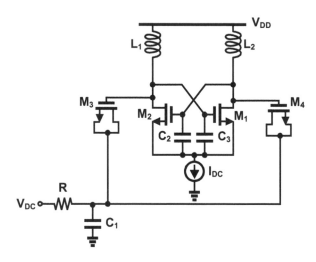

Figure 3.28: Cross-coupled oscillator.

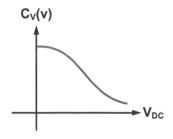

Figure 3.29: Characteristics of the nonlinear MOSFET varactor.

resonant circuit terminals. The frequency of the oscillation then can be obtained as

$$f = \frac{1}{2\pi} \sqrt{\frac{1}{L\left(C_N + C_V\left(v\right)\right)}} \tag{3.98}$$

where in Equation 3.98, C_N is the total capacitance at the output node and $C_v(v)$ is the nonlinear bias dependent varactor capacitance. The nonlinearity of varactor results in changing the VCO gain which in fact changes the closed-loop gain and the phase margin.

In the previous sections, we discussed the frequency demodulation with PLL. Now, we focus on FM with a PLL. Figure 3.30 illustrates both frequency modulation and demodulation with PLL.

Figure 3.30 shows the system-level structure of a frequency modulator. We have seen that by varying the varactor voltage, we are able to make a frequency modulator. The varactor was the MOS device which was biased in the reverse region. As an example, consider a 100 mV single-tone input signal in the control voltage of the VCO with the frequency of 10 Hz as

$$V_{MOD} = 0.1 \sin\left(2\pi \times 10\,\text{Hz}\right) \tag{3.99}$$

Moreover, suppose that the VCO is locked to 10.7 MHz. Depending on the bandwidth or speed of the loop, different outputs can be achieved. If the loop is faster than the input signal of the oscillator, it doesn't let the VCO to change its frequency (maintains the frequency of the loop as stable). On the other hand, for slow loops, the FM will

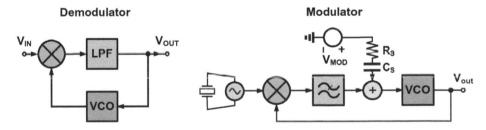

Figure 3.30: Frequency modulation and demodulation with PLL.

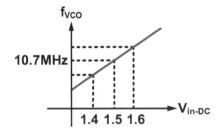

Figure 3.31: VCO characteristics for a 10.7 MHz carrier.

be materialized. The PLL here plays a main role to hold the intermediate frequency (the carrier frequency) as constant. From the quantitative analysis point of view, we remember that loop's f_{-3dB} is selected near ω_n to have complex poles with proper settling time. Now assume that the natural frequency of the loop is 500 Hz. Thus, the loop is fast enough not to let the frequency change. Now, if the input signal to the VCO changes its frequency to 5 kHz, the VCO changes its frequency with a rate of 5 kHz. The frequency deviation in the oscillator is merely dependent on the variations of its control voltage. It is clear that the larger-signal input to the VCO will result in more frequency deviation from the center frequency of the oscillator. Figure 3.31 depicts the characteristics of the oscillator for this example.

In an ordinary PLL, the output follows the input to find the same frequency. However, in FM with PLL, the loop resists against the carrier frequency variation. In fact, the loop has an output with the average frequency of 10.7 MHz and will find a frequency deviation corresponding to the input signal. It can be stated that in the frequency modulator, the loop should be designed as a slow loop, and in the frequency demodulator, the loop should be designed as a fast loop. Thus, the lower limit in frequency modulator is ω_n and the upper limit is specified by the low-pass filter for the modulating signal.

> **Example 3.10** Is is possible to feed the baseband signal to the VCO for the sake of FM generation without a PLL?
>
> **Answer:**
> Although an FM modulator with a simple VCO is conceivable, practically it is not possible, because of the requirement for the carrier frequency stability. The frequency stability of the PLL is then necessary for correct operation of the FM generation which is guaranteed by means of the negative feedback in the PLL loop. Moreover, PLL shapes the phase noise of the oscillator which is of great importance as well. ∎

3.4.2 PLL Application in Frequency Synthesizers and Its Transfer function

Consider Figure 3.32.

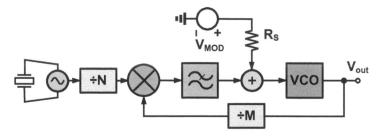

Figure 3.32: Frequency synthesizer block diagram for a frequency modulation scheme.

Figure 3.32 is the usual frequency synthesizer which is used to generate a frequency-modulated signal. The divider shown in Figure 3.32 is a digital counter which after M input pulses, generates one pulse. Note in any counter, the value of M can be selected digitally. In the steady-state condition, both the inputs of the phase detector will have the same frequency and as a result, we will have

$$\frac{f_{\text{Crystal}}}{N} = \frac{f_{\text{VCO}}}{M} \tag{3.100}$$

Suppose the crystal frequency is equal to 10.7 MHz, then for the VCO frequency, one may obtain

$$f_{\text{VCO}} = \frac{M}{N} f_{\text{Crystal}} = \frac{M}{N} 10.7 \,\text{MHz} \tag{3.101}$$

Moreover, assume that for the input divider, we have $N = 107$. Thus, the comparison frequency will be equal to 100 kHz. Now, assume that $M = 9000$. As a result, the output signal will be at 900 MHz and the channel spacing could be 100 kHz. The channel selection can be achieved by changing M, and thus $\frac{10.7\,\text{MHz}}{N}$ will be the minimum channel step. Assuming an input sinusoidal signal with 100 mV for the modulation signal, for the frequency of the VCO, we will have

$$f_{\text{VCO}} = \frac{M}{N} f_{\text{Crystal}} + 100 \,\text{mV} \times K_{\text{VCO}} \sin\left(\omega_{\text{BB}} t\right) \tag{3.102}$$

M is changed by the digital circuitry, and therefore one may hop from one channel to another. With respect to different standards, we can change the comparison frequency to change the channel spacing. In the high-frequency applications (e.g., higher than 5 GHz), we should break the divider into several stages and design a special counter for the first stage which operates at high frequency.

Till now, we have learned how to demodulate a frequency-modulated signal by a quadrature resonator or a PLL. Suppose the input signal frequency to the PLL is 10.7 MHz $\pm$ 70 kHz (in other words, the frequency deviation is 70 kHz), therefore the VCO follows the input frequency variations and the output of the phase detector through the low-pass filter gives in the detected FM baseband. However, in the quadrature FM detector, if the transmitted signal carrier frequency is changed, the detector could not detect thoroughly the input FM because the phase shift in quadrature component will be no longer about 90° and the detector would not perform correctly.

We have also shown that, using a PLL, we are able to generate a frequency-modulated signal which is shown in Figure 3.33. As stated earlier, the bandwidth of the FM signal at the PLL output in Figure 3.33 is

$$BW = 2\left(f_{\text{dev}} + f_{\text{m}}\right) \tag{3.103}$$

where f_{dev} is the maximum frequency deviation and f_{m} is the maximum frequency of the baseband signal. However, the bandwidth of the PLLs is far less than the above bandwidth. The frequency deviation is proportional to the amplitude of the modulating signal. As discussed earlier, the correct operation of the frequency modulator has two

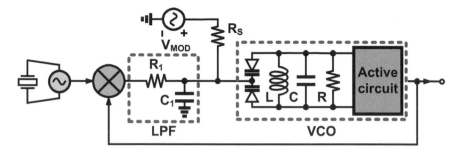

Figure 3.33: System- and circuit-level implementation of frequency modulation using a PLL.

margins which are specified by the natural frequency of the loop and the maximum frequency component of the modulating signal. In fact, the time constant of the loop characterizes the upper margin of the PLL. For increasing the time constant, making the loop slow, we can increase the capacitance in the loop filter.

Now, we derive equations for the dynamic behavior of the synthesizer in Figure 3.32. We stated that in the steady-state condition, both inputs of the phase detector will have the same frequency. It can be asserted that the phase detector is somehow a frequency detector as well and we can employ frequency modulator system as a phase modulator block too. Now, if we write the relation between the output phase and the input phase in Figure 3.32, we reach to

$$\frac{\phi_o}{\phi_i} = \frac{f_o}{f_i} = \frac{\left(1 + \frac{s}{\omega_{LPF}}\right)^{-1} K_{PD} \frac{K_{VCO}}{s}}{1 + \left(1 + \frac{s}{\omega_{LPF}}\right)^{-1} K_{PD} \frac{K_{VCO}}{s} \frac{1}{M}} \tag{3.104}$$

Thus, Equation 3.104 gives the transfer function of the frequency synthesizer. One of the important parameters in this loop is the transition time to shift from one channel frequency to another channel frequency which can be calculated through the inverse Laplace transform of Equation 3.104 which yields the settling time as

$$T_S = \frac{4}{\zeta \omega_n} \tag{3.105}$$

Note that the settling time is defined as the lapse of time required for the output frequency to reach 98% of its final value. Here ζ is the damping factor and it is expressed by

$$\zeta = \frac{1}{2Q} \tag{3.106}$$

Now, suppose the oscillator in Figure 3.32 has a frequency equal to 900 MHz and the channel spacing is 30 kHz. We can obtain channel spacing as follows

$$\text{Channel Spacing} = \frac{f_{Crystal}}{N} \tag{3.107}$$

Here, N determines the channel spacing. Thus, we now have implemented a frequency synthesizer which is used for frequency generation for both the receiver and the transmitter. Moreover, by putting a baseband signal in the control voltage of the latter, we will have a frequency modulator. As an example, for a crystal oscillator of 15 MHz frequency, one may obtain the value of N as

$$N = \frac{15000}{30} = 500 \qquad (3.108)$$

To change the channel frequency, we are able to change the value of M. Another usual method for frequency synthesis is the direct digital synthesis (DDS) which is very precise with the precision of hundredth of hertz (at IF frequency). The DDS-based design is out of the scope of this text.

Nowadays, the FM is not used in high-speed and high-performance transceivers. It is mainly used in commercial broadcast systems which depend on great number of conventional FM receivers. However, digital modulations such as $M - QAM$ and $QPSK$ are common in data communication which we discuss in the following chapters.

3.5 Advanced Topic: PLL Type II

The problems that the type I PLLs introduce have driven the designers to find a second type of PLL structure which does not have those imperfections. The very first problem of the type I PLL is its tradeoff between the stability and the output distortion. To mitigate the spur level at the output of the VCO, we can bring the pole of the LPF near the origin; however, this scheme will decrease the phase margin and as a result, there will be a greater possibility of instability. Another drawback of the type I PLL is its limited locking range due to the phase detector characteristics. These drawbacks were the incentive for the designers to propose a structure named phase-frequency-detector (PFD) which is also able to detect the difference of the input frequencies and consequently increase the locking range. We now introduce the basic behaviors of the PFD and the charge pump which are crucial in type II PLL. Using two D flip-flops and an AND gate, the system-level implementation of the PFD is shown in Figure 3.34.

As Figure 3.34 suggests, applying two signals with the same frequency and slight phase difference results in periodic output which is proportional to the phase difference of the inputs. This behavior is the same as that of the phase detector. The operation of the PFD is as follows: while signal A goes high, the output Q_A goes high till the input B goes high as well and both outputs Q_A and Q_B are applied to the AND gate and the output of it goes high. Then, the output of the flip-flops will be reset. Now, suppose the inputs have different frequencies then PFD will generate a signal proportional to the frequency difference which finally makes the oscillator to lock to the input frequency. In type II PLL, the charge-pump circuit alleviates the tradeoff between the stability and the spur level by introducing a new parameter. Figure 3.35 shows the charge-pump circuit.

The output signals of the PFD drive the up and the down inputs of the charge pump and as a result, the current sources may charge or discharge the capacitance charge level. Figure 3.36 depicts the operation of the charge pump alongside the PFD.

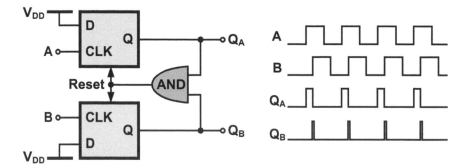

Figure 3.34: System-level implementation of PFD using two D flip-flops and an AND gate.

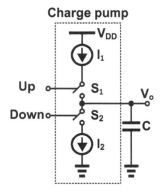

Figure 3.35: A typical charge-pump circuit using two current sources and two switches.

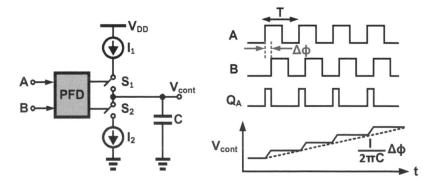

Figure 3.36: Operation of the charge-pump circuit under the excitation of the PFD.

Figure 3.36 suggests that when the switch is on, the capacitor is charged linearly and the circuit can be assumed as an integrator and when the switch is off the capacitor holds its value. The output voltage increment in Figure 3.36 can be approximated as

$$\Delta V_{\text{cont}} = \frac{\Delta \varphi}{2\pi} T \frac{I}{C} \tag{3.109}$$

Equation 3.109 can be rewritten for the control voltage of the oscillator as

$$V_{\text{cont}}(t) = \frac{\Delta \varphi}{2\pi} \frac{I}{C} t u(t) \tag{3.110}$$

By taking the Laplace transform of Equation 3.110, we reach to

$$\frac{V_{\text{cont}}(s)}{\Delta \varphi} = \frac{I}{2\pi C} \frac{1}{s} \tag{3.111}$$

Equation 3.111 shows the integration behavior of the circuit explicitly. Finally, by placing the charge-pump circuit subsequent to the PFD, and applying a unity feedback, the type II PLL can be achieved as in Figure 3.37.

The reason that we call this architecture type-II is that it has two poles at the origin in the open-loop transfer function (one for the charge pump and another for the VCO). The two poles at the origin make the instability of great concern. Thus, for the stability issues, we place a series resistor with the capacitor and rewrite the charge-pump equation as (this brings a zero in the open-loop as well as the closed-loop transfer function)

$$\frac{V_{\text{cont}}}{\Delta \varphi}(s) = \frac{I}{2\pi} \left(\frac{1}{Cs} + R \right) \tag{3.112}$$

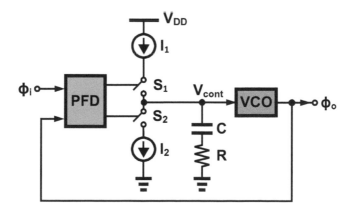

Figure 3.37: Type-II PLL block diagram including a PFD, a charge pump, and a VCO.

Thus, we can write the overall transfer function for type-II PLL as

$$H(s) = \frac{\varphi_0}{\varphi_i} = \frac{\frac{IK_{VCO}}{2\pi C}(RCs+1)}{s^2 + \frac{I}{2\pi}K_{VCO}Rs + \frac{I}{2\pi C}K_{VCO}} = \frac{\omega_n^2\left(1 + \frac{2\xi}{\omega_n}s\right)}{s^2 + 2\xi\omega_n s + \omega_n^2} \qquad (3.113)$$

where the parameters of the loop are

$$\xi = \frac{R}{2}\sqrt{\frac{ICK_{VCO}}{2\pi}} \qquad (3.114a)$$

$$\omega_n = \sqrt{\frac{IK_{VCO}}{2\pi C}} \qquad (3.114b)$$

$$Q = \frac{1}{2\xi} \qquad (3.114c)$$

Furthermore, the poles and the zero for the transfer function, $H(s)$, are

$$s_{p_{1,2}} = \left(-\xi \pm \sqrt{\xi^2 - 1}\right)\omega_n \qquad (3.115a)$$

$$s_z = \frac{-\omega_n}{2\xi} = -\frac{1}{RC} \qquad (3.115b)$$

As Equation 3.114 suggests, to mitigate the spur level, we can increase the value of C, and therefore ζ will be increased which now does not pose any problem for the instability. Thus, the drawbacks of the type-I PLL are now resolved at the cost of lower phase margin and consideration for stability due to increased order of the transfer function. To increase the locking speed, one should increase ω_n, and therefore, IK_{VCO} should be increased, or C could be decreased. Regarding the stability check of the type-II PLL, further reading in the given references is recommended.

3.6 Conclusion

In this chapter, the general configuration of the PLLs was studied. Care should be taken that in a PLL, the parameter of the study whose stability and response should be considered is the phase (and consequently, the frequency), so here we are considering the frequency response of the phase (or the frequency) in the loop. The phase detector was one of the main components of the PLL whose implementation using a Gilbert cell or an XOR was introduced. FM using a varactor-tuned oscillator was introduced alongside an FM demodulator using a quadrature resonator. The FM demodulation is possible using a sufficiently high-speed PLL. This concept was introduced as well. FM is possible using a low-speed PLL whose concept was described in this chapter. Frequency synthesizers are one of the basic building blocks of the modern transceivers. The basic structure of a frequency synthesizer using a crystal oscillator, a frequency divider, and a PLL including a second frequency divider was introduced as well. Type I PLLs are based on a phase detector, a low-pass filter, and a VCO. This type of PLL suffers from the problem of instability, and limited locking range. Type II PLLs were introduced to mitigate the problem of instability and the locking range. The type II PLL is based on a phase-frequency-detector, a charge pump, and a VCO. In sum, the building blocks described in this chapter can be used as frequency modulators, frequency demodulators, synthesizers, and eventually phase modulators.

3.7 References and Further Reading

1. F.M. Gardner, *Phaselock Techniques*, second edition, New York, NY: J. Wiley & Sons, 1979.
2. P.R. Gray, P.J. Hurst, S.H. Lewis, R.G. Meyer, *Analysis and Design of Analog Integrated Circuits*, fifth edition, Hoboken, NJ: J. Wiley & Sons, Inc., 2009.
3. D.H. Wolaver, *Phase-Locked Loop Circuit Design*, United Kingdom: Prentice Hall, 1991.
4. B. Razavi, *RF Microelectronics*, second edition, Castleton, NY: Prentice-Hall, 2011.
5. T.C. Carusone, D.A. Johns, K.W. Martin, *Analog Integrated Circuit Design*, Singapore: J. Wiley & Sons, 2013.
6. K.K. Clarke, D.T. Hess, *Communication Circuits, Analysis and Design*, United States: Krieger Publishing Company, 1994.
7. D.O. Pederson, K. Marayam, *Analog Integrated Circuits for Communications*, Boston, MA: Kluwer Academic Publishers, 1990.
8. L.W. Couch, *Digital and Analog Communication systems*, eighth edition, New Jersey: Prentice-Hall, 2013.

3.8 Problems

Problem 3.1 Figure 3.38 depicts a simplified frequency synthesizer.

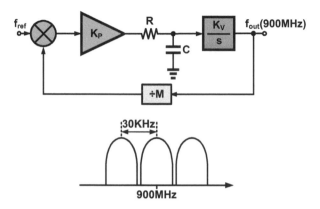

Figure 3.38: Type I frequency synthesizer.

In the transfer function of the loop, the value of ζ is taken as 0.707 and $\omega_n = 500$ rad/sec, and $K_P = 10$ V/rad.

1. First show that in this loop $\omega_n = \sqrt{\frac{\omega_{LPF} K_V K_P}{M}}$, $\zeta = \frac{1}{2}\sqrt{\frac{M\omega_{LPF}}{K_V K_P}}$.
2. For 30 kHz reference frequency, design the synthesizer for the channel spacing of 30 kHz and a center frequency of 900 MHz. (Find the divider's modulus M, the low-pass filter's RC time constant, and the VCO gain, K_V).
3. Find the settling time of the loop when it hops from the current channel to the adjacent channel.
4. If we replace the phase detector with a bipolar Gilbert cell, find the value of R_L for a bias current of 5 mA to obtain $K_P = 10$ V/rad.

Problem 3.2 It is possible to make an FM modulator out of a PLL as in Figure 3.39.

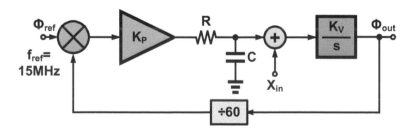

Figure 3.39: Frequency modulator using a PLL.

Suppose that we have $f_{3dBLPF} = 100$ Hz and $\omega_n = 2\pi \times 50$ Hz.
1. Determine ϕ_{out}/ϕ_{in} in the Laplace domain.

2. If the signal x_{in} is injected to the input of the VCO through R_S, determine the minimum and maximum frequency of the baseband input. Suppose that $R = 10\,k\Omega$, $C = 159.2\,nF$, $R_S = 100\,k\Omega$, and the average capacitance seen through the VCO is $C_{in,0} = 160\,pF$. You may use the equivalent circuit shown in Figure 3.40 for this purpose.

3. If $R_1 = 400\,\Omega$, and $K_V = 2\pi \times 100\,kHz/V$ determine the required K_P and consequently the tail current of the Gilbert cell phase detector.

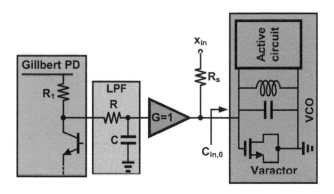

Figure 3.40: The equivalent circuit of the part of the PLL used as the FM modulator.

Problem 3.3 In the synthesizer depicted in Figure 3.41, we have $\omega_n = 2\pi \times 45\,kHz$, and $Q = 0.5$,

1. Find the loop filter's cut-off frequency and the phase-detector gain if the VCO gain is $K_{VCO} = 2\pi \times 1\,MHz/V$.
2. If the value of M changes from 1000 and 1001, draw the control voltage waveform.

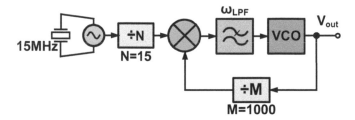

Figure 3.41: Frequency synthesizer using type I PLL.

Problem 3.4 **FM Modulator;** In the MOS oscillator stage depicted in Figure 3.42, the right-hand section acts as a variable reactance which loads the left-hand oscillator stage. Here, assume that $r \ll \frac{1}{C\omega_0}$. Determine an expression for the variable reactance seen through the right-hand section and from there obtain an expression for the frequency of oscillations (the carrier frequency and the frequency deviation) in terms of the circuit parameters. Assume that both MOS transistors operate in the square-law active region. Secondly, write an expression for the oscillation condition which determines the amplitude of oscillations. Here, $f(t)$ is the low frequency baseband signal varying between $+1$ and -1.

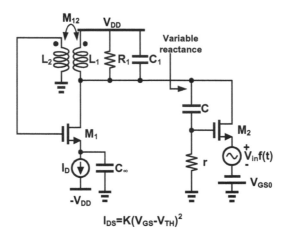

$$I_{DS}=K(V_{GS}-V_{TH})^2$$

Figure 3.42: MOS-based FM modulator/VCO.

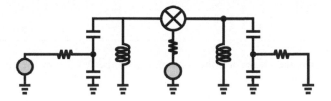

4. Mixers

4.1 Mixer Concept

Mixers are of integral parts of radio systems. Due to large-signal input, this block is usually quite nonlinear. This three-port block is used in receivers to downconvert the RF signal and in transmitters to upconvert the modulated signal. Due to their intrinsic nonlinear behavior and port-to-port leakage, these blocks mandate specific analysis for their operation. Moreover, taking into account their ever-existing harmonics in transceivers, it is necessary to achieve a high-performance system. In normal mixer operation, there is a large signal which is the local oscillator and two other small signals which are the IF and the RF signals. Upon driving a mixer's input toward large-signal regime (either of IF or RF signals) depending on the fact that the mixer is an upconverter or a downconverter), the output can pass through saturation.

4.1.1 The Conceptual Behavior of Single-Diode Mixers

Consider a simple mixer schematic as depicted in Figure 4.1.

Regarding the thermal voltage ($V_T = 26$ mV at $300°$ K), one can roughly consider the signals with an amplitude of less than 15 mV as small signal, and the signals in excess of 100 mV as large signal. A silicon diode will turn on by the threshold voltage

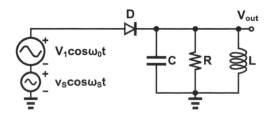

Figure 4.1: Basic mixer schematic with a single diode.

of, say, 700 mV and shows a finite turn on resistance. We know the on-resistance of the diode is equal to

$$r_{\text{on}} = \frac{V_T}{I_D} \tag{4.1}$$

Normally, the turn-on resistance of the diode is in order of the few ohms which could be considered as a short circuit compared to the load resistance (R). As such, once the diode is on, a whole RF voltage would appear at the output. Once the diode is turned off (has a large series impedance) in the negative half cycle of the LO signal, the output voltage goes to zero. As such, the input RF signal is sampled at the rate of the LO signal. The output voltage can be expressed as

$$v_{\text{out}} = v_s \cos{(\omega_s t)} . S{(\omega_0 t)} \tag{4.2}$$

where $S(\omega_0 t)$ is a square-wave signal that toggles between one and zero with the period of the LO. Its Fourier expansion is expressed in Equation 4.40. The small-signal output waveform is shown in Figure 4.2. It is obvious that within the right-hand product of Equation 4.2, there exists the sum and difference frequency components of the RF and the LO terms. As such, if the RF is at the input, the difference component gives in the IF signal and if the IF was at the input, the sum component would give in the RF signal.

For now, we have shown that the large-signal input makes diode to be on and off and when the diode is on, the input small signal appears at the output and when the diode is off, there would be no signal at the output. Moreover, by virtue of the tuned circuit, the desired frequency component of the signal would appear at the output. In the next section, we delve into the nonlinear transconductance which is approximated by a polynomial expansion.

4.1.2 A Nonlinear Circuit as a Mixer

Consider Figure 4.3. As Figure 4.3 suggests two input signal sources with finite resistance generate a voltage v at the input of our nonlinear device. Then, the output current passes through a resonant circuit. This voltage-dependent current source can be assumed as a nonlinear transconductance. Here, we have assumed the characteristic polynomial of the third order; however, in reality, this polynomial might be a more

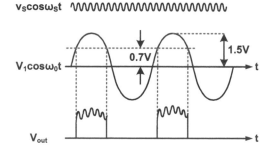

Figure 4.2: Input and output signals for Figure 4.1.

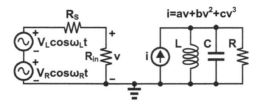

Figure 4.3: Polynomial model for a nonlinear current source.

complex function such as an exponential one. In radio communication, the weak signal is received by the antenna which is noted in Figure 4.3 as $(V_R \cos \omega_R t)$ and then this signal is mixed with the local oscillator signal noted as $(V_L \cos \omega_L t)$. Therefore, one may express the output current as

$$i = a\left(V_R \cos(\omega_R t) + V_L \cos(\omega_L t)\right) + b\left(V_R \cos(\omega_R t) + V_L \cos(\omega_L t)\right)^2 + \qquad (4.3)$$
$$c\left(V_R \cos(\omega_R t) + V_L \cos(\omega_L t)\right)^3$$

Our objective is to find the product term of RF and LO frequencies in Equation 4.3. We can expand Equation 4.3 to arrive at Equation 4.4:

$$i = aV_R \cos(\omega_R t) + aV_L \cos(\omega_L t) + bV_R^2 \cos^2(\omega_R t) + bV_L^2 \cos(\omega_L t) \qquad (4.4)$$
$$+ 2bV_R V_L \cos(\omega_L t)\cos(\omega_R t) + cV_R^3 \cos^3(\omega_R t) + cV_L^3 \cos^3(\omega_L t)$$
$$+ 3cV_R^2 \cos^2(\omega_R t)V_L \cos(\omega_L t) + 3cV_R \cos(\omega_R t)V_L^2 \cos^2(\omega_L t)$$

Each nonlinear circuit is capable of receiving both large and small signals, and by virtue of its nonlinearity generates the harmonics of the inputs and their products. We can define each component of Equation 4.4 as "RF," "LO" themselves, and "RF 2nd harmonic and a DC component," "LO 2nd harmonic and a DC component," "desired component of IF," "3rd harmonic of RF," and finally "3rd harmonic of LO." With the following trigonometric equations

$$\cos^2(\omega t) = \frac{1 + \cos(2\omega t)}{2} \qquad (4.5)$$

$$\cos^3(\omega t) = \frac{3}{4}\cos(\omega t) + \frac{1}{4}\cos(3\omega t) \qquad (4.6)$$

In real design, however, the large signal is the signal of local oscillator which can degrade the performance of the circuit due to nonlinear characteristic of diodes. Moreover, this signal can leak to other points of the circuit through the supply voltage line and the ground line, and cause undesirable effects. This leaked signal upon a nonlinear element can generate unwanted harmonics and mixing products. Finally, the main drawback of a nonlinear system is the handling of strong interferes and intermodulation products. This unfavorable mixing occurs in any nonlinear circuit with large-signal input. Tuning circuit may be useful to mitigate the effect of harmonic generation. For instance, if the LO frequency resides at 945 MHz and the RF frequency is at 900 MHz (as in the GSM case), by tuning the resonant circuit at 45 MHz, we can suppress the unwanted mixing products.

4.2 Third Order Intermodulation Concept in a Nonlinear Amplifier

Consider Figure 4.3 where the amplifier had a third order polynomial characteristics, here the input signals consist of two adjacent channels which we call father and mother signals. If we write the $I - V$ equation for the nonlinear amplifier, we arrive at Equation 4.7:

$$V_O = a\left(V_f\cos(\omega_f t) + V_m\cos(\omega_m t)\right) \tag{4.7}$$
$$+ b(V_f\cos(\omega_f t) + V_m\cos(\omega_m t))^2$$
$$+ c(V_f\cos(\omega_f t) + V_m\cos(\omega_m t))^3$$

If rewrite Equation 4.7, we can reach to

$$V_O = a\left(V_f\cos(\omega_f t) + V_m\cos(\omega_m t)\right) \tag{4.8}$$
$$+ b\left((V_f\cos(\omega_f t))^2 + (V_m\cos(\omega_m t))^2\right)$$
$$+ 2b\left(V_f\cos(\omega_f t)\right)\left(V_m\cos(\omega_m t)\right)$$
$$+ c\left((V_f\cos(\omega_f t))^3 + (V_m\cos(\omega_m t))^3\right)$$
$$+ 3c\left((V_f\cos(\omega_f t))^2\left(V_m\cos(\omega_m t)\right) + (V_f\cos(\omega_f t))\left(V_m\cos(\omega_m t)\right)^2\right)$$

Now, we can expand Equation 4.8 to obtain all the harmonic at the output. Until now, we have carried out equations for the output harmonics of a nonlinear circuit. Another important issue in a nonlinear amplifier is named as intermodulation (IM). Our IM of interest is IM3 which is the intermodulation product caused by third-order nonlinearity. Regarding two inputs as $v_1 = V_f\cos(\omega_f t)$ and $v_2 = V_m\cos(\omega_m t)$ as to adjacent channels, with respect to Equation 4.4, we then reach to

$$V_O = aV_m\cos(\omega_m t) + aV_f\cos(\omega_f t) + bV_m^2\cos^2(\omega_m t) + bV_f^2\cos^2(\omega_f t) \tag{4.9}$$
$$+ 2bV_fV_m\cos(\omega_f t)\cos(\omega_m t) + cV_m^3\cos^3(\omega_m t) + cV_f^3\cos^3(\omega_f t)$$
$$+ 3cV_m^2V_f\cos^2(\omega_m t)\cos(\omega_f t) + 3cV_mV_f^2\cos(\omega_m t)\cos^2(\omega_f t)$$

Then the third order IM components will be obtained as

$$V_{IM} = \frac{3}{4}cV_m^2V_f\cos((2\omega_m - \omega_f)t) + \frac{3}{4}cV_mV_f^2\cos((2\omega_f - \omega_m)t) \tag{4.10}$$

Figure 4.4 depicts the signal spectra at the input and the output of the nonlinear amplifier.

As Figure 4.4 suggests by the virtue of nonlinearity in the amplifier, different mixing products of the two input signals are generated at the output. However, in this derivation, we have merely taken into account a polynomial of third order. Magnitude of each component in Figure 4.4 can be easily computed by Equation 4.9. The green component in Figure 4.4 is called the IM product of third order, because this

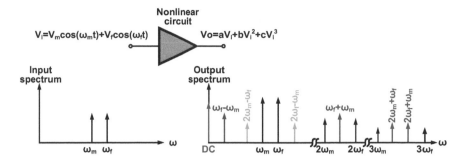

Figure 4.4: Representation of mixing products for two input adjacent channels.

term is generated due to the cubic term of the polynomial. This component can be troublesome in wide-band receivers and we then linearize the amplifier to mitigate this effect. As an example, suppose we have two adjacent channels with the frequency of $\omega_f = 2\pi \times 100.2\,\mathrm{MHz}$ and $\omega_m = 2\pi \times 100\,\mathrm{MHz}$. Thus, IM3 components reside at $2\omega_f - \omega_m = 2\pi \times 100.4\,\mathrm{MHz}$ and $2\omega_m - \omega_f = 2\pi \times 99.8\,\mathrm{MHz}$. As each channel is normally modulated by a random signal, the intermodulation products (IM3) could be considered as a random noise for either of the channels. Thus, this might be a drawback in receivers which can degrade signal-to-noise ratio (SNR) of the alternative channel. Assuming the magnitude of adjacent channel equal to V, Equation 4.9 suggests that the IM3 competent grows by V^3 and each channel power grows by V. This is an important point which exacerbates more the SNR. Note, if the power of each channel is added by 1 dB, IM3 component power will be added by 3 dB. This concern is mitigated by linearizing nonlinear circuit.

4.2.1 Characteristic of Third-Order IM and Measurement Method

The evolution of the aforementioned concept is shown in Figure 4.5. The upper nonlinear curve demonstrates the compression of the output signal with respect to the increase of the input signal. Its slope for lower values of the input signal is

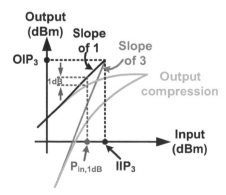

Figure 4.5: Intercept point of first harmonic and third-order intermodulation.

approximately equal to one (or 10 dB/dec). The lower nonlinear curve shows the evolution of the IM products level with respect to the input level. Its slope at the lower values of the input is about three times of that of the main output (30 dB/dec). Both of these curves saturate (experience a decrease in their respective slopes) at the high levels of the input signals.

If one draws the tangents at the two curves at lower signal levels and extends them far enough towards the higher levels, the two lines would intersect at a point which we call the third input-intercept-point (IIP_3 on the abscissa). Moreover, the output point is called oip_3. Figure 4.5 shows a real compression of the output signal which is denoted by the green line. In fact, the IIP_3 point is a practical indication of the nonlinearity of the amplifier. The higher it is, the more linear is the amplifier. The lower it is, the more nonlinear is the amplifier. Another point of interest is the saturation point of the amplifier and that point is where the difference between the linear input/output (tangent line) characteristic and the nonlinear (the real) input/output characteristic comes to 1 dB difference value, is called the compression point. It is another indication of the linearity of the amplifier. The higher the compression point, the more linear is the amplifier. Regarding the compression point refer to (equ compression),

$$v_{mo} = \left(aV_m + \frac{3}{4}cV_m{}^3 \right) \cos(\omega_m t) \tag{4.11}$$

$$v_{fo} = \left(aV_f + \frac{3}{4}cV_f{}^3 \right) \cos(\omega_f t) \tag{4.12}$$

Normally, in physical electronic devices, c/a is negative. As such, while increasing the input, the slope of the output signal level decreases. This phenomenon is called the compression in amplifier gain.

4.3 Basic Concept of Third-Order IM in a Basic Mixer

Till now, we have carried out computations for a nonlinear amplifier with two inputs. However, in a mixer, we may have two adjacent channels at one port and the large-signal LO at another port. Here, for the sake of simplicity, we assume that all the signal components are added up at a single input port. Furthermore, we assume that the device's input capacitance is small enough such that its reactance is much larger than the source resistance of the input signal. The nonlinear $I - V$ characteristics for the mixer are assumed to be a polynomial of the fourth order. Furthermore, a tuned circuit is employed at the output to select the desirable components. Here we try to demonstrate the same concepts of IM and compression for a mixer.

As it is obvious from Figure 4.6, the LO signal with two adjacent channels (namely, the mother (v_m) and the father signal (v_f)) are applied at the input of the nonlinear mixer. We now compute the frequency content at the output. Using the mentioned $I - V$ characteristics, we arrive at Equation 4.13:

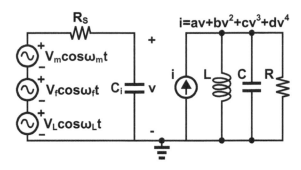

Figure 4.6: An approximation of a mixer with three input signals.

$$i = a(V_f \cos(\omega_f t) + V_m \cos(\omega_m t) + V_L \cos(\omega_L t)) \qquad (4.13)$$
$$+ b(V_f \cos(\omega_f t) + V_m \cos(\omega_m t) + V_L \cos(\omega_L t))^2$$
$$+ c(V_f \cos(\omega_f t) + V_m \cos(\omega_m t) + V_L \cos(\omega_L t))^3$$
$$+ d(V_f \cos(\omega_f t) + V_m \cos(\omega_m t) + V_L \cos(\omega_L t))^4$$

Equation 4.13 shows a large number of mixing products at the output. However, with enough suppression of unwanted products achieved by the high-Q output tuned circuit, most of these products are eliminated. Finally, at the output, two desired downconverted signals plus two third-order IM products remain at the output. To better understand the effect of IM3 product in mixers, suppose two adjacent channels residing at 901 MHz and 902 MHz with an LO frequency of 945 MHz. The desired IF signals would be at 43 MHz and 44 MHz and the undesired IM3 components will be at 42 MHz and 45 MHz. Equation 4.14 describes the desired components of the downconverted signal in the mixer. The first term in each of these equations stands for the linearly converted signal and the second terms describe the compressive components of the desired output signal. Note that in physical electronic devices normally $\frac{d}{b} < 0$ so that the second term in this equation is a compressive one:

$$v_{IF_m} = bV_m V_L R \cos(\omega_L - \omega_m)t + \frac{3}{2}dV_m^3 V_L R \cos(\omega_L - \omega_m)t \qquad (4.14)$$

$$v_{IF_f} = bV_f V_L R \cos(\omega_L - \omega_f)t + \frac{3}{2}dV_f^3 V_L R \cos(\omega_L - \omega_f)t \qquad (4.15)$$

Equations 4.16 and 4.17 describe the third-order IM product in this mixer. As it is obvious, both of them increase with a slope of 30 dB/dec with respect to the input signals:

$$v_{IM_m} = \frac{3}{2}dV_m^2 V_f V_L R \cos(\omega_L - 2\omega_m + \omega_f)t \qquad (4.16)$$

$$v_{IM_f} = \frac{3}{2}dV_m V_f^2 V_L R \cos(\omega_L - 2\omega_f + \omega_m)t \qquad (4.17)$$

Figure 4.7 shows adjacent channels and LO frequency spectra.

This configuration of input signals may be troublesome in receivers and degrade the performance of the receiver. The signals in Figure 4.7 are shown to be incident at the receiver of Figure 4.8 of a desired channel with a frequency of 900 MHz.

In Figure 4.8, the desired channel is at 900 MHz. The desired signal at the output of the first mixer will be at 45 MHz. The subsequent filter has suppressed the other products and passes the signal with a 2 MHz bandwidth. Next, the downconverted signal mixes again with 45.455 MHz LO and is downconverted to 455 kHz for final filtering. Here, the problem arises from two alternative strong channels which are at 901 MHz and 902 MHz. These two channels give in mixing products at the mixer output at 44 MHz and 43 MHz, as well as IM3 products at 42 MHz and 45 MHz. The second IM3 product is atop of our desired downconverted signal and corrupts its SNR. Figure 4.9 depicts this problem clearly.

4.3.1 The Desired Channel Blocking with the Third-Order IM Component

Since mixer circuits generate lots of IM components, we should take into account the effect of those affecting our desired channel SNR. This component is the last term of Equation 4.13. If we expand this term, we then arrive at Equation 4.18,

$$d(\cdots + 12V_f^2\cos^2(\omega_f t)V_m\cos(\omega_m t)V_L\cos(\omega_L t) \tag{4.18}$$
$$+ 12V_f\cos(\omega_f t)V_m^2\cos^2(\omega_m t)V_L\cos(\omega_L t) + \cdots)$$

Now, if we just look at the low-pass signal components of Equation 4.18, we can obtain

$$\omega_1 = -2\omega_f + \omega_m + \omega_L, \omega_2 = -2\omega_m + \omega_f + \omega_L \tag{4.19}$$

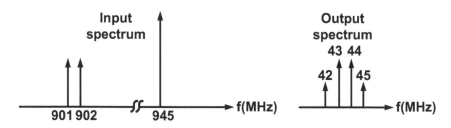

Figure 4.7: Input/output signal spectrums at the mixer ports.

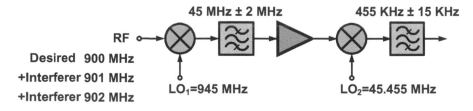

Figure 4.8: IM3 product problem in DAMPS.

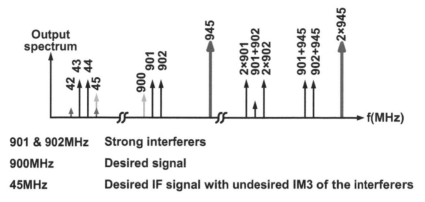

901 & 902MHz   **Strong interferers**

900MHz   **Desired signal**

45MHz   **Desired IF signal with undesired IM3 of the interferers**

Figure 4.9: The spectrum of signals for the mixer of Fig. 4.8.

The frequency components in Equation 4.19 are IM_3 which should be taken into account from linearity perspective. Whenever we record the input–output characteristic of a linear system, we reach to a line with the slope of one which shows the small-signal constant gain. In other words, if the input grows with just 1 dB, the output will be added by the same value. However, in nonlinear systems, IM_3 component will experience 3 dB growth with 1 dB input increase. For a highly linear mixer, the IIP_3 value is high. The problem that may arise is in the fading case of the desired signal and the presence of high level adjacent interfering (blocker) channels. The IM_3 components of the strong adjacent channels might fall within the reception bandwidth of the receiver. This may be troublesome in radio systems. In the real world, however, this issue can be alleviated by frequency hopping and the use of error-correcting codes. The presence of a strong blocker (interferer) signal in a nonlinear mixer is a challenge. One way to handle this challenge is to linearize the mixer.

Another IM component is IM_5 that increases by 50 dB/dec of input increase and has emerged by virtue of a term with the sixth order in the nonlinear model of the transconductance. As stated earlier, IIP_3 is the parameter which gives a measure of linearity in a system, thus we intend to find an easy method to compute it through input/output measurement. It can be proved that this value can be written as

$$IIP_{3\,dBm} = input_{dBm} + \frac{\Delta dB}{2} \tag{4.20}$$

where in Equation 4.20, ΔdB is the difference between lines of output signal (slope one) and output intermodulation (slope three), please refer to Figure 4.11. This equation is proved in the next subsection.

4.3.2 Special Content: IM with Any Nonlinear Circuit as a Mixer

Consider Figure 4.10 which depicts how two signals are added at the input of the mixer.

In the mixing mode, the output merely has the components of sum and difference frequencies of the input signal with the LO. To take into account just the IM_3 component, consider Figure 4.11.

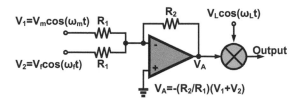

Figure 4.10: Implementation of signals' sum to be applied to a mixer input.

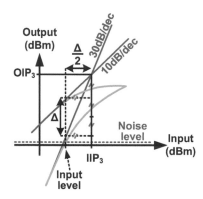

Figure 4.11: Interpolation of the output signal and the intermodulation curves to obtain third-order intercept point (IIP$_3$).

Our goal is to derive a simple equation for IIP$_3$. To obtain IIP$_3$, first of all choose a point which is in the low input power region for both lines. This is done for a better approximation of the slopes of the tangents to those curves. Then, by a simple subtraction of the dB levels recorded on those two lines, divide it by two, and adding this value to the selected point operating value, we reach to IIP$_3$. That is

$$\text{IIP}_3 \, (\text{dBm}) = P_{\text{in}} \, (\text{dBm}) + \frac{\Delta P \, (\text{dB})}{2} \tag{4.21}$$

where ΔP is the signal to IM ratio in dB.

IIP$_3$ could be roughly estimated at 7 dBm for a silicon diode mixer in the 50 Ω system. In a silicon bipolar transistor Gilbert cell, it varies between -20 dBm and -12 dBm at the input, and for its MOS counterpart, this value is in the range of -15 dBm to -5 dBm. For applications which necessitate highly linear mixers, IIP$_3$ can be up to 14 dBm. High IIP$_3$ mixer is of great importance in radio systems. In nonlinear systems and in the presence of interfering channels, signal detection is somehow tough. In reality, the lines in Figure 4.11 never reach to one another due to the compression phenomenon; however, the tangent lines give us the measure of nonlinearity. If one decreases the level of the input signal, such that the desired output component goes under the noise floor, that point determines the mixer sensitivity. On the other hand, if we increase our input signal such that the output goes beyond the compression point, and the resulting distortion in the signal causes error in the received bits, that point is considered as the saturation point of the mixer. The difference in dB between those mentioned levels defines the dynamic range of the mixer. To check the accuracy of ones measurement, one may increase the signal by 1 dB and check the IM3 component to increase by 3 dB. Another important point in Figure 4.11 is the 1 dB compression point which is noted by $p_{1\,\mathrm{dB}}$. Due to nonlinearity, the gain of the mixer will be decreased, and the point where the gain drops by 1 dB is of great importance. In practical system design, we usually work at a back-off (at a level 6 dB–10 dB lower than the compression point to assure the required linearity) of roughly between 6 dB and 10 dB with respect to compression point to prevent compression. In modern applications, we need new techniques to manipulate IM component for better signal detection. Figure 4.12 depicts a conventional receiver example.

We can also use a mixer to upconvert the signal, in a transmitter which is shown in Figure 4.13.

In transmitters, both the LO and IF signal are large signals. 900 MHz band-pass filter is placed to attenuate the other component of mixing residing at 990 MHz

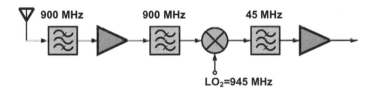

Figure 4.12: Typical heterodyne conventional receiver block diagram.

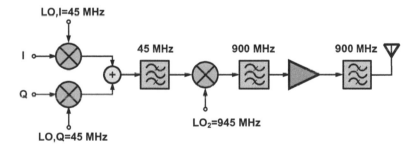

Figure 4.13: Typical conventional quadrature transmitter block diagram.

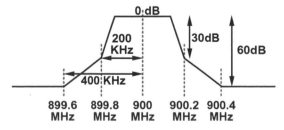

Figure 4.14: Wide-band spectrum standard for GSM and ACPR effect.

alongside other components made up due to mixing process. In radio regulation specifications, there are exact power versus frequency transmission windows which are specified by the regulatory organizations in which a transmitter should fit its own signal. Figure 4.14 illustrates the standard specification for one channel of the GSM. In this standard, the base station may have seven different power levels which are adjusted with respect to distance of the users.

   In far distances, due to high-power transmission, the device battery will be discharged fastly. The transmitted signal has a finite skirt in the frequency domain. By virtue of nonlinearity in the receivers, the leaked spectrum may be troublesome for adjacent channels. This effect is called adjacent channel power ratio (ACPR). Thus, we have more complex consideration in transmitter design than the receiver, because the leaked signal may act as an interferer for the other channel. It can be stated that in both receiver and transmitter, mixers are crucial. If one desires to have a highly linear mixer, they can use a diode mixer at the cost of lower gain. However, nowadays given the availability of good MOS switches, we are able to design highly linear active and passive switching mixers. In the next section, we discuss simple methods to analyze those kinds of mixers.

4.4   Bipolar Transistor Active Mixer

A typical bipolar transistor mixer is depicted in Figure 4.15. The transistor is biased in its active region. As it is seen in this figure, the input RF or IF signal is applied to the base of the transistor and the local oscillator signal is applied to the emitter of the transistor.

   Assuming an exponential nonlinear characteristics for the emitter–base junction, one can write

$$i_e(t) = I_{ES} e^{qV_{BE}/kT} \tag{4.22}$$

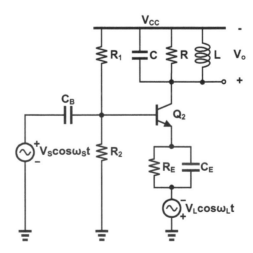

Figure 4.15: A typical bipolar transistor mixer in its active region.

Here, the total base–emitter voltage consists of a DC voltage, a local oscillator voltage, and an input signal voltage:

$$v_{BE} = V_{BEQ} + v_S + v_L \qquad (4.23)$$

By substituting the corresponding values of DC, LO, and input signal voltages in Equation 4.22, we obtain

$$i_e(t) = I_{ES} e^{qV_{BEQ}/kT} e^{q(V_S \cos(\omega_S t))/kT} e^{q(V_L \cos(\omega_L t))/kT} \qquad (4.24)$$

Expanding Equation 4.24, we obtain

$$i_c(t) = \alpha I_{ES} e^{qV_{BEQ}/kT} \left[I_0(y) + 2 \sum_{m=1}^{\infty} I_m(y) \cos(m\omega_S t) \right]$$
$$\left[I_0(x) + 2 \sum_{n=1}^{\infty} I_n(x) \cos(n\omega_L t) \right] \qquad (4.25)$$

Here, $I_n(x)$ or $I_m(y)$ are modified Bessel functions of the first kind which exponentially increase with respect to their argument. It is noteworthy that $I_0(x)$ tends to unity when its argument tends to zero. $I_n(x)$ for $n \geq 0$ tends to zero when its argument approaches zero. Furthermore, $I_1(x) \approx x/2$ for $x < 1$. It should be added that

$$\frac{I_{n+1}(x)}{I_n(x)} < 1, \; for \; x > 0, n \geq 0 \qquad (4.26)$$

Equation 4.25 can be simplified to

$$i_c(t) = \alpha I_{ES} e^{qV_{BEQ}/kT} I_0(y) I_0(x) \left[1 + 2 \sum_{m=1}^{\infty} \frac{I_m(y)}{I_0(y)} \cos(m\omega_S t) \right]$$
$$\left[1 + 2 \sum_{n=1}^{\infty} \frac{I_n(x)}{I_0(x)} \cos(n\omega_L t) \right] \qquad (4.27)$$

Let's denote the DC current, I_{E0}, by

$$I_{E0} = I_{ES} e^{qV_{BEQ}/kT} I_0(y) I_0(x) \qquad (4.28)$$

The collector current can be expressed as

$$i_c(t) = \alpha I_{E0} \left[1 + 2\frac{I_1(y)}{I_0(y)} \cos(\omega_S t) + \cdots \right] \left[1 + 2\frac{I_1(x)}{I_0(x)} \cos(\omega_L t) + \cdots \right] \quad (4.29)$$

Or

$$i_c(t) = \alpha I_{E0} \left[1 + 2\frac{I_1(y)}{I_0(y)} \cos(\omega_S t) + 2\frac{I_1(x)}{I_0(x)} \cos(\omega_L t) \right.$$
$$\left. + 4\frac{I_1(y)}{I_0(y)} \frac{I_1(x)}{I_0(x)} \cos(\omega_S t) \cos(\omega_L t) + \cdots \right] \qquad (4.30)$$

Considering small-signal input and a large-signal LO, and given the fact that $I_0(y) \cong 1$, $\frac{I_1(y)}{I_0(y)} \approx \frac{y}{2}$, one can rewrite the collector current expression as

$$i_c(t) = \alpha I_{E0}\left[1 + y\cos(\omega_S t) + 2\frac{I_1(x)}{I_0(x)}\cos(\omega_L t) + 2y\frac{I_1(x)}{I_0(x)}\cos(\omega_S t)\cos(\omega_L t) + \cdots\right]$$

(4.31)

By separating the different components of the collector current, one can deduce from Equation 4.31 that each component of the collector current appears through a certain transconductance as it is followed.

The input signal frequency component would appear in the collector through a small-signal transconductance, namely, g_m:

$$I_S = \alpha I_{E0} y = g_m v_S$$

(4.32)

The local oscillator frequency component would appear in the collector through a large-signal transconductance, namely, $G_m(x)$:

$$I_L = \alpha I_{E0}\frac{2I_1(x)}{I_0(x)} = g_m\frac{2I_1(x)}{xI_0(x)}v_L = G_m(x)v_L$$

(4.33)

As it is seen in Figure 4.16, the large-signal transconductance of a bipolar transistor decreases monotonically with the input large-signal voltage. So the large-signal transconductance is generally smaller than the small-signal operating transconductance. The large-signal transconductance goes from a normalized unity value, for the small signal case, toward zero for very large values of the input LO signal.

The mixing products (sum or difference frequencies) would appear in the collector through a conversion conductance, namely, g_C:

$$I_{\omega_L \pm \omega_S} = \alpha I_{E0} y\frac{I_1(x)}{I_0(x)} = g_m\frac{I_1(x)}{I_0(x)}v_S = g_C v_S$$

(4.34)

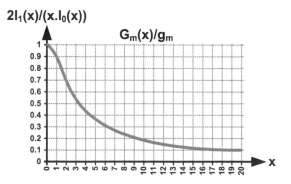

Figure 4.16: The normalized large-signal transconductance of a single bipolar transistor stage as a function of the normalized local oscillator voltage.

The conversion conductance, g_C, is

$$g_C = \frac{I_{IF}}{V_{RF}} = \frac{I_{RF}}{V_{IF}} = g_m \frac{I_1(x)}{I_0(x)} \tag{4.35}$$

As it is seen in Figure 4.17, the conversion conductance of a bipolar mixer increases monotonically with the input large-signal local oscillator amplitude. It goes from zero value for the small-signal LO to a saturating normalized value of unity with respect to the operating point transconductance.

If the output RLC circuit is a high-Q one and it is tuned to the corresponding mixing product (the sum or the difference frequency), the output voltage would have either of the following forms

$$v_O = V_{CC} - g_C V_S R_L \cos((\omega_S + \omega_L)t) \tag{4.36}$$

which is used for an upconverting mixer.

$$v_O = V_{CC} - g_C V_S R_L \cos((\omega_S - \omega_L)t) \tag{4.37}$$

which is used for a downconverting mixer. As such, our active mixer would have a gain of $g_C R_L$.

It is noteworthy that other unwanted signal components would appear at the output if the Q factor of the RLC circuit is not sufficiently high. As an example, the unwanted LO component and the unwanted RF signal component would have the following values

$$V_{L,out} = -G_m(x) V_L Z_L(j\omega_L) \tag{4.38}$$

$$V_{S,out} = -g_m V_S Z_L(j\omega_S) \tag{4.39}$$

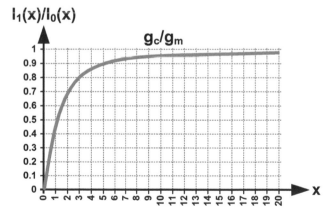

Figure 4.17: The normalized conversion conductance of a single bipolar transistor stage as a function of normalized input LO voltage.

4.5   Mixer types Based on Switching Circuits

In this part, we introduce very useful methods to analyze mixers. This method is based on the assumption of complete switching in the transistors that are driven by the large LO signal. In our analysis, we use a Fourier series expansion of the switching signal. Consider Figure 4.18 which shows three configurations of mixer circuits for both bipolar and MOS implementation. Note that in all these figures, the required DC bias of the LO and the RF signals is not shown.

For Figure 4.18, mixers (a) and (d) which are called unbalanced mixers, both RF and LO signal leak to the output. Mixers (b) and (e) are called single-balanced mixers from which RF signal leakage is removed (the LO signal appears at the output in addition to the mixing signals). Finally, mixers (e) and (f) are called double-balanced mixers where the RF and the LO components are nonexistent at the output of the mixer. We will return back to this point later.

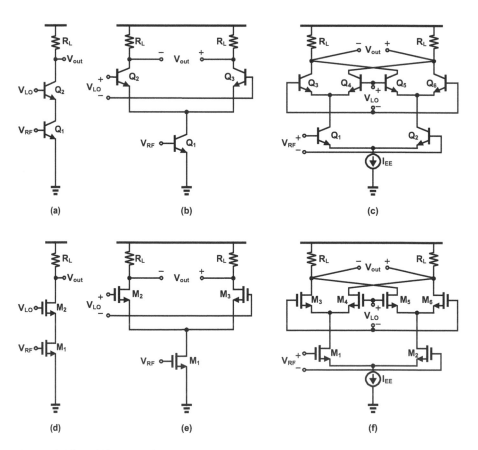

Figure 4.18: Different mixer circuit topologies, (a) bipolar unbalanced, (b) bipolar balanced, (c) bipolar double-balanced, (d) MOS unbalanced, (e) MOS balanced, and (f) MOS double-balanced.

4.5.1 Conversion Gain and Local Oscillator Leakage

To analyze the operation of the mixers, we first need to know Fourier series coefficients of a square-wave signal. It can be shown that for a pulse signal toggling between 1 and -1 with the period of T_{LO}, the signal and its corresponding Fourier series coefficients can be expressed as Equation 4.40:

$$s(\omega_0 t) = \sum_{n=1}^{\infty} a_n \cos(n\omega_0 t) \tag{4.40}$$

$$a_n = \frac{\sin\left(\frac{n\pi}{2}\right)}{\frac{n\pi}{4}}$$

where Equation 4.40 is valid for odd values of n; however, for even values of n, coefficients are zero. Now, with respect to Fourier series coefficients, we perform our computations for three different mixer configurations. Note for a pulse signal toggling between 1 and 0 the Fourier series will have a $\frac{1}{2}$ as the DC component and the AC components of its Fourier series will be half of the AC component of the bipolar pulse signal.

Unbalanced Mixer

One may obtain the output signal of mixers (a) and (d), considering a nonlinear power series transconductance for the lower transistor switched by the LO driven upper transistor in Figure 4.18 as

$$V_{out} = (a + bV_{RF} + cV_{RF}^2 + \cdots)\left(\frac{1}{2} + \frac{2}{\pi}\cos(\omega_{LO}t) - \frac{2}{3\pi}\cos(3\omega_{LO}t) + \cdots\right) \tag{4.41}$$

Equation 4.41 is written using Equation 4.40. To simplify the operation of the mixers, we can state that Q_2 turns on and off by the LO signal. This implies that when this transistor is on, the RF signal appears at the output; otherwise, the output is tied to the supply voltage. Thus, we can assume that the RF signal is multiplied by a square wave with the amplitude of 0 and 1 by the LO period. Therefore, we can attain a new set of coefficients as

$$b_0 = \frac{1}{2} \tag{4.42a}$$

$$b_n = \frac{a_n}{2} \tag{4.42b}$$

Equation 4.41 suggests that there will be lots of mixing products at the output of the mixer. Thus, we usually employ a low-pass filter at the output to suppress the unwanted products. Moreover, note that the leakage of RF and LO signals to the output has come from the DC within the parenthesis terms in Equation 4.42.

Single-Balanced Mixer

We can derive the output signal of mixers (b) and (e) in Figure 4.18 as for Equation 4.43,

$$V_{out} = (a + bV_{RF} + cV_{RF}^2 + \cdots)\left(\frac{4}{\pi}\cos(\omega_{LO}t) - \frac{4}{3\pi}\cos(3\omega_{LO}t) + \cdots\right) \tag{4.43}$$

The important point in Equation 4.43 is the effect of differential circuit on the Fourier series of the LO frequency. It seems that the RF signal now is multiplied by a square wave with the alternative amplitudes 1 and -1. Thus, no DC component at LO Fourier series coefficients suggests no RF feedthrough at the output.

Double-Balanced Mixer

Finally, the output of mixers (c) and (f) in Figure 4.18 can be written as

$$V_{\text{out}} = (bV_{\text{RF}} + dV_{\text{RF}}^3 + \cdots)\left(\frac{4}{\pi}\cos(\omega_{\text{LO}}t) - \frac{4}{3\pi}\cos(3\omega_{\text{LO}}t) + \cdots\right) \tag{4.44}$$

Equation 4.44 introduces no DC components at both LO and RF sides, thus the concept of double-balanced mixer which doesn't permit these signals to appear at the output mixer is obvious. In fact, with this powerful analysis, we are able to compute any mixing product gain and moreover understand the port-to-port leakages. Nonetheless, with inevitable mismatches and offset voltages, a finite leakage signals would be present at the output of the mixer. Today, MOS process offers very fast switches due to lower capacitances and on-resistances which can operate for high frequencies. One of the most important specifications of mixers is their linearity issue which has come from the nonlinear transconductance of the input transistor which converts the input RF voltage to the current that passes through the switch loads. LO signal applied to the other transistors just turns them on and off and roughly doesn't affect the linearity issues. Another type of mixer which is called a passive switching mixer is shown in Figure 4.19.

These circuits manifest better linearity because of no transconductance device between the switch and the load. In other words, the signal itself is chopped by means of switches and reaches the output. Figure 4.20 depicts a differential implementation of a passive mixer which is somehow alike active ones without transconductance.

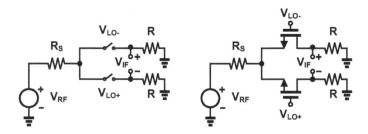

Figure 4.19: Passive switching mixer circuits.

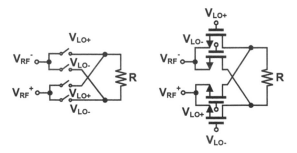

Figure 4.20: Differential implementation of passive mixer circuits.

Now, we present a set of relations to compute the small-signal conversion gains of the mixers in Figure 4.18 (here, we have considered the transconductance for the transistors is linear). For mixers (a) and (d) we may obtain

$$V_{out} = V_{CC} - R_L(I_{Bias} + g_m V_{in} \cos(\omega_R t)) \left(\frac{1}{2} + \frac{2}{\pi} \cos(\omega_L t) \right.$$

$$\left. - \frac{2}{3\pi} \cos(3\omega_L t) + \frac{2}{5\pi} \cos(5\omega_L t) - \cdots \right) \tag{4.45}$$

Equation 4.45 confirms the previously mentioned port-to-port leakage in an unbalanced mixer and we can compute LO and RF leakage amplitudes as $2/\pi R_L I_{bias}$ and $0.5 g_m R_L$, respectively. We can also attain the same equation for single-balanced mixers (b) and (e) as

$$V_{out} = R_L(I_{Bias} + g_m V_{in} \cos(\omega_R t)) \left(\frac{4}{\pi} \cos(\omega_L t) - \frac{4}{3\pi} \cos(3\omega_L t) + \frac{4}{5\pi} \cos(5\omega_L t) - \cdots \right)$$

$$\tag{4.46}$$

Equation 4.46 shows that input signal does not appear at the output and the LO leakage is equal to $4/\pi R_L I_{bias}$. Finally, the double-balanced mixer output signal for mixers (c) and (f) can be calculated as

$$V_{out} = g_m R_L V_{in} \cos(\omega_R t) \left(\frac{4}{\pi} \cos(\omega_L t) - \frac{4}{3\pi} \cos(3\omega_L t) + \frac{4}{5\pi} \cos(5\omega_L t) - \cdots \right)$$

$$\tag{4.47}$$

where it shows there is no leakage to the output. However, with the definition of conversion gain, i.e., the gain from IF signal to RF can be carried out as

$$\frac{V_{out}(IF)}{V_{in}(RF)} = \frac{1}{\pi} g_m R_L \quad \text{for unbalanced} \tag{4.48a}$$

$$\frac{V_{out}(IF)}{V_{in}(RF)} = \frac{2}{\pi} g_m R_L \quad \text{for single-balanced} \tag{4.48b}$$

$$\frac{V_{out}(IF)}{V_{in}(RF)} = \frac{2}{\pi} g_m R_L \quad \text{for double-balanced} \tag{4.48c}$$

Note that the coefficient $4/\pi$ has come from the Fourier series expansion and $1/2$ is due to one of the sum or difference components obtained out of the multiplication of cosines. Nowadays, double-balanced mixers are more frequently applied due to suppression of port-to-port leakages. MOS devices present proper switches for mixing purposes; however, their quadratic $I - V$ characteristics are such that for a given bias current, MOS devices have lower transconductance than their bipolar counterparts. Moreover, note that their output impedance is lower than those of bipolar devices which is not a merit. It is instructive to note that the main parameter in mixers is their linearity issue rather than their conversion gain. Moreover, to alleviate the linearity issue, we should linearize the input active device, because the upper side in the aforementioned mixers is just switches. Figure 4.21 shows a bipolar unbalanced mixer with a tuned circuit load.

In Figure 4.21, LO signal is connected to the base of Q_1 and RF signal is applied to Q_2. LO signal is large and might have the amplitude of a few hundred millivolts or more and the RF signal is small. Transistor Q_1 will roughly be on and off within each LO period. When this device is on, it let the current flow to reach the resonant load and the output voltage appears across the tuned circuit load. However, when Q_1 is off, the current passing through the Q_2 collector is nearly zero and the output will be tied to V_{CC}. Figure 4.22 illustrates the concept of mixing in the mixer in Figure 4.21.

In each cycle, the following happens:

1. Q_1 is off (negative half cycle of LO): in this case $I_C = 0$.

2. Q_1 is on (positive half cycle of LO): in this case $I_C = I_{E0} + g_m V_R$, where I_{E0} is

$$I_{E0} = (1+\beta)\frac{V_{BB2} - V_{BEQ}}{R_2} \tag{4.49}$$

Then, in sum, the collector current of Q_2 can be expressed as

$$i_C(t) \simeq [I_{E0} + g_m V_R \cos(\omega_R t)] S(\omega_L t) \tag{4.50}$$

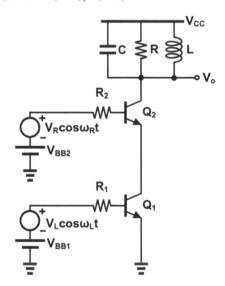

Figure 4.21: Bipolar unbalanced mixer functioning on the LO switching basis (downconverter).

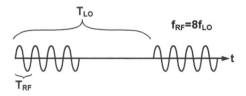

Figure 4.22: A rough approximation of the output signal of Figure 4.21.

where $S(\omega_L t)$ is a monopolar square wave varying between zero and one at the rate of LO. Then

$$
\begin{aligned}
i_C(t) &\simeq \left[I_{E0} + \frac{qI_{E0}}{kT} V_R \cos(\omega_R t) \right] S(\omega_L t) \\
&= \left[I_{E0} + \frac{qI_{E0}}{kT} V_R \cos(\omega_R t) \right] \left[\frac{1}{2} + \frac{2}{\pi} \cos(\omega_L t) \right. \\
&\qquad\qquad \left. - \frac{2}{3\pi} \cos(3\omega_L t) + \frac{2}{5\pi} \cos(5\omega_L t) + \cdots \right] \quad (4.51)
\end{aligned}
$$

Finally, if the RLC circuit is tuned to the difference frequency, the output AC voltage becomes

$$
v_{out} \simeq \frac{R_L}{\pi} \frac{qI_{E0}}{kT} V_R \cos((\omega_R - \omega_L)t) \tag{4.52}
$$

In another mode of operation, a similar circuit topology can be used as an upconverting mixer. Here a bypass capacitor C_E is used between the Q_2 emitter and the ground (Figure 4.23). This capacitor should be sufficiently large to be short at the LO frequency and adequately small to be open at the IF frequency. As such, the transistor Q_1 acts as a time-varying current source biasing Q_2 at the rate of IF. Here, we have

$$
i_E = \frac{V_{BB1} - V_{BEQ} + V_{IF} \cos(\omega_{IF} t)}{R_E} \tag{4.53}
$$

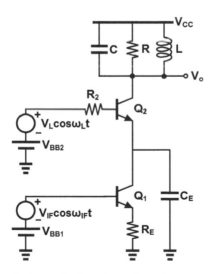

Figure 4.23: Bipolar unbalanced mixer based on time-varying transconductance (upconverter).

As the LO signal is considered to be a large signal, we should use the large-signal transconductance of the bipolar transistor. That is

$$G_m(x) = g_m \frac{2I_1(x)}{x I_0(x)} \tag{4.54}$$

where $x = \frac{q V_L}{kT}$, and

$$g_m = \frac{q}{kT}(I_{E0} + I_{E_{IF}} \cos(\omega_{IF} t)) \tag{4.55}$$

and

$$I_{E_{IF}} = \frac{V_{IF}}{R_E} \tag{4.56}$$

The Q_2 collector current becomes

$$i_C(t) \simeq \frac{q}{kT}(I_{E0} + I_{E_{IF}} \cos(\omega_{IF} t)) \frac{2I_1(x)}{x I_0(x)} V_L \cos(\omega_L t) \tag{4.57}$$

Finally, the output voltage of the mixer (if the RLC circuit is tuned to the sum frequency) becomes

$$v_{out} \simeq \frac{R_L V_{IF}}{R_E} \frac{I_1(x)}{I_0(x)} \cos((\omega_L + \omega_{IF}) t) \tag{4.58}$$

Mixers introduce a large amount of mixing products within the frequency spectrum of the output current by virtue of device nonlinearity. However, the desired signal is usually $\omega_{RF} - \omega_{LO}$ component which is selected by the tuned band-pass filter. This nonlinearity generates mixing products by two main sources. First, two adjacent interferers may cause an undesired signal atop the desired signal due to IM3 component as described before. Secondly, considerable leakage of LO and RF at the output causes difficulties in extracting the desired signal. Another way to mix the two signals can be implemented by applying both LO and RF signals to the base of a bipolar transistor. Similarly, we can apply the LO signal at the emitter of a bipolar transistor and the RF signal to its base. Finally, the exponential $I - V$ characteristics of the device will produce our desired mixing product. Figure 4.24 depicts a differential implementation of a bipolar mixer which is single-balanced.

It should be noted that these mixers can also be implemented by MOS devices. The important point in Figure 4.24 is the need for lower signal amplitude to achieve the switching of Q_2 and Q_3. In fact, in these devices, the RF current is applied to each branch with a rate of LO signal. It can be roughly with a voltage of (V_{LO}) between 100 mV and 500 mV, the upper tree can be switched efficiently. In Figure 4.24, the RF signal in each cycle appears at either of output terminals, while the other terminal is grounded. Thus, we can assert the RF signal is multiplied by +1 or −1 alternatively. Our objective is to obtain an equation for the output of the single-balanced mixer in which the RF signal appears in common mode in the differential output. Since the

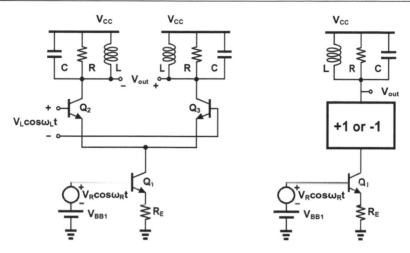

Figure 4.24: Single-balanced bipolar mixer.

LO drive is 180° out of phase, with respect to the base terminals, its leakage will be present at the differential output. One may obtain the output of the single-balanced mixer of Figure 4.24 as follows

$$v_{C2} = V_{CC} - i_{C2} * Z_L \tag{4.59}$$

and

$$v_{C3} = V_{CC} - i_{C3} * Z_L \tag{4.60}$$

The output voltage becomes

$$v_{out} = v_{C3} - v_{C2} = (i_{C2} - i_{C3}) * Z_L \tag{4.61}$$

Here $*$ sign stands for the convolution in the time domain or equivalently multiplication of the corresponding impedances and current harmonics in the frequency domain. The currents in each branch of the upper tree can be described as

$$i_{C2,3} = \frac{I_C}{2}\left(1 \pm \tanh\left(\frac{V_L \cos(\omega_L t)}{2V_T}\right)\right) \tag{4.62}$$

The bias current of the upper tree is

$$I_C = \frac{V_{BB1} - V_{BEQ} + V_R \cos(\omega_R t)}{R_E} \tag{4.63}$$

The differential output current becomes

$$\Delta I_C = i_{C2} - i_{C3} = \frac{1}{R_E}[V_{BB1} - V_{BEQ} + V_R \cos(\omega_R t)]\tanh\left(\frac{V_L \cos(\omega_L t)}{2V_T}\right) \tag{4.64}$$

For $V_L \gg V_T$, one can write

$$\Delta I_C \simeq \frac{1}{R_E} \left[V_{BB1} - V_{BEQ} + V_R \cos\left(\omega_R t\right)\right] S\left(\omega_L t\right) \tag{4.65}$$

where $S\left(\omega_L t\right)$ is the bipolar switching function or the bipolar square wave toggling between $+1$ and -1 with a rate of LO, where its Fourier expansion becomes as follows

$$\Delta I_C \simeq \frac{1}{R_E} \left[V_{BB1} - V_{BEQ} + V_R \cos\left(\omega_R t\right)\right] \left[\frac{4}{\pi} \cos\left(\omega_L t\right) - \frac{4}{3\pi} \cos\left(3\omega_L t\right)\right.$$
$$\left. + \frac{4}{5\pi} \cos\left(5\omega_L t\right) + \cdots\right] \tag{4.66}$$

Finally, the output voltage, if the RLC circuits are tuned to the difference frequency, becomes as follows

$$v_{out} \simeq \frac{2}{\pi} \frac{R_L}{R_E} V_R \cos\left(\left(\omega_R - \omega_L\right)t\right) \tag{4.67}$$

At last, we introduce the well-known Gilbert cell as a possible candidate for double-balanced mixer. Figure 4.25 depicts the Gilbert cell. The output of this mixer can be described in the same way as Equation 4.47. By the fact that the output is loaded by C_L, the low-pass term of Equation 4.47 would appear at the output of the mixer.

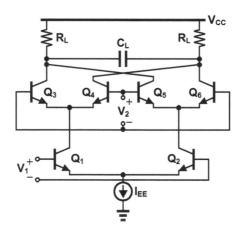

Figure 4.25: Double-balanced mixer.

Example 4.1 Since the mixer circuit has three ports, how can we define power in its ports? Moreover, discuss the IM$_3$ component for two alternate interfering channels with different spacings.

Solution:
Consider Figure 4.27.

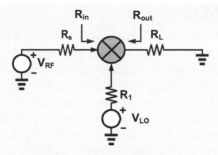

Figure 4.26: Mixer circuit.

If we have matching at the input and at the output, we have $R_{in} = R_s$, $R_{out} = R_L$, and then we may write the conversion power gain

$$G_P = \frac{\frac{V_{IF}^2}{2R_L}}{\frac{V_{RF}^2}{2R_{in}}} \tag{4.68}$$

Suppose that the desired channel resides at 900 MHz and the LO signal is at 945 MHz. Therefore, the IF signal will be at 45 MHz.

(a) Consider the interfering channels are at $f_1 = 901$ MHz and $f_2 = 902$ MHz, and (b) imagine the two interfering channels are at $f_1 = 900.03$ MHz and $f_2 = 900.06$ MHz. Both of these channels could make IM_3 components (e.g., $\omega_{LO} - (2\omega_1 - \omega_2)$) atop of the desired signal. One way to mitigate this issue is the implementation of a band-pass filter at the mixer's input to eliminate those interfering channels. Figure 4.27 depicts the structure in this case. Using a band-pass filter with 1 MHz bandwidth, it is possible to eliminate the interfering channels in the case (a). But having a band-pass filter of 60 kHz bandwidth at 900 MHz is practically impossible and eliminating the interfering signals would become impossible in case (b) at this stage (in this case, either the linearity of the mixer should be improved or the wireless standard should require the levels of the adjacent channels to be less than a predetermined value).

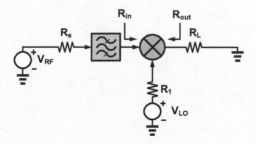

Figure 4.27: Employing a band-pass filter to suppress input blockers in order to avoid the resulting IM_3 component.

4.6 Matching in Mixers

As in any RF circuit for the maximum power transfer, it is required that the input RF/IF port and the output IF/RF port of the mixer to be matched to the source and to the load, respectively. For this purpose, standard step-up or step-down LC matching circuits could be used. Consider Figure 4.28 as an example of the input and output matching.

As Figure 4.28 suggests, a capacitive matching circuit along with an inductor is placed at the input of the mixer to transform 1 kΩ input impedance of the mixer, to 50 Ω source impedance value. Likewise at the output, another matching has been realized to transform 1500 Ω output impedance of the mixer to 50 Ω value of load. Here the mixer block could be replaced by a Gilbert cell, for example.

4.7 Calculating IIP₃ in Nonlinear Amplifier/Mixer

In this section, we investigate the nonlinear behavior of active devices to obtain the input intercept point level. Suppose Figure 4.29 that shows a transistor which has a bias of V_{BB} and two input signals.

The $I - V$ characteristic of the amplifier is approximated by a polynomial of third order to obtain a IIP$_3$ level. Let

$$i(t) = \alpha_0 + \alpha_1 V + \alpha_2 V^2 + \alpha_3 V^3 \tag{4.69}$$

Applying two input signals as in Figure 4.29 and substituting in Equation 4.69, we then obtain

$$i(t) = \alpha_0 + \alpha_1 \left(A_1 \cos\left(\omega_1 t\right) + A_2 \cos\left(\omega_2 t\right)\right) \tag{4.70}$$
$$+ \alpha_2 (A_1 \cos\left(\omega_1 t\right) + A_2 \cos\left(\omega_2 t\right))^2 + \alpha_3 (A_1 \cos\left(\omega_1 t\right) + A_2 \cos\left(\omega_2 t\right))^3$$

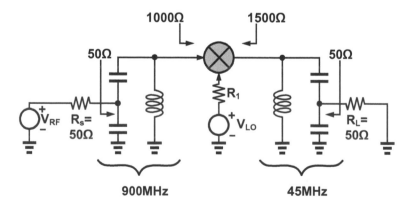

Figure 4.28: Typical matching circuit for a mixer, step-up capacitive input matching, and step-down capacitive output matching.

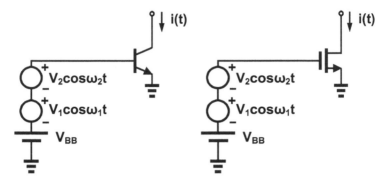

Figure 4.29: Applying two large signals to a nonlinear device (bipolar or MOS transistors) to compute the compression point and the third-order intercept point.

If we expand Equation 4.70, it results in

$$i(t) = I_{\text{Bias}} + \left(\alpha_1 A_1 + \overbrace{\frac{3}{4}\alpha_3 A_1^3 + \frac{3}{2}\alpha_3 A_1 A_2^2}^{\text{signal compression}} \right) \cos(\omega_1 t) \qquad (4.71)$$

$$+ \left(\alpha_1 A_2 + \frac{3}{4}\alpha_3 A_2^3 + \frac{3}{2}\alpha_3 A_2 A_1^2 \right) \cos(\omega_2 t)$$

$$+ \frac{3\alpha_3}{4} A_2 A_1^2 \left(\cos((2\omega_1 + \omega_2)t) + \cos((2\omega_1 - \omega_2)t) \right)$$

$$+ \frac{3\alpha_3}{4} A_1 A_2^2 \left(\cos((2\omega_2 + \omega_1)t) + \cos((2\omega_2 - \omega_1)t) \right)$$

$$+ \frac{\alpha_3}{4} A_1^3 \cos(3\omega_1)t + \frac{\alpha_3}{4} A_2^3 \cos(3\omega_2)t$$

As stated earlier, one of the important parameters in the nonlinear amplifiers is their measure of linearity which is obtained by means of a two-tone test. In this test, by increasing the amplitude of tones, the output will be compressed and the low-level slopes of the first- and the third-order terms will intersect at a point which we call the intercept point. The term shown in Equation 4.71. is called signal amplitude compression term which has a nonlinear relation with the input level and causes the decrease in the amplifier gain as the input level is increased. Figure 4.30 illustrates two different curves, one traces the fundamental harmonic term at the output as a function of input level, and the other illustrates the output third-order intermodulation amplitudes as a function of the input too, both on the log–log scale.

The -1 dB compression point is a point where the output fundamental level is 1 dB less than the presumed linear fundamental output level. Given $A_1 = A_2 = A$, the compression point is computed as

$$\alpha_1 A + \frac{3}{4}\alpha_3 A^3 + \frac{3}{2}\alpha_3 A^3 = 10^{\frac{-1}{20}}.\alpha_1 A \qquad (4.72)$$

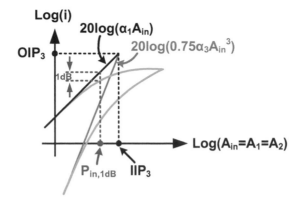

Figure 4.30: Output current of the amplifier versus its input signals' amplitudes.

or

$$\frac{9}{4}\frac{\alpha_3}{\alpha_1}A^2 = -0.11 \tag{4.73}$$

Note that for the nonlinear amplifier to be compressive, we should have

$$\frac{\alpha_3}{\alpha_1} < 0 \tag{4.74}$$

Otherwise, for $\frac{\alpha_3}{\alpha_1} > 0$, the amplifier would be expansive which is generally a nonphysical amplifier. Therefore, in the case compressive (physical) amplifier, we would have

$$A_{1\,\text{dB}} = 0.22\sqrt{-\frac{\alpha_1}{\alpha_3}} \tag{4.75}$$

Note that this is the "two-tone" compression point. It can be easily shown, by putting $A_2 = 0$, that a single-tone compression point can be expressed as

$$A_{1\,\text{dB}} = 0.38\sqrt{-\frac{\alpha_1}{\alpha_3}} \tag{4.76}$$

Verification of the above equation is left to the reader. From Figure 4.30, we are able to compute IIP$_3$ by the intersection of the two linear terms (tangents) as

$$20\log\left(\alpha_1 A_{\text{in}}\right) = 20\log\left(-\frac{3\alpha_3}{4}A_{\text{in}}^3\right) \tag{4.77}$$

which finally gives the corresponding amplitude as

$$A_{\text{IIP}_3} = \sqrt{\frac{4}{3}\left|\frac{\alpha_1}{\alpha_3}\right|} = \sqrt{-\frac{4}{3}\frac{\alpha_1}{\alpha_3}} \tag{4.78}$$

Note that in the above computations, we have assumed that $A_1 = A_2 = A$. Furthermore, notice that in a physical compressive amplifier, α_3/α_1 is always negative.

Equation 4.78 states that an amplifier with high fundamental harmonic content (large α_1) and a low third harmonic content (small absolute value of α_3) results in high level of IIP₃, that is, an amplifier with better linearity.

4.7.1 Compression Point and IIP₃ in a Nonlinear Transconductance Mixer

Consider a nonlinear transconductance mixer with a fourth-order nonlinearity:

$$i(t) = \alpha_0 + \alpha_1 v + \alpha_2 v^2 + \alpha_3 v^3 + \alpha_4 v^4 \tag{4.79}$$

If we consider an input with the following form

$$v = V_1 \cos(\omega_0 t) + V_S \cos(\omega_S t) \tag{4.80}$$

Then the output current will have the following expression

$$\begin{aligned}
i(t) = &\alpha_0 + \alpha_1 \left(V_1 \cos(\omega_0 t) + V_S \cos(\omega_S t)\right) \\
&+ \alpha_2 \left(V_1 \cos(\omega_0 t) + V_S \cos(\omega_S t)\right)^2 \\
&+ \alpha_3 \left(V_1 \cos(\omega_0 t) + V_S \cos(\omega_S t)\right)^3 \\
&+ \alpha_4 \left(V_1 \cos(\omega_0 t) + V_S \cos(\omega_S t)\right)^4
\end{aligned} \tag{4.81}$$

By sorting out only the desired output components at $\omega_0 - \omega_S$, we will have

$$\begin{aligned}
i(t) = &\alpha_2 V_1 V_S \cos\left((\omega_0 - \omega_S)t\right) \\
&+ \frac{3}{2} \alpha_4 V_1 V_S^3 \cos\left((\omega_0 - \omega_S)t\right) \\
&+ \frac{3}{2} \alpha_4 V_1^3 V_S \cos\left((\omega_0 - \omega_S)t\right) \\
&+ \cdots
\end{aligned} \tag{4.82}$$

Therefore, the desired mixer output will have the following form

$$v_{\text{out}} = \alpha_2 V_1 V_S R_L \left[1 + \frac{3}{2}\frac{\alpha_4}{\alpha_2} V_S^2 + \frac{3}{2}\frac{\alpha_4}{\alpha_2} V_1^2\right] \cos\left((\omega_0 - \omega_S)t\right) \tag{4.83}$$

As it is obvious from the above equation, the output signal is compressed both with respect to V_S and with respect to V_1. So we define the -1 dB compression point as a two-variable equation as follows

$$1 + \frac{3}{2}\frac{\alpha_4}{\alpha_2} V_S^2 + \frac{3}{2}\frac{\alpha_4}{\alpha_2} V_1^2 = 10^{\frac{-1}{20}} = 0.891 \tag{4.84}$$

Or

$$V_S^2 + V_1^2 = -0.11 \times \frac{2}{3}\frac{\alpha_2}{\alpha_4} \tag{4.85}$$

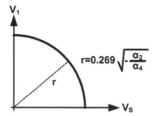

Figure 4.31: The locus of the saturation voltages in the V_S-V_1 plane.

Or in another form

$$\left(V_S^2 + V_1^2\right)^{\frac{1}{2}} = 0.269\sqrt{-\frac{\alpha_2}{\alpha_4}} \tag{4.86}$$

This describes the compression effect in a mixer, which depends both on the LO level and the signal level. This equation also describes a circle in the V_1,V_S plane (Fig. 4.31). For example, if one considers the signal as a small input, he/she would obtain the compression point by putting $V_S = 0$ in the above equation, and obtain the compression point in terms of V_1. Otherwise, if one considers the LO as a small signal, he/she would obtain the compression point by putting $V_1 = 0$ in the above equation, and obtain the compression point in terms of V_S. In a more general manner, one can consider any proportion between V_1 and V_S, and compute the compression point through Equation 4.86.

Two-tone −1 dB compression point in a nonlinear mixer

If we consider an input with the following form

$$v = V_1 \cos\left(\omega_0 t\right) + V_{S1} \cos\left(\omega_{S1} t\right) + V_{S2} \cos\left(\omega_{S2} t\right) \tag{4.87}$$

Then the output current will have the following expression

$$\begin{aligned}
i(t) =&\, \alpha_0 + \alpha_1 \left(V_1 \cos\left(\omega_0 t\right) + V_{S1} \cos\left(\omega_{S1} t\right) + V_{S2} \cos\left(\omega_{S2} t\right)\right) \\
&+ \alpha_2 \left(V_1 \cos\left(\omega_0 t\right) + V_{S1} \cos\left(\omega_{S1} t\right) + V_{S2} \cos\left(\omega_{S2} t\right)\right)^2 \\
&+ \alpha_3 \left(V_1 \cos\left(\omega_0 t\right) + V_{S1} \cos\left(\omega_{S1} t\right) + V_{S2} \cos\left(\omega_{S2} t\right)\right)^3 \\
&+ \alpha_4 \left(V_1 \cos\left(\omega_0 t\right) + V_{S1} \cos\left(\omega_{S1} t\right) + V_{S2} \cos\left(\omega_{S2} t\right)\right)^4
\end{aligned} \tag{4.88}$$

By sorting out only the desired output components at $\omega_0 - \omega_{S1}$ and at $\omega_0 - \omega_{S2}$, we will have

$$\begin{aligned}
v_{out} =&\, \alpha_2 V_1 V_{S1} R_L \left[1 + \frac{3}{2}\frac{\alpha_4}{\alpha_2}V_{S1}^2 + 3\frac{\alpha_4}{\alpha_2}V_{S2}^2 + \frac{3}{2}\frac{\alpha_4}{\alpha_2}V_1^2\right] \cos\left(\left(\omega_0 - \omega_{S1}\right)t\right) \\
\end{aligned} \tag{4.89}$$

$$+ \alpha_2 V_1 V_{S2} R_L \left[1 + \frac{3}{2}\frac{\alpha_4}{\alpha_2}V_{S2}^2 + 3\frac{\alpha_4}{\alpha_2}V_{S1}^2 + \frac{3}{2}\frac{\alpha_4}{\alpha_2}V_1^2\right] \cos\left(\left(\omega_0 - \omega_{S2}\right)t\right)$$

In a similar manner, as described in the computation of the -1 dB compression point for a single-tone, one can show that the -1 dB compression point can be computed through the following equations for either of the two tones

$$V_{S1}^2 + 2V_{S2}^2 + V_1^2 = -0.11 \times \frac{2}{3}\frac{\alpha_2}{\alpha_4} \tag{4.90}$$

and

$$V_{S2}^2 + 2V_{S1}^2 + V_1^2 = -0.11 \times \frac{2}{3}\frac{\alpha_2}{\alpha_4} \tag{4.91}$$

These two equations describe the compression phenomenon in a nonlinear mixer for two-tone excitation. If one considers $V_{S1} = V_{S2} = V_S$, the above equations simplify to the following

$$3V_S^2 + V_1^2 = -0.11 \times \frac{2}{3}\frac{\alpha_2}{\alpha_4} \tag{4.92}$$

Or in another form

$$\left(3V_S^2 + V_1^2\right)^{\frac{1}{2}} = 0.269\sqrt{-\frac{\alpha_2}{\alpha_4}} \tag{4.93}$$

This describes an elliptical contour in the $V_1 - V_S$ plane. That is the contour which describes a predetermined compression value (here, 1 dB) in the $V_1 - V_S$ plane.

IIP$_3$ calculation in a nonlinear mixer

If we consider an input with the following form

$$v = V_1 \cos(\omega_0 t) + V_{S1}\cos(\omega_{S1} t) + V_{S2}\cos(\omega_{S2} t) \tag{4.94}$$

For a fourth-order nonlinear transconductance, the desired downconverted components of the signal would be approximately (ignoring the saturating components of the signals)

$$v_{IF} = \alpha_2 V_{S1} V_1 R_L \cos((\omega_0 - \omega_{S1})t) + \alpha_2 V_{S2} V_1 R_L \cos((\omega_0 - \omega_{S2})t) \tag{4.95}$$

The third-order IM components would be then

$$v_{IM} = \frac{3}{2}\alpha_4 V_{S1}{}^2 V_{S2} V_1 R_L \cos(\Omega_0 - 2\omega_{S1} + \omega_{S2})t +$$
$$\frac{3}{2}\alpha_4 V_{S1} V_{S2}{}^2 V_1 R_L \cos(\omega_0 - 2\omega_{S2} + \omega_{S1})t \tag{4.96}$$

With the same reasoning as in section 4.7, and considering $V_{S1} = V_{S2} = V_S$, the IIP$_3$ for a mixer is deduced

$$A_{IIP_3} = V_S = \sqrt{-\frac{2}{3}\frac{\alpha_2}{\alpha_4}} \tag{4.97}$$

It is noteworthy that in a physical mixer whose characteristics is compressive, $\frac{\alpha_2}{\alpha_4}$ is always negative.

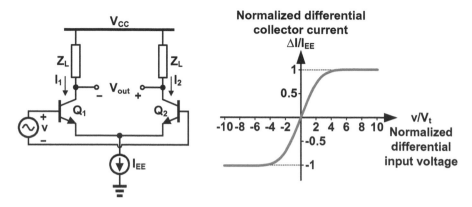

Figure 4.32: A typical differential bipolar stage and its corresponding nonlinear transfer characteristics.

4.7.2 IIP$_3$ of Differential Pair Amplifiers

As an example, we investigate the linearity of a differential pair bipolar transistor. We can write for each branch of this cell's current

$$I_1 = \frac{I_{EE}}{1 + \exp\left(-\frac{V\cos(\omega t)}{V_t}\right)} \tag{4.98a}$$

$$I_2 = \frac{I_{EE}}{1 + \exp\left(\frac{V\cos(\omega t)}{V_t}\right)} \tag{4.98b}$$

and subtracting the first equation from the second equation in Equation 4.98, we obtain the differential current as

$$\Delta I = I_{EE} \tanh\left(\frac{V\cos(\omega t)}{2V_t}\right) \tag{4.99}$$

If we employ the Taylor expansion of *tanh* {.} as

$$\tanh(x) = x - \frac{x^3}{3} + \cdots \tag{4.100}$$

Now, the current–voltage characteristic becomes

$$\Delta I = \frac{V}{2V_t}\cos(\omega t) - \frac{1}{3}\left(\frac{V}{2V_t}\cos(\omega t)\right)^3 + \cdots \tag{4.101}$$

which results in

$$A_{IP3} = \sqrt{\frac{4}{3}\frac{\frac{1}{2V_t}}{\frac{1}{24V_t^3}}} = 4V_t \approx 100\,\text{mV} \tag{4.102}$$

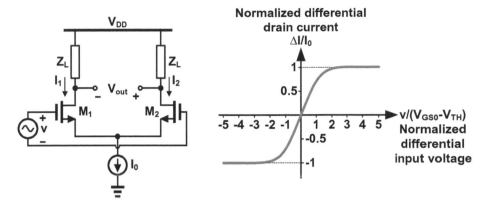

Figure 4.33: A typical differential MOS stage and its corresponding nonlinear transfer characteristics.

As another example for a MOS differential pair, one can write with a good approximation

$$I_1 = k\left(V_{GS0} + \frac{v}{2} - V_{TH}\right)^2 \quad \text{for} \quad \left|\frac{v}{2}\right| < V_{GS0} - V_{TH} \tag{4.103}$$

and

$$I_2 = k\left(V_{GS0} + \frac{-v}{2} - V_{TH}\right)^2 \quad \text{for} \quad \left|\frac{v}{2}\right| < V_{GS0} - V_{TH} \tag{4.104}$$

Note for $\left|\frac{v}{2}\right| > V_{GS0} - V_{TH}$, one of the transistors goes to saturation and the other one goes to cut-off. Here V_{GS0} is the common DC bias voltage of either of the transistors whose value is obtained by the following

$$V_{GS0} = V_{TH} + \sqrt{\frac{I_0}{2k}} \tag{4.105}$$

Then

$$\frac{I_1}{I_2} = \frac{\left(V_{GS0} + \frac{v}{2} - V_{TH}\right)^2}{\left(V_{GS0} - \frac{v}{2} - V_{TH}\right)^2} \tag{4.106}$$

Given

$$I_1 + I_2 = I_0 \tag{4.107}$$

Then

$$I_1\left[1 + \left(\frac{V_{GS0} - v/2 - V_{TH}}{V_{GS0} + v/2 - V_{TH}}\right)^2\right] = I_0 \tag{4.108}$$

and

$$I_1 = \frac{I_0}{1 + \left(\frac{V_{GS0} - v/2 - V_{TH}}{V_{GS0} + v/2 - V_{TH}}\right)^2} \tag{4.109}$$

similarly

$$I_2 = \frac{I_0}{1 + \left(\frac{V_{GS0} + v/2 - V_{TH}}{V_{GS0} - v/2 - V_{TH}}\right)^2} \tag{4.110}$$

Finally, the differential output current would have the following form, for $\left|\frac{v}{2}\right| < V_{GS0} - V_{TH}$

$$\Delta I = I_1 - I_2 = I_0 \left(\frac{\frac{v}{V_{GS0} - V_{TH}}}{1 + \frac{v^2}{4(V_{GS0} - V_{TH})^2}}\right) \tag{4.111}$$

Otherwise

$$\begin{cases} I_1 = I_0 \\ I_2 = 0 \end{cases} \quad \text{for} \quad \frac{v}{2} > V_{GS0} - V_{TH} \tag{4.112}$$

This means $\Delta I = I_0$ for $v > 2(V_{GS0} - V_{TH})$.
    and

$$\begin{cases} I_1 = 0 \\ I_2 = I_0 \end{cases} \quad \text{for} \quad \frac{v}{2} < -(V_{GS0} - V_{TH}) \tag{4.113}$$

This means $\Delta I = -I_0$ for $v < -2(V_{GS0} - V_{TH})$.
    The current in Equation 4.111 attains its maximum value of I_0 once $v/2 = V_{GS0} - V_{TH}$. If $v/2 > V_{GS0} - V_{TH}$, the transistor M_1 goes to saturation and transistor M_2 goes to cut-off and the differential current remains at I_0. If $-v/2 > V_{GS0} - V_{TH}$ the transistor M_2 goes to saturation and transistor M_1 goes to cut-off and the differential current remains at $-I_0$. For a rough estimate of the IIP$_3$, one can write

$$\Delta I \approx \frac{v I_0}{(V_{GS0} - V_{TH})} \left[1 - \left(\frac{v}{2(V_{GS0} - V_{TH})}\right)^2\right] \tag{4.114}$$

Consequently, the third-order input intercept point amplitude is calculated as

$$A_{IIP_3} = \sqrt{\frac{4}{3}\left(\frac{\frac{1}{(V_{GS0} - V_{TH})}}{\frac{1}{4(V_{GS0} - V_{TH})^3}}\right)} = \frac{4(V_{GS0} - V_{TH})}{\sqrt{3}} = 2.31(V_{GS0} - V_{TH}) \tag{4.115}$$

4.8 Linearization Methods in Mixers

For now, we have learned that the nonlinearity is one of the most important issues in the mixer design. In this section, we introduce methods to increase IIP$_3$ in mixers. One practical method is merely adding a small resistor in the emitter (or the source) of the input transistors. This resistor is called the degeneration resistor due to decrease in the conversion of gain of the mixers. This, however, alleviates linearity problems in the mixers. Nevertheless, these resistors' drawbacks are the worse noise figure and lower conversion gain. Figure 4.34 compares the linearity of a bipolar transistor and a MOS transistor pairs.

Not surprisingly, however, MOS devices versus bipolar show better linearity due to quadratic $I-V$ characteristics of the former versus the exponential characteristics of the latter. Moreover, Figure 4.34 shows that the acceptable peak-to-peak range for linear operation in bipolar devices is roughly $4V_t$ and this value for MOS is $2.3\,(V_{GS0} - V_{TH})$ peak-to-peak, which is normally larger, given the fact that the bias voltage above the threshold of the currently used MOSFET's is much larger than the thermal voltage of the bipolar transistors. Thus, we can employ MOS device for the input RF signal to achieve a better IIP$_3$. As stated earlier, one method to mitigate the nonlinearity is to add degeneration resistor in the emitter/source which is shown in Figure 4.35. As Figure 4.35 suggests, resistors R decrease the signal on the base emitter junction of

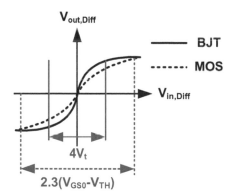

Figure 4.34: Input–output voltage characteristic of MOS and bipolar devices.

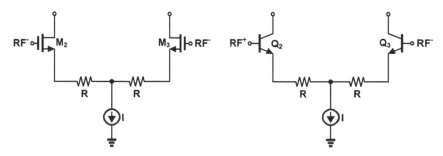

Figure 4.35: Degeneration resistor implementation to achieve linear behavior.

Q_2 and Q_3 transistors or gate–source terminal of M_2 and M_3 MOS transistors, thus reducing the nonlinearity. The required condition to linearize the differential stage is that the series resistance should be much larger than the inverse of the transconductance of each transistor as the following

$$g_m R_E \gg 1 \quad \text{or} \quad \frac{I_{EE} R_E}{2V_t} \gg 1 \tag{4.116}$$

To determine the large-signal characteristics of the differential stage with the degeneration resistors, given the above condition, one can write

$$i_{e1} = \frac{I_{EE}}{2} + \frac{v}{2R_E} \tag{4.117}$$

$$i_{e2} = \frac{I_{EE}}{2} - \frac{v}{2R_E} \tag{4.118}$$

Then

$$\Delta i_{ee} = i_{e1} - i_{e2} = \frac{v}{R_E} \quad \text{for } |v| \leq R_E I_{EE} \tag{4.119}$$

$$\Delta i_{ee} = i_{e1} - i_{e2} = I_{EE} \quad \text{for } |v| > R_E I_{EE} \tag{4.120}$$

The overall transfer characteristics of the differential stage are depicted in Figure 4.36.

The required condition to linearize the differential MOS stage is that the series degeneration resistance should be much larger than the inverse of the transconductance of each transistor, as follows

$$g_m R_S \gg 1 \quad \text{or} \quad R_S \sqrt{2kI_0} \gg 1 \tag{4.121}$$

To determine the large-signal characteristics of the differential MOS stage with the degeneration resistors, given the above condition, one can write

$$i_{d1} = \frac{I_0}{2} + \frac{v}{2R_S} \tag{4.122}$$

$$i_{d2} = \frac{I_0}{2} - \frac{v}{2R_S} \tag{4.123}$$

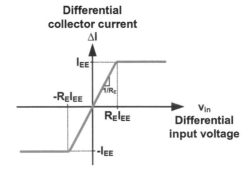

Figure 4.36: The transfer characteristics of a differential stage with degeneration resistors.

Then

$$\Delta i_{dd} = i_{d1} - i_{d2} = \frac{v}{R_S} \quad \text{for} \quad |v| \leq R_S I_0 \tag{4.124}$$

$$\Delta i_{dd} = i_{d1} - i_{d2} = I_0 \quad \text{for} \quad |v| > R_S I_0 \tag{4.125}$$

The overall transfer characteristics of the differential MOS stage are depicted in Figure 4.37.

A drawback of the structure depicted in Figure 4.35 is reduction in voltage head-room which is wasted on these resistors. Therefore, we can modify Figure 4.35 to solve this problem which is shown in Figure 4.38.

This structure doesn't consume DC power in the resistor and it is a good prototype for a more linear mixer. Note that in any case the conversion gain of the mixer with degenerative resistors would be reduced with respect to nondegenerative mixer. That is the conversion gain of the mixer goes from a maximum value obtained for $R_E = 0$ or $R_S = 0$ to zero for $R_E \gg 1/g_m$ or $R_S \gg 1/g_m$.

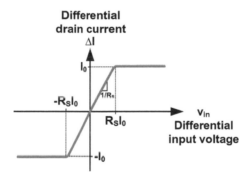

Figure 4.37: The transfer characteristics of a differential MOS stage with degeneration resistors.

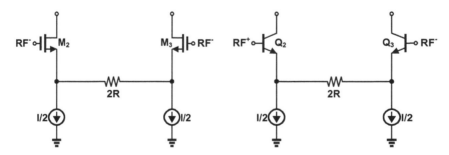

Figure 4.38: Degeneration resistor implementation without voltage-headroom usage.

Example 4.2 Find the input and output power alongside conversion voltage gain for the single-balanced downconverting mixer. Here the capacitors, C, are considered to be short-circuit at RF and the LO frequencies, and they are considered as open-circuit at the IF frequency. Furthermore, R_C is small compared to R_L.

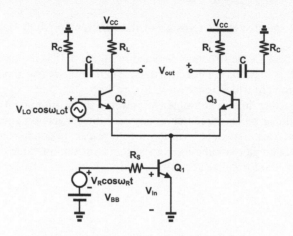

Figure 4.39: Single-balanced mixer.

Solution:
For computing the input power, we can write

$$V_{\text{in}} = \frac{1}{1 + g_{\text{in}}R_S} V_R \tag{4.126}$$

where g_{in} is the input transconductance of the transistor Q_1. Then

$$P_{\text{in}} = g_{\text{in}}V_{\text{in}_{\text{rms}}}^2 = \frac{V_{\text{in}_{\text{rms}}}^2}{R_{\text{in}}} \tag{4.127}$$

Given the fact that C in open at the IF frequency, for the output power, we can write

$$P_{\text{out}_{\text{IF}}} = \frac{\left(\frac{V_{\text{out}}}{2\sqrt{2}}\right)^2}{R_L} \times 2 = \frac{V_{\text{out}}^2}{4R_L} \tag{4.128}$$

Finally, for the conversion gain, given the fact that the upper tree is switched at the LO rate,one may write

$$A_V = \left(\frac{4}{\pi}\right)\frac{1}{2}g_m R_L = \frac{2}{\pi}g_m R_L \tag{4.129}$$

Example 4.3 In the given mixer circuit, the LO is at 2.4 GHz with 1 V differential for the upper tree and the RF frequency is at 2.41 GHz.
(a) With 1 mV signal for RF amplitude, find the IF component amplitude.
(b) If the input amplitude for RF signal is 1 mV, find the capacitor C in order to attenuate the adjacent channel with the same amplitude residing at 2.45 GHz by 6 dB. The desired channel bandwidth is 2 MHz.
(c) Calculate the amplitude of LO signal without the capacitor C at the output.
(d) Suppose the double-balanced Gilbert cell and write its advantage.
(e) While in single-balanced given circuit only one branch has the RF current, why the RF leakage is zero?

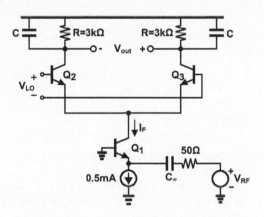

Figure 4.40: Single-balanced mixer.

Solution:
(a) Considering complete switching of the differential pair, we have

$$g_C = \frac{2}{\pi} g_m \tag{4.130}$$

$$r_{in} = \frac{KT}{qI_E} = 52\,\Omega \tag{4.131}$$

$$V_{in} = V_{BE}\frac{r_{in}}{r_{in}+R_S} \approx \frac{1}{2}V_{RF} \quad V_{IF} = \frac{2}{\pi}g_m R \times \frac{1\,\text{mV}}{2} = \frac{2}{\pi}\frac{0.5\,\text{mA}}{25\,\text{mV}}3\text{k}\times 0.5\,\text{mV}$$

$$= 19.1\,\text{mV} \tag{4.132}$$

(b) To have 6 dB attenuation for the adjacent channel, we should have

$$-20\log\frac{\omega_2}{\omega_c} = -6 \Rightarrow \frac{\omega_2}{\omega_c} = 2 \Rightarrow \omega_c = \frac{1}{RC} = \frac{\omega_2}{2} \tag{4.133}$$

where ω_c is the cut-off frequency of the output filter. Consequently, the value for capacitor will be

$$C = \frac{2}{R\omega_2} = \frac{2}{3000\,(2\pi)\,50\,\left(10^6\right)} = 2.12\,\text{pF} \tag{4.134}$$

(c) For the LO leakage (without consideration of load capacitance, the LO will have a square-wave form, then), we have

$$V_{\text{out}} = R\left(I_{\text{EDC}} + I_{\text{RF}}\cos\omega_0 t\right) S\left(\omega_0 t\right) \tag{4.135}$$

$$V_{\text{out}} = R\left(I_{\text{EDC}} + I_{\text{RF}}\cos\omega_0 t\right)\left[\frac{4}{\pi}\cos\omega_0 t - \frac{4}{3\pi}\cos 3\omega_0 t + \cdots\right] \tag{4.136}$$

where

$$I_{\text{RF}} = \frac{V_{\text{RF}}}{50 + r_e} \tag{4.137}$$

Then

$$V_{\text{LO}} = \frac{4}{\pi}RI_{\text{EDC}} = 1.9\,\text{V} \tag{4.138}$$

(d) An important feature of double-balanced Gilbert cell is removing the LO and RF leakage to the output.

(e) As it is seen from the above equations, only the LO and the mixing components appear at the output and because the RF signal is common mode, no RF signal will emerge at the output (even without C). ∎

One of the most practically applied mixers is the MOS Gilbert cell shown in Figure 4.41. This mixer doesn't show second-order nonlinearity due to its symmetry. Moreover,

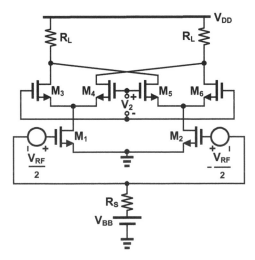

Figure 4.41: MOS double-balanced mixer.

since it has no current source at the source of input devices, it provides a larger linear range operation for the RF signal.

4.9 Calculating Third-Order Input Intercept Point in Cascaded Stages

Till now, we have introduced useful parameters to understand the nonlinearity of an amplifier or a mixer as a black box. In this section, we discuss the cascaded nonlinear blocks' behavior. The input–output relation of the two nonlinear blocks could be assumed as

$$y_1(t) = \alpha_1 x(t) + \alpha_2 x^2(t) + \alpha_3 x^3(t) \tag{4.139a}$$

$$y_2(t) = \beta_1 y_1(t) + \beta_2 y_1{}^2(t) + \beta_3 y_1{}^3(t) \tag{4.139b}$$

Replacing $x(t)$ by a two-tone input signal, and computing $y_1(t)$ as before, and then replacing $y_1(t)$ by the computed result, we would obtain a sinusoidal expansion for $y_2(t)$. Then, by following the same procedure as described in section 4.7, it can be shown that IIP$_3$ voltage will be

$$A_{IP3} = \sqrt{\frac{4}{3} \left| \frac{\alpha_1 \beta_1}{\alpha_3 \beta_1 + 2\alpha_1 \alpha_2 \beta_2 + \alpha_1{}^3 \beta_3} \right|} \tag{4.140}$$

4.9.1 Third-Order Input Intercept Voltage of Cascaded stages in Terms of Single-Stage Intercept Voltage

One can rewrite Equation 4.140 to obtain

$$\frac{1}{A^2{}_{IP3}} \approx \frac{1}{A^2{}_{IP3,1}} + \frac{\alpha_1{}^2}{A^2{}_{IP3,2}} + \frac{3\alpha_2 \beta_2}{2\beta_1} \tag{4.141}$$

Note that, given the fact that most of practical mixers/amplifiers use differential pairs which have odd symmetry in their transfer characteristic, chances are that α_2 and β_2 are nearly zero. So, the third term in Equation 4.141 could be neglected with respect to the first two terms. Equation 4.141 gives us an explicit equation to obtain IIP$_3$ of two cascaded stages. An important point is the effect of nonlinearity in subsequent stages which will be more severe. We can compare Equation 4.141 with equivalent resistance of parallel resistors, and by the assumption that the third term is neglected, we can generalize Equation 4.141 to give the equivalent IP$_3$ point for multiple stages (here for three stages or more) as Equation 4.142:

$$\frac{1}{A^2{}_{IP3}} \approx \frac{1}{A^2{}_{IP3,1}} + \frac{\alpha_1{}^2}{A^2{}_{IP3,2}} + \frac{\alpha_1{}^2 \beta_1{}^2}{A^2{}_{IP3,3}} + \cdots \tag{4.142}$$

Here, it could be seen that the total IIP$_3$ of cascaded stages is lower than each of them in Equation 4.142. In other words, by the assumption of sufficient gain for previous stages, the total IIP$_3$ will be lower than the third stage IIP$_3$ divided by previous gains.

In the above equation A_{IIP_3} being the signal (voltage) amplitude and α_1 and β_1 being the voltage gains, for a fixed input impedance system (e.g., $50\,\Omega$), one can reexpress Equation 4.142 in another form in terms of IIP_3 powers and the power gains of the stages as follows

$$P_{\text{IIP}_3,\text{total}}^{-1} = P_{\text{IIP}_3,1}^{-1} + P_{\text{IIP}_3,2}^{-1}G_{P,1} + P_{\text{IIP}_3,3}^{-1}G_{P,1}G_{P,2} + \dots \qquad (4.143)$$

4.9.2 Combination of Amplifier and Mixer

In many receivers' applications, we use a structure of cascaded low-noise amplifier and a mixer. As Equation 4.142 suggests, the IIP_3 of total chain will be less than the IIP_3 of the mixer divided by the gain of LNA.

Example 4.4 The given architecture is for a global positioning system receiver. The received signal frequency is 1575 MHz and has a 2 MHz bandwidth and its power is $-130\,$dBm. The LO frequency is 1579 MHZ which downconverts the RF signal to a 4 MHz carrier. With the given specifications, find
(a) The overall noise figure and overall IIP_3.
(b) If the mixer is linear, with the given two interferer signals, how much the interferers' IM_3 component is lower than the desired GPS signal.
(c) We consider the effect of IIP_3 of the mixers and the LNA, what is the input interferer signal level which results in an IM_3 component with $-140\,$dBm power level at the output.
The system impedance is 50 ohms.

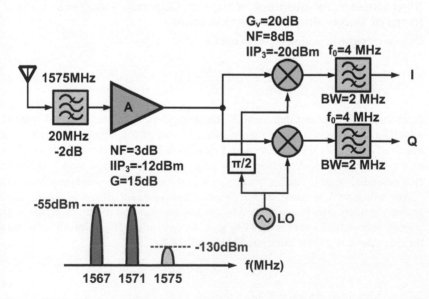

Figure 4.42: Typical GPS receiver architecture and the neighboring interfering signals.

Solution:

(a) For the noise figure of cascaded stages one can write (see cascaded noise figure expression in section 9.5)

$$F = L \times \left(F_1 + \frac{F_2 - 1}{G_1} \right) = 10^{0.2} \times \left(10^{0.3} + \frac{10^{0.8} - 1}{10^{1.5}} \right)$$

$$= 1.58 \times (1.99 + 0.16) = 3.42 \tag{4.144}$$

$$F_{dB} = 10 \log(3.42) = 5.35 \, dB$$

To calculate $IIP_{3,tot}$, we have

$$IIP_3 = 10 \log \left(\frac{\frac{A_{IIP_3}^2}{2R}}{1 \, mW} \right) \tag{4.145}$$

Converting dBm level into Volts, in a $50 \, \Omega$ system we can write

$$A_{IIP_3} = \sqrt{10^{\frac{IIP_3}{10}} \times 1 \, mW \times 2R} = \sqrt{10^{\frac{IIP_3}{10} - 1}} \tag{4.146}$$

For overall IIP_3 we have

$$\frac{1}{A_{IIP_{3tot}}^2} = \frac{1}{A_{IIP_{3_1}}^2} + \frac{G_1^2}{A_{IIP_{3_2}}^2} \tag{4.147}$$

Given the fact that there is a 2 dB loss ahead of the LNA, then the IIP_3 of the combined filter and LNA would be

$$IIP_{3_1} = 2 - 12 = -10 \, dBm \tag{4.148}$$

Finally, for the overall IIP_3, we will have

$$\frac{1}{A_{IIP_{3tot}}^2} = \frac{1}{10^{\frac{IIP_{3_1}}{10} - 1}} + \frac{G_1^2}{10^{\frac{IIP_{3_2}}{10} - 1}} = \frac{1}{10^{\frac{-10}{10} - 1}} + \frac{10^{\frac{26}{20}}}{10^{\frac{-20}{10} - 1}} \Rightarrow A_{IIP_{3tot}} = 7.06 \, mV \tag{4.149}$$

$$IIP_{3tot} = 10 \log \left(\frac{\frac{A_{IIP_{3tot}}^2}{2R}}{1 \, mW} \right) = -33 \, dBm \tag{4.150}$$

(b) At the LNA input, we have a pair of $-57\,\text{dBm}$ interfering signals. From Equation 4.77, it can be seen that

$$A_{\text{IM3}} = \frac{3}{4}\alpha_3 A^3 \tag{4.151}$$

or it can be rewritten as

$$A_{\text{IM3}} = \frac{3}{4}\frac{\alpha_3}{\alpha_1}\alpha_1 A^3 = \frac{1}{A_{\text{IIP3,LNA}}^2}\alpha_1 A^3 \tag{4.152}$$

Now, the amplitude of the IM_3 component, at the LNA output, can be computed as

$$A_{\text{IM3}} = \frac{1}{A_{\text{IIP3,LNA}}^2}\alpha_1 A^3 = \frac{1}{10^{\frac{-12}{10}}-1}10^{\frac{15}{20}} \times \left(\sqrt{10^{\frac{-57}{10}}-1}\right)^3 = 79.4\,\text{nV} \tag{4.153}$$

The desired GPS signal level at the output of the LNA (with $13\,\text{dB}$ total gain) would be

$$A_{\text{GPS}} = \sqrt{10^{\frac{-117}{10}}-1} = 447\,\text{nV} \tag{4.154}$$

Since

$$20\log\left(\frac{447}{79.4}\right) = 15\,\text{dB} \tag{4.155}$$

Therefore, the desired signal is $15\,\text{dB}$ higher than the third-order IM of the interfering signal.

(c) To obtain an IM level of $-140\,\text{dBm}$ at the output, we should have

$$A_{\text{IM3}} = \frac{3}{4}\frac{\alpha_3}{\alpha_1}\alpha_1 A_{\text{in}}^3 = \frac{1}{A_{\text{IIP3,tot}}^2}\alpha_1 A_{\text{in}}^3 = \frac{1}{10^{\frac{-33}{10}}-1}10^{\frac{33}{20}} \times A_{\text{in}}^3 = 31.6\,\text{nV} \Rightarrow$$

$$\tag{4.156}$$

$$A_{\text{in}} = 32.9\,\mu\text{V}$$

$$P_{\text{in}} = 10\log\left(\frac{\frac{A_{\text{in}}^2}{2R}}{1\,\text{mW}}\right) = -79.7\,\text{dBm}$$

■

4.10 Important Point in RF Circuit Simulation

One of the most important parameters in any circuit simulation is the computation time. It is quite clear that more points of simulation in each period result in much more computation time. The important point is to decrease the ratio of the highest operating frequency of the signal to its lowest frequency component. Once we have two exciting tones with small frequency difference, their beat frequency will be very small. As such in simulations related to IIP_3, it is recommended to increase the frequency distance between the exciting tones. By this, we will avoid the requirement of too many points, in the simulation, to differentiate between the frequencies of the exciting tones (and consequently the beat frequency).

4.11 Conclusion

In this chapter, we have studied the different mixer topologies normally used in RF circuits. The main application of the mixer block is to downconvert RF signal to IF for detection in the receiver, or upconvert the IF signal to the RF in the transmitter.

It was shown that each mixer operates by virtue of its nonlinearity or switching characteristics at the LO rate; however, evaluating the generated components needs careful considerations to suppress unwanted mixing products. Moreover, a parameter was introduced which is a measure of nonlinearity and was named 1 dB compression point. Furthermore, in the presence of multiple input signals, another quantity was introduced as the IIP_3, for computation of which analytical relations were presented. Three different mixer topologies were studied, namely, unbalanced, single balanced, and double balanced, and relations for port-to-port signal conversion were carried out. Finally, methods to improve linearity in mixers by means of degeneration resistors were investigated.

4.12 References and Further Reading

1. P.R. Gray, P.J. Hurst, S.H. Lewis, R.G. Meyer, *Analysis and Design of Analog Integrated Circuits*, fifth edition, Hoboken, NJ: J. Wiley & Sons, Inc., 2009.
2. K.K. Clarke, D.T. Hess, *Communication Circuits, Analysis and Design*, United States: Krieger Publishing Company, 1994.
3. F. Farzaneh, *RF Communication Circuits* (in Persian), Tehran: Sharif University Press, 2005.
4. B. Razavi, *RF Microelectronics*, second edition, Castleton, NY: Prentice-Hall, 2011.
5. R. Chi-Hsi Li, *RF Circuit Design*, Hoboken, NJ: J. Wiley & Sons, Inc., 2009.
6. J.R. Smith, *Modern Communication Circuits*, second edition, New York, NY: McGraw Hill, 1997.
7. D.O. Pederson, K. Marayam, *Analog Integrated Circuits for Communications*, Boston, MA: Kluwer Academic Publishers, 1990.
8. R. Dehghani, *Design of CMOS Operational Amplifiers*, Norwood, MA: Artech House, 2013.
9. P. Wambacq, W. Sansen, *Distortion Analysis of Analog Integrated Circuits*, Norwell, MA: Kluwer Academic, 1998.
10. J. Everard, *Fundamentals of RF Circuit Design with Low Noise Oscillators*, United Kingdom: J. Wiley & Sons, Inc., 2000.

4.13 Problems

Problem 4.1 In the mixer circuit depicted in Figure 4.43,

1. Determine the output IF signal amplitude at 10.7 MHz for the case where the
   RF current is $i_{RF} = 100\,\mu A \sin(2\pi \times 100\,MHz \times t)$ and the LO signal is
   $V_{LO} = 300\,mV \sin(2\pi \times 89.3\,MHz \times t)$.
2. If the input signal has two components of the same amplitude one residing at
   100 MHz and the parasitic one at 111.3 MHz, find the required 3 dB bandwidth
   of the output low-pass filter. In order that the downconverted component of
   the parasitic signal is 6 dB lower than the desired IF component, in this case,
   calculate the appropriate value of C.
3. If the output low-pass filter has a 3 dB bandwidth of 11 MHz, find the parasitic
   LO component at the output.
4. If the input has two components of 100 MHz and 100.1 MHz, compare the
   conversion gain in dB for the output components at 10.7 MHz and 10.8 MHz
   in comparison with the gain of IM3 components residing at 10.6 MHz and
   10.9 MHz. In this case, we have
   $V_{RF} = 50\,mV \sin(2\pi \times 100\,MHz \times t) + 50\,mV \sin(2\pi \times 100.1\,MHz \times t)$ and the
   bias current is $500\,\mu A$.

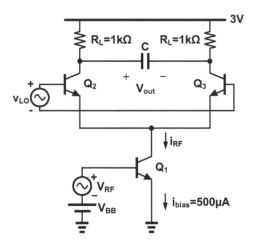

Figure 4.43: Single-balanced differential pair mixer.

Problem 4.2 In the circuit in Figure 4.44, a single-ended mixer with the input and
output matching networks is depicted.

1. Determine the values of L_1 and C_1 in order to match the RF input to $50\,\Omega$, assume
   that the input impedance of the transistor is $2\,k\Omega$. Furthermore, determine the
   values of L_2 and C_2 in order to match the output to $50\,\Omega$. Suppose that the
   output impedance of the transistor is $200\,\Omega$. The RF frequency is 1900 MHz
   and the IF frequency is 200 MHz. Assume that the capacitances C_{B1} and C_{B2}
   are RF short-circuit.
2. If the signals at the base of the transistor are $V_{RF} < V_T$ and $V_{LO} > 10V_T$, calculate
   the IF output current alongside the RF leakage current at the output in terms of

transistor's g_m and find an expression for the IF output voltage and the output leakage voltage in this case.

Note that the RF trap circuit is open circuit at the RF frequency and short circuit at other frequencies, and the the LO trap circuit is open circuit at the LO frequency and short circuit at other frequencies.

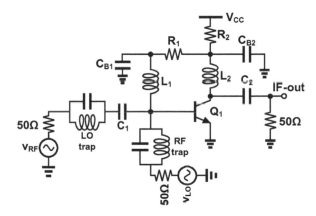

Figure 4.44: Single-transistor mixer with corresponding matching circuits.

Problem 4.3 In the given mixer circuit depicted in Figure 4.45, assume the $I - V$ characteristics are described by $i_1 - i_2 = \left(0.4v_{RF} - 0.01v_{RF}^3\right)\tanh\left(\frac{v_{LO}}{2V_T}\right)$,

1. Considering two signals of 50 mV amplitude at 104 MHz and 104.1 MHz at the RF input, obtain the output components at 10.7 MHz and 10.6 MHz, considering a rectangular LO voltage in the upper tree (how?). The LO frequency is 114.7 MHz.
2. Compute the IM3 components at the output, and obtain the IIP₃ point.

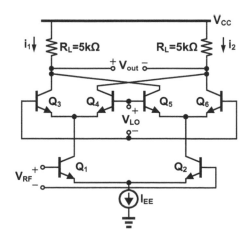

Figure 4.45: Gilbert cell double-balanced mixer.

Problem 4.4 In the circuit depicted in Figure 4.46, determine the IIP$_3$ through ADS computer simulation. To save the computation time, use two distant tones with 150 MHz and 155 MHz frequencies as an example. The LO frequency is 225 MHz. Note that the output low-pass filter has a cut-off frequency of 75 MHz which affects the outputs. If one employs two close tones with 150 MHz and 150.1 MHz frequencies, for example, he/she might obtain the same results with a much larger computation time (why?). Choose bipolar transistors with a f_T greater than 5 GHz in this simulation.

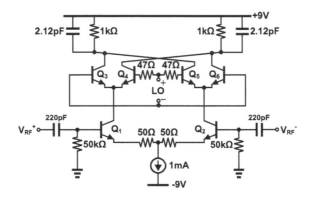

Figure 4.46: Gilbert cell double-balanced mixer with a degenerative resistor.

Problem 4.5 Consider the differential pair MOS mixer circuit depicted in Figure 4.47 where the RF input signal is applied to the gate of M_1, the LO signal is applied between the gates of the differential pair, and the output-tuned circuits are tuned to the difference frequency. Considering the RF signal as $V_S \cos(\omega_S t)$ and the LO signal as $V_0 \cos(\omega_0 t)$. Find an expression for the output IF signal. Here consider that the RF signal is a small signal and the LO signal is a large signal with $V_0 \leq \frac{1}{4}(V_{GS0} - V_{TH})$ where V_{GS0} is the bias voltage of the MOS differential pair transistors. All the MOS transistors are biased in the active region.

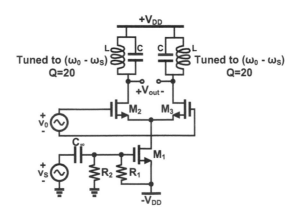

Figure 4.47: Differential pair MOS balanced mixer.

Problem 4.6 In the downconverting mixer depicted in Figure 4.48, consider β of the transistors is sufficiently large such that $2r_\pi \geq 25\,\mathrm{k\Omega}$, and $L = 765\,\mathrm{nH}$ with $Q_L = 10$. Find the appropriate values of C_1 and C_2 for matching the $50\,\Omega$ RF source to the input of the mixer. Furthermore, determine the required value of I to achieve a conversion power gain $G_{PC} = \frac{P_{\mathrm{out}}(10.7\,\mathrm{MHz})}{P_{\mathrm{in}}(104\,\mathrm{MHz})} = 12\,\mathrm{dB}$. Note that the RF signal at the input of the center tapped capacitor is $v_{RF} = 1\,\mathrm{mV}\cos\left(2\pi \times 104\,\mathrm{MHz} \times t\right)$. Moreover, the quality factor of the capacitors is assumed to be large.

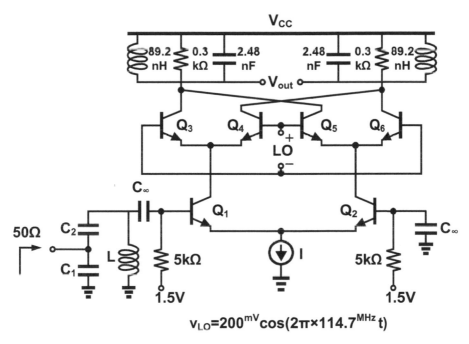

Figure 4.48: Gilbert cell double-balanced mixer with input matching.

Problem 4.7 In this problem, we try to learn how to simulate the IIP$_3$ point determination. In order to have a lower simulation time, we should choose two-tones far enough for more time-efficient simulation. To begin the simulation (e.g., by ADS or Cadence IC design), run the transient simulation for two different low-power signals to obtain points A and B depicted in Figure 4.51. At the same time, find points C and D through the IM3 component computation. By extrapolating the two straight lines, you may find the IIP$_3$ point. Note that a common-mode 2.5 V bias is applied to the LO port. Furthermore, that you can use the following relations in your simulations:
$$v_a = V_a \cos 2\pi\left(103\,\mathrm{MHz}\right)t, v_b = V_b \cos 2\pi\left(102\,\mathrm{MHz}\right)t, P_{\mathrm{in}} = \frac{(V_a)^2}{8(R_{\mathrm{in}})} = \frac{(V_b)^2}{8(R_{\mathrm{in}})}, P_{\mathrm{out}} = \frac{(V_{\mathrm{out}})^2}{2(R_{\mathrm{out}})}$$
where $R_{\mathrm{in}} = 50\,\Omega$ and $R_{\mathrm{out}} = 1\,\mathrm{k\Omega}$.
Hint: choose a small value for V_a and V_b while $V_a = V_b$, and increase both of them gradually.

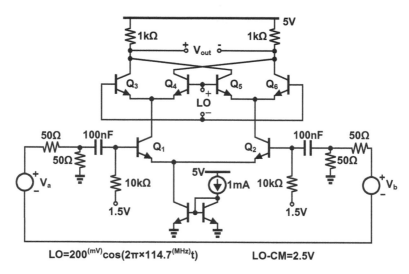

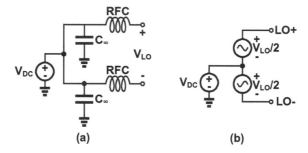

Figure 4.49: A matched Gilbert cell double-balanced mixer to test the IIP$_3$.

Figure 4.50: LO bias implementation, (a) practical implementation, and (b) LO and bias application in the simulations.

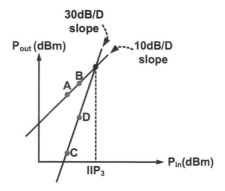

Figure 4.51: IIP$_3$ point determination through computer simulation.

Problem 4.8 If in an amplifier for the input signal pair of -70 dBm level, we obtain output signals of -50 dBm and output IM3 components of -80 dBm,

1. Determine the IIP$_3$ of a single-stage amplifier in dBm and its gain in dB (Hint: use formula $\text{IIP}_3 = \frac{\Delta p}{2} + p_{\text{in}}$).
2. If two similar stages of the same amplifier are cascaded, determine the overall IIP$_3$ in dBm, and the output IM3 components in dBm in case of -70 dBm input signals.

Problem 4.9 If the output spectrum of the first mixer in the receiver chain depicted in Figure 4.52 is like what is shown in the figure,

1. Determine the IIP$_3$ for the first mixer, if it has a conversion gain of 10 dB. (Input signals reside at 899.97 MHz and 899.94 MHz. Moreover, $f_{\text{LO},1} = 945$ MHz, and $f_{\text{LO},2} = 45.455$ MHz).
2. Assuming the second mixer having the same nonlinear characteristics as the first mixer, find the spectrum of the output of the third filter and then given unequal input signals, find the IM3 components at the output of the second mixer (the transfer function of the third filter is also shown).
3. Find the output spectrum of the fourth filter.

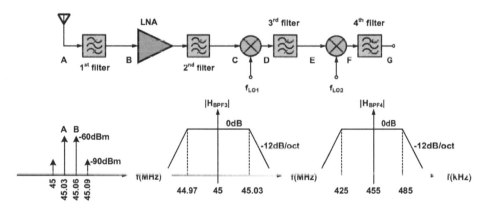

Figure 4.52: The receiver chain intended for IM3 computation.

Problem 4.10 In the given receiver depicted in Figure 4.53, determine the overall noise figure and the overall IIP$_3$ (see section 9.5 for the expression for the noise figure of the cascaded stages). You may determine the noise figure and the IIP$_3$ of the first four blocks, then those of the last two blocks, and consequently the overall noise figure and the overall IIP$_3$. Then, suppose that the desired signal power is -60 dBm at the input. If two interferer signals both with a power of -50 dBm at 60 kHz off the desired signal and 120 kHz off the desired signal, respectively, appear at the receiver input, calculate their effect at the output of the IF amplifier. What kinds of unwanted signals appear at the output of the IF amplifier? How much the unwanted signals are lower than the desired signal?

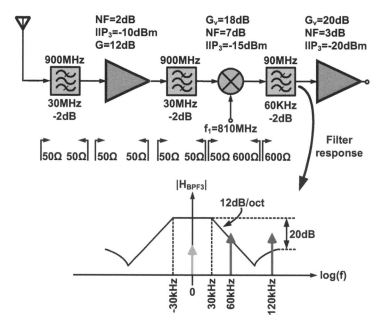

Figure 4.53: Receiver chain for determination of the overall noise figure and overall IIP$_3$.

Problem 4.11 Considering a modulated IF input signal v_s and a carrier signal v_1 as follows

$v_s = 50^{mV} \left[1 + 0.8\cos\left(2\pi \times 10^4 t\right)\right]\cos\left(2\pi \times 10.7 \times 10^6 t\right)$

$v_1 = 1.5^V \cos\left(2\pi \times 10.7 \times 10^6 t\right)$

determine the output voltage as shown in Figure 4.54.

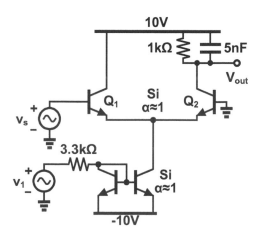

Figure 4.54: A mixer used as a synchronous detector.

Problem 4.12 Determine the main mixing component at the output. Furthermore, calculate the LO leakage signal at the output of the mixer circuit depicted in Figure 4.55.

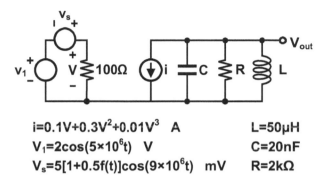

$i=0.1V+0.3V^2+0.01V^3$ A	$L=50\mu H$
$V_1=2\cos(5\times10^6 t)$ V	$C=20nF$
$V_s=5[1+0.5f(t)]\cos(9\times10^6 t)$ mV	$R=2k\Omega$

Figure 4.55: A transconductance harmonic mixer.

Problem 4.13 In the mixer circuit depicted in Figure 4.56, the LO signal is a square-wave pulse train as depicted in the figure. Determine the output voltage at the sum frequency and the unwanted component at the difference frequency.

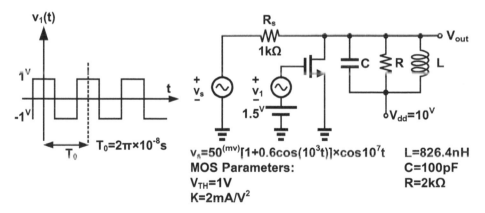

$v_s=50^{(mv)}[1+0.6\cos(10^3 t)]\times\cos10^7 t$ $L=826.4nH$
MOS Parameters: $C=100pF$
$V_{TH}=1V$ $R=2k\Omega$
$K=2mA/V^2$

Figure 4.56: A switching upconverting mixer.

Problem 4.14 A transconductance mixer of Figure 4.57 has the $I-V$ characteristics as follows

$$i = \alpha V + \beta V^2 + \gamma V^3 \tag{4.157a}$$

where

$$\alpha = 2\,\text{mA/V} \tag{4.157b}$$

$$\beta = 0.5\,\text{mA/V}^2 \tag{4.157c}$$

$$\gamma = -0.2\,\text{mA/V}^3 \tag{4.157d}$$

Find the downconverted IF component and the parasitic component at $\omega = 1.4 \times 10^8$ rad/s. Note that there is an inherent negative feedback in this mixer (how?), so you should first find the time-variant transconductance of the nonlinear device.

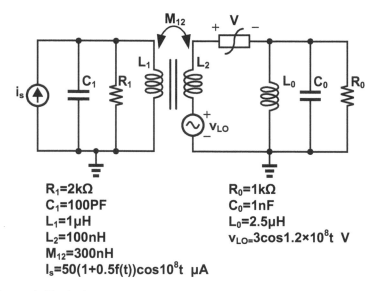

R₁=2kΩ   R₀=1kΩ
C₁=100PF   C₀=1nF
L₁=1µH   L₀=2.5µH
L₂=100nH   v_LO=3cos1.2×10⁸t V
M₁₂=300nH
I_s=50(1+0.5f(t))cos10⁸t µA

$R_1=2k\Omega$
$C_1=100\text{PF}$
$L_1=1\mu H$
$L_2=100nH$
$M_{12}=300nH$
$I_s=50(1+0.5f(t))\cos10^8t \ \mu A$

$R_0=1k\Omega$
$C_0=1nF$
$L_0=2.5\mu H$
$v_{LO}=3\cos1.2\times10^8t \ V$

Figure 4.57: A downconverting mixer using a series nonlinear device.

Problem 4.15 Consider the nonlinear transconductance with the specified $I-V$ characteristic in Figure 4.58. First, find the large-signal time-varying transconductance and then determine an expression for the output signal of the mixer. Furthermore, find an expression for the parasitic component at $\omega_0 + \omega_s$ in this scenario.

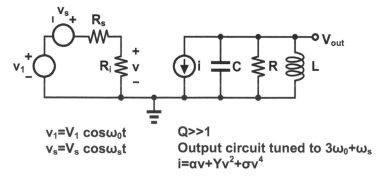

$v_1=V_1 \cos\omega_0 t$ $Q\gg1$
$v_s=V_s \cos\omega_s t$ Output circuit tuned to $3\omega_0+\omega_s$
                                   $i=\alpha v+Yv^2+\sigma v^4$

Figure 4.58: A harmonic upconverting mixer.

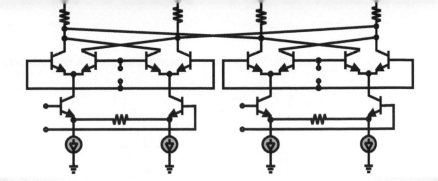

5. Modulation/Demodulation of Amplitude/Phase

Baseband signals are generally band-limited low-pass signals which cannot be directly transmitted over the transmission medium. Furthermore, a huge number of different signals should be transmitted simultaneously in a transmission medium (apparently the air or other transmission medium), so it is imperative that the baseband signals to be modulated over the radio carriers at different frequencies before being transmitted over the air, with antennas of limited sizes. This allows different modulating signals to be differentiated or to be distinct in the frequency domain. In this chapter, we discuss the conventional modulation schemes alongside modern digital modulations with high bandwidth efficiency. Moreover, we investigate the receiver structures for the signal demodulation. One of the long-existing modulation schemes which is used even today is the amplitude modulation (AM) for long-distance broadcasting. In this method, the data are embedded on the amplitude of the signal, therefore it is sensitive to amplitude noise. Furthermore, we introduce circuits to demodulate AM signals. One of the applicable modulations is the phase modulation (PM). In this modulation, the baseband data are embedded in the phase of the radio signal enabling high data rates. We will then discuss quadrature amplitude modulation (QAM) which is one of the most applied modulation schemes in modern radios, phase modulator and demodulator in this chapter. Finally, a few special modulation schemes are investigated.

5.1 AM Modulation

We can represent a sinusoidal AM-modulated signal as follows

$$V_{AM} = A\cos(\omega_C t)\left(1 + m\cos(\omega_m t)\right) \tag{5.1}$$

where $A\cos(\omega_C t)$ denotes the carrier signal, m is the modulation index, and $\cos(\omega_m t)$ is the baseband signal. For AM modulation signal, modulation index is equal to or less than unity. Figure 5.1 depicts a typical AM-modulated signal. As it is obvious from Figure 5.1, the baseband signal is embedded as the envelope of the signal. Moreover,

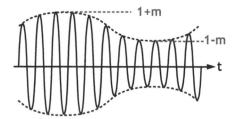

Figure 5.1: Sinusoidal AM-modulated signal.

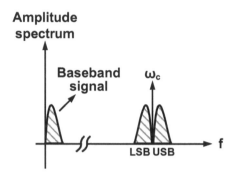

Figure 5.2: Typical baseband and AM modulator output spectrum.

signal amplitude is limited between $1 + m$ and $1 - m$ values. Notice that we can expand Equation 5.1 to achieve

$$V_{AM} = A\cos(\omega_C t) + \frac{m}{2}A\cos(\omega_C t - \omega_m t) + \frac{m}{2}A\cos(\omega_C t + \omega_m t) \qquad (5.2)$$

Equation 5.2 shows that three frequency components appear at the output which are $\omega_C \pm \omega_m$, and ω_C. If we assume that the baseband signal has a finite spectrum (low pass spectrum), the output will be similar to what is shown in Figure 5.2.

Figure 5.2 illustrates the fact that the baseband signal is upconverted around the carrier signal, and also subscripts *LSB* and *USB* refer to the lower sideband and the upper sideband, respectively. This shows that the baseband signal is present at both sides of the carrier.

5.2 AM Demodulation

The easiest method to demodulate the AM signal is to extract the baseband signal from the envelope of the received RF signal. This can be done using a diode and a capacitor which is shown in Figure 5.3.

In Figure 5.3, the antenna receives the AM-modulated signal, and develops an AM voltage at the diode input. The $R - C$ low-pass filter extracts the low-pass components of the rectified signal which is the envelope of the RF signal. Figure 5.4 illustrates the behavior of the circuit presented in Figure 5.3. As in Figure 5.4, the envelope of the

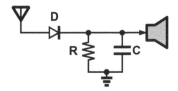

Figure 5.3: Simple AM demodulator.

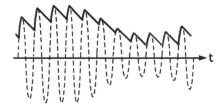

Figure 5.4: The concept of AM demodulation using a diode with R–C circuit.

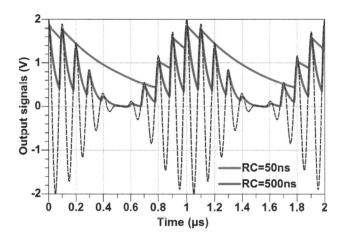

Figure 5.5: A failure-to-follow distortion once the modulation index is at its maximum.

RF signal is proportional to the baseband data. As such, the nonlinearity might affect the demodulation process. Figure 5.5 illustrates an AM signal with the modulation index of unity.

In Figure 5.5, the carrier frequency is selected to be 10 MHz and the baseband signal is assumed to be 1 MHz. A problem arises once the R-C time constant of the output circuit is too long with respect to the period of the modulating signal. In this case, a failure-to-follow distortion is caused once the R-C low-pass bandwidth is insufficient. That is

$$\frac{1}{RC} \leq \omega_m \tag{5.3}$$

the output distortion will occur, so normally one should choose $\omega_c \gg \frac{1}{RC} \geq \omega_m$ to have less distortion.

5.3 Generating AM Signals

In this section, we introduce circuits which generate AM signals. Consider Figure 5.6 where the baseband signal is injected to the base of Q_1 and generates a corresponding baseband current which flows through the upper differential tree. If we assume that the operation of the differential pair is switching (i.e., $V_1 \gg V_t$), the baseband signal is upconverted to the carrier frequency. Care should be taken such that the input transistor does not enter the nonlinear region, so that the modulation process is held linear. On the other hand, the RF signal imposed at the upper tree's differential pair input can be large enough to bring the pair into the nonlinear region. In this case, the carrier and its harmonics will be modulated by the baseband signal. By virtue of the output band-pass filters, the harmonics of the carrier can be suppressed. We now present a detailed analysis of the modulation phenomenon. The time-varying bias current of the upper tree, I_E (the collector current of Q1), is expressed as

$$I_E = \frac{V_{BB} + V_m \cos \omega_m t - V_{BEQ}}{R_E} \tag{5.4}$$

Defining the normalized RF input voltage amplitude of the upper tree as x

$$x = \frac{qV_1}{kT} = \frac{V_1}{V_t} \tag{5.5}$$

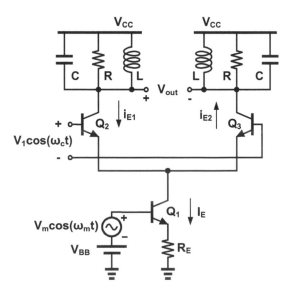

Figure 5.6: Implementation of the AM modulator.

The collector currents of the transistor pair (Q2 and Q3) can be expressed as

$$i_{E1} = \frac{I_E}{2}\left(1 + \tanh\left(\frac{x}{2}\cos(\omega_c t)\right)\right) \tag{5.6}$$

$$i_{E2} = \frac{I_E}{2}\left(1 - \tanh\left(\frac{x}{2}\cos(\omega_c t)\right)\right) \tag{5.7}$$

The time-domain differential output voltage can be described as

$$V_{out} = V_{CC} - i_{E1} * z_L(t) - (V_{CC} - i_{E2} * z_L(t)) \tag{5.8}$$

where $z_L(t)$ is the impulse response of the load impedance, and $*$ stands for the convolution process. The output voltage reduces to

$$V_{out} = (i_{E2} - i_{E1}) * z_L(t) \tag{5.9}$$

or

$$V_{out} = I_E \tanh\left(\frac{x}{2}\cos(\omega_c t)\right) * z_L(t) \tag{5.10}$$

The load impedance in the frequency domain can be expressed as

$$Z_L(j\omega) = \frac{R}{1 + jQ\left(\frac{\omega}{\omega_c} - \frac{\omega_c}{\omega}\right)} \tag{5.11}$$

For harmonic components of the input frequency, the load impedance can be described in terms of harmonic frequencies (i.e., for the nth harmonic). Note that the output tuned circuits should be tuned to ω_c

$$Z_L(jn\omega_c) = \frac{R}{1 + jQ\left(n - \frac{1}{n}\right)} = \frac{nR}{n + jQ(n^2 - 1)} \approx \frac{-jnR}{Q(n^2 - 1)} \tag{5.12}$$

The output voltage in terms of harmonic frequencies can be expressed as

$$V_{out} = \sum_{n=0}^{\infty} [I_{E2}(n\omega_c) - I_{E1}(n\omega_c)] \times |Z_L(jn\omega_c)| \cos(n\omega_c t + \angle Z_L(jn\omega_c)) \tag{5.13}$$

Once the input RF voltage is sufficiently small (less than 50 mV for a bipolar differential pair), the harmonics become negligible and the above expression is reduced to

$$V_{out} = -\frac{I_E(t)}{2}x\cos(\omega_c t) \times R \tag{5.14}$$

And the output in this case becomes the following which is apparently an AM-modulated signal

$$V_{out} = -\frac{R(V_{BB} - V_{BEQ})}{2R_E}\left(1 + \frac{V_m}{V_{BB} - V_{BEQ}}\cos(\omega_m t)\right)\frac{qV_1}{kT}\cos(\omega_c t) \tag{5.15}$$

The modulation index of the above amplitude-modulated signal is

$$m = \frac{V_{\mathrm{m}}}{V_{\mathrm{BB}} - V_{\mathrm{BEQ}}} \tag{5.16}$$

In the case where the RF input signal is a large signal, the transistors Q_2 and Q_3 will be switched on and off sequentially. The differential output current can be written as

$$\Delta i_{\mathrm{out}} = \frac{I_{\mathrm{E}}(t)}{2} S(\omega_{\mathrm{c}} t) \tag{5.17}$$

where $s(\omega_{\mathrm{c}} t)$ is a bipolar RF square wave with a radian frequency, ω_{c}. Given the tuned circuit loads, the output voltage will have the following form

$$V_{\mathrm{out}} = -\frac{2}{\pi} I_{\mathrm{E}}(t) R \cos(\omega_{\mathrm{c}} t) \tag{5.18}$$

Or

$$V_{\mathrm{out}} = -\frac{1}{\pi} \frac{R(V_{\mathrm{BB}} - V_{\mathrm{BEQ}})}{R_{\mathrm{E}}} \left(1 + \frac{V_{\mathrm{m}}}{V_{\mathrm{BB}} - V_{\mathrm{BEQ}}} \cos(\omega_{\mathrm{m}} t)\right) \cos(\omega_{\mathrm{c}} t) \tag{5.19}$$

The harmonics of the carrier can be modulated by the baseband signal as well if the output-tuned circuit is tuned to the either of the harmonics (3rd, 5th, 7th, etc.). This phenomenon is shown in Figure 5.7. However, once the bandpass filter is tuned to the carrier frequency, its harmonics will be attenuated by the output band-pass filters. In any case, it is possible to compute the harmonic components at the output. Note that in AM modulation we have always a carrier component, ω_{c}, and two sidebands at $\omega_{\mathrm{c}} + \omega_{\mathrm{m}}$ and $\omega_{\mathrm{c}} - \omega_{\mathrm{m}}$.

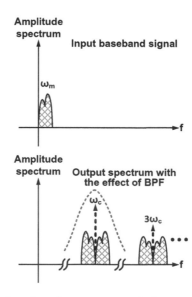

Figure 5.7: The input baseband spectrum and the output spectrum of an AM modulator (the two sidebands and the carrier are distinct in the output spectrum).

5.4 Double-Sideband and Single-Sideband Suppressed Carrier Generation

We are able to remove the carrier signal at the output of the AM modulator by just multiplying the baseband and the carrier signal in a balanced modulator. This can be achieved by a balanced pair of AM modulators. Since the carrier itself is absent at the output, this scheme is called double-sideband suppressed carrier (DSBSC). For this case, the AM signal can be expressed as

$$V_{AM} = kV_c \cos(\omega_c t) \cdot (V_m \cos(\omega_m t)) \tag{5.20}$$

where k is the mixer's output proportionality factor. As Equation 5.20 suggests at the output, we have two frequency components as $\omega_c \pm \omega_m$ and there is no effect of the carrier signal itself. Since the AM signal has components at both sides of the carrier (it has two sidebands) which transmit the same amount of information, the idea of removing one of the sidebands comes to mind in order to have a more bandwidth-efficient modulation. This modulation is called single-sideband suppressed carrier (SSBSC). We can implement SSBSC by the block diagram shown in Figure 5.8.

Assuming sinusoidal baseband, for the sake of simplicity, the outputs of each of the mixers become

$$V_1 = kV_c V_m \cos\left(\omega_m t + \frac{\pi}{4}\right) \cos(\omega_c t) \tag{5.21}$$

$$V_2 = kV_c V_m \cos\left(\omega_m t - \frac{\pi}{4}\right) \sin(\omega_c t) \tag{5.22}$$

and the total output after the summer becomes

$$V_{out} = kV_c V_m \left[\cos\left(\omega_m t + \frac{\pi}{4}\right)\cos(\omega_c t) + \sin\left(\omega_m t + \frac{\pi}{4}\right)\sin(\omega_c t)\right] \tag{5.23}$$

or

$$V_{out} = kV_c V_m \cos\left((\omega_c - \omega_m)t - \frac{\pi}{4}\right) \tag{5.24}$$

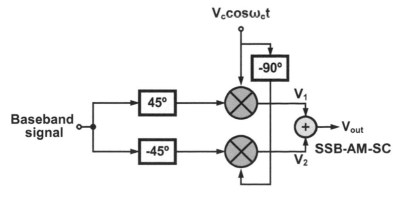

Figure 5.8: Block diagram implementation of the SSBSC AM signal.

which is apparently an SSBSC signal. The structure in Figure 5.8 occupies half of the bandwidth of the DSBSC AM signal; however, its drawback is the implementation of precise wideband phase shifter circuits which should have more than two decades of bandwidth, for example, from 100 Hz to 12 kHz. To mitigate this problem, we can implement the phase shifters after the upconverters. Finally, Figure 5.9 can be presented as a modified version of Figure 5.8.

As Equation 5.24 suggests, the carrier and the upper sideband are not present at the output and as a result, a better spectral efficiency is achieved.

Figure 5.10 illustrates an AM modulator where the LO signal is not present at the output (suppressed carrier) because of the symmetry of the circuit at the baseband.

Here, the bias current of the upper tree becomes a function of the carrier voltage:

$$I_E = \frac{V_{BB} + V_c \cos(\omega_c t) - V_{BEQ}}{R_E} \tag{5.25}$$

The output of the upper tree becomes

$$V_{out} = (i_{E2} - i_{E1}) * z_L(t) \tag{5.26}$$

Assuming $V_m \leq V_t$, we can write

$$V_{out} = -\frac{I_E}{2} \frac{qV_m}{kT} \cos(\omega_m t) \times R \tag{5.27}$$

In another form

$$V_{out} = -\frac{V_c}{2R_E} \frac{qV_m R}{kT} \cos(\omega_m t) \cos(\omega_c t) = -\frac{V_c}{2R_E} \frac{V_m R}{V_t} \cos(\omega_m t) \cos(\omega_c t) \tag{5.28}$$

The carrier signal is suppressed at the output because it is the common mode at the upper tree as stated in Chapter 4.

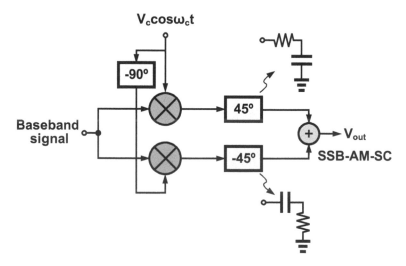

Figure 5.9: Block diagram implementation of the SSBSC AM signal with phase shifters at the carrier frequency.

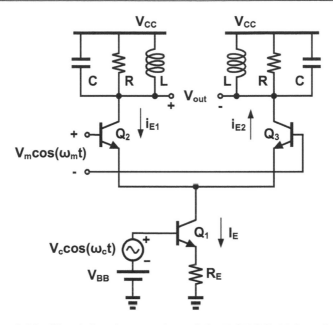

Figure 5.10: Circuit implementation of the DSBSC AM modulator.

5.5 Synchronous AM Detection

In this section, we introduce a number of circuits to detect the AM signal. Consider Figure 5.11.

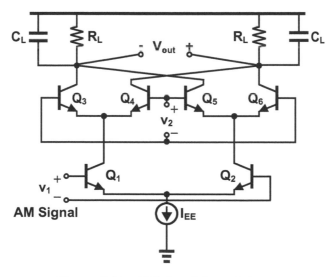

Figure 5.11: AM signal demodulator.

The circuit in Figure 5.11 is a synchronous AM demodulator based on a Gilbert cell multiplier. In this circuit, the received AM signal is converted to RF current at collector outputs of the lower tree. Using the Gilbert cell, these currents are multiplied by the differential input of the upper tree. An important assumption here is that the generated carrier frequency at the receiver, v_2, is phase locked to the carrier of the transmitter. The analysis of the detection process is as follows.

The differential output current of the Gilbert cell is described as

$$\Delta I_E = I_{EE} \tanh\left(\frac{qv_1}{2kT}\right) \tanh\left(\frac{qv_2}{2kT}\right) \tag{5.29}$$

The output voltage will be in general form as

$$V_{out} = \Delta I_E * z_L(t) \tag{5.30}$$

Assuming small-signal inputs, such that the Gilbert cell functions in the linear range, the output simplifies to

$$V_{out} = I_{EE}\left(\frac{qv_1}{2kT}\right)\left(\frac{qv_2}{2kT}\right) * z_L(t) = I_{EE}\left(\frac{v_1}{2V_t}\right)\left(\frac{v_2}{2V_t}\right) * z_L(t) \tag{5.31}$$

Assuming the input voltages as

$$v_1 = V_1\left(1 + m\cos\left(\omega_m t\right)\right)\cos\left(\omega_c t\right) \tag{5.32}$$

and

$$v_2 = V_2 \cos\left(\omega_c t\right) \tag{5.33}$$

The low-pass output voltage simplifies to

$$V_{out} = I_{EE}R_L \frac{V_1 V_2}{8V_t^2} m\cos\left(\omega_m t\right) \tag{5.34}$$

If we assume a large signal for V_2 (i.e., hard switching of the upper tree), the AM signal is in effect multiplied by a square wave of the carrier frequency, and as such, the carrier harmonics are multiplied by the AM signal and the low-pass component will appear at the output with a form as follows

$$V_{out} = \frac{2}{\pi} I_{EE}R_L \frac{V_1}{V_t} m\cos\left(\omega_m t\right) \tag{5.35}$$

The output is clearly proportional to the modulating signal. Figure 5.12 illustrates the demodulation process.

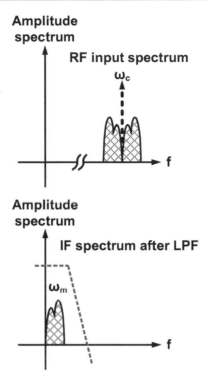

Figure 5.12: AM demodulation alongside low-pass filtering of the output.

5.5.1 A Synchronous AM Detection (with carrier extraction)

While AM demodulation needs the carrier signal, we are able to use the input RF signal to generate a proper LO (carrier) signal. Now, consider Figure 5.13.

We can amplify the input RF signal and limit it to obtain the LO (carrier) signal to downconvert AM signal. The red amplifier is a limiting amplifier which suppresses the modulating data and acts as a proper LO, carrier generator (limiters are covered in Chapter 6). This receiver architecture is a low power consuming one because it does not need a PLL circuit. The degeneration resistor R_s is placed for the sake of linearity to decrease the level of baseband distortion at the output of the demodulator.

The upper tree's differential pair current can be expressed as

$$\Delta I_D = k v_1 v_2 \tag{5.36}$$

The input AM voltage at the lower tree is

$$v_1 = V_1 \left(1 + m \cos(\omega_m t)\right) \cos(\omega_c t) \tag{5.37}$$

The AM signal voltage is limited to V_L, by the limiting amplifier, so the input voltage of the upper tree becomes

$$v_2 = V_L \cos(\omega_c t) \tag{5.38}$$

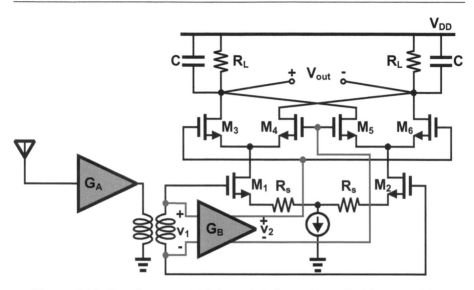

Figure 5.13: Synchronous AM demodulation using a limiting amplifier.

The low-pass component of the product of the above two voltages at the output becomes

$$V_{out} = \frac{k}{2} V_1 V_L \left(1 + m \cos\left(\omega_m t\right)\right) \tag{5.39}$$

5.6 Gilbert Cell Applications

Gilbert cell is one of the most widely used circuits in RF communications. Table 5.1 shows its application with respect to input signal magnitude.

The same application could be implemented for the MOS Gilbert cell if we replace V_t for the BJT by $V_{eff} = V_{GS0} - V_{th}$ for MOS devices. In Gilbert cell, as a phase detector, the upper tree is operating as a switch and the lower tree is fed by a large signal. In multiplier mode, both lower and upper trees of the circuit are fed by small signals.

Table 5.1: Gilbert cell applications (BJT).

Mixer	Multiplier	Phase Detector
Upper tree large-signal	Upper tree small-signal	Upper tree large-signal
Lower tree small-signal	Lower tree small-signal	Lower tree large-signal
$V_{RF} < V_t, V_{LO} \gg V_t$	$V_{RF} < V_t, V_{LO} < V_t$	$V_{RF} \gg V_t, V_{LO} \gg V_t$

Finally, the most applicable usage of Gilbert cell which is a mixer is achieved by prefect switching of the upper tree and small-signal injection at the lower tree.

5.7 Modern Practical Modulations

The aforementioned modulation schemes are not power efficient, and therefore they don't have further usage in today's RF communications except for legacy radio and television broadcasting. Today, the widespread modulations are *QAM*, *M − QAM*, *PSK*, *QPSK*, and *GMSK*. These modulation techniques are visualized by means of signal constellation. In signal constellation, each point has a unique amplitude and phase which is corresponding to its transmitted or received signal. Figure 5.14 shows signal constellation of 64-*QAM* which represents 64 points on the *I − Q* plane.

Moreover, we are able to demodulate these signals by the received constellation point. If we know the amplitude and the phase of the received signal, based on the constellation, we are able to understand the baseband transmitted data. In the following subsections, we focus on these modulation schemes with their corresponding constellation.

5.7.1 Binary Phase Shift Keying

This modulation has two points on its constellation. Figure 5.15 depicts the BPSK signals and its own constellation.

This modulation scheme is capable of transmitting one bit per symbol on the quadrature constellation. Its main drawback is hopping from $0°$ to $180°$ which is not bandwidth efficient. It occupies more bandwidth in comparison with other high data rate modulation schemes. In this scheme, we can say that bit "1" is transmitted by $A\cos(\omega_c t)$ and bit "0" by $-A\cos(\omega_c t)$.

5.7.2 Quadrature Phase Shift Keying

This modulation scheme employs 4 symbols in its constellation. Figure 5.16 depicts the QPSK constellation.

This scheme can send two bits per symbol, thus it has a higher bit rate than BPSK in the same bandwidth. In QPSK, each unique two bits are transmitted by a signal as mentioned in Equation 5.40a.

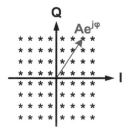

Figure 5.14: 64-*QAM* signal constellation.

$$11 : A\cos\left(\omega_c t + \frac{\pi}{4}\right) \tag{5.40a}$$

$$10 : A\cos\left(\omega_c t + \frac{3\pi}{4}\right) \tag{5.40b}$$

$$00 : A\cos\left(\omega_c t - \frac{\pi}{4}\right) \tag{5.40c}$$

$$01 : A\cos\left(\omega_c t - \frac{3\pi}{4}\right) \tag{5.40d}$$

Figure 5.16 depicts the RF symbols in QPSK modulation for a 4 MHz carrier (as an example) and 1 M symbol per second (2 Mbit/s) data rate, where all of them have the same amplitude; however, all adjacent symbols are orthogonal to each other.

5.7.3 Quadrature Amplitude Modulation (16 – QAM)

In this modulation, the constellation has 16 symbols. Figure 5.18 illustrates its 16-*QAM* constellation.

Generally, the relation between the symbol rate and the bit rate is

$$R_b = S_R \log_2 m \tag{5.41}$$

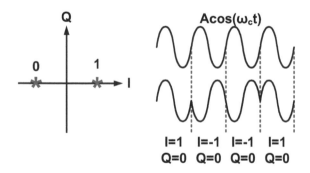

Figure 5.15: Signal constellation of BPSK modulation and its corresponding time-domain waveform.

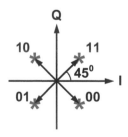

Figure 5.16: Constellation of QPSK modulation.

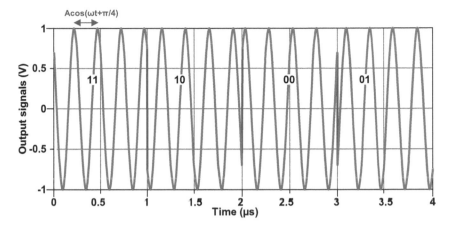

Figure 5.17: Signal waveform in QPSK modulation and the corresponding assigned bits.

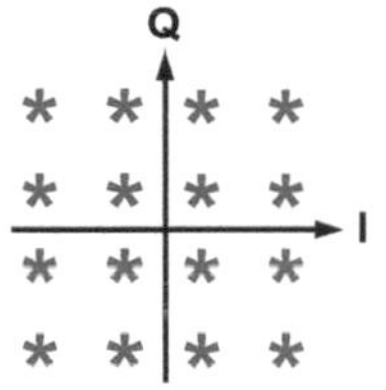

Figure 5.18: $16 - QAM$ constellation.

where R_b is the bit rate, S_R is the symbol rate, and m is the number of symbol levels (normally, the transmission bandwidth, BW, is chosen about the symbol rate). This modulation transmits four bits per symbol. Thus, 16-QAM has a higher bit rate with respect to the $QPSK$ within the same bandwidth. In this scheme, each four bits are transmitted by a single symbol. For the symbols, there are 3 different amplitude levels and 12 different phase levels as shown in Figure 5.18. In this case there are 16 distinct symbols, and therefore, $R_b = 4S_R$.

5.7.4 Quadrature Amplitude Modulation (64 – QAM)

In this modulation, there are 64 symbols in the constellation. Figure 5.19 illustrates 64-QAM constellation.

This modulation transmits 6 bits per symbol. Thus, it has a better bandwidth efficiency than 16-QAM. Symbols in this scheme have 10 different amplitude levels and 52 different phase levels as shown in Figure 5.19.

It is noteworthy that the more bits per symbol are transmitted (more complex modulation), the larger amount of SNR is needed to discern a symbol from its adjacent one. That is to say, higher modulation levels need higher levels of transmission power.

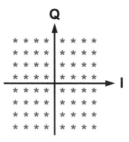

Figure 5.19: $64 - QAM$ constellation.

5.7.5 Generating Binary Phase Shift Keying Signal

Consider Figure 5.20.

The feed current of the differential pair in this figure is a function of the Q_1 base's input RF signal:

$$I_E \approx \frac{V_{BB} - V_{BEQ} + V_c \cos(\omega_c t)}{R_E} \tag{5.42}$$

The input baseband signal of the differential pair is expressed as

$$V_{BB} = V_m f(t) \tag{5.43}$$

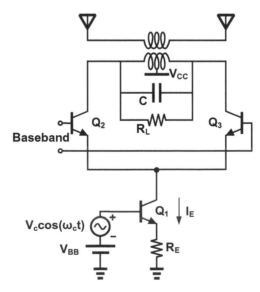

Figure 5.20: Circuit implementation of a *BPSK* modulator using a differential pair.

where $f(t)$ is a random binary signal which varies between $+1$ and -1, and V_m is the logic amplitude. The difference current of the differential pair (if $\frac{V_m}{V_t} < 1$) becomes

$$\Delta i_{EE} = \frac{I_E}{2} \frac{V_m f(t)}{2V_t} \tag{5.44}$$

The analog output voltage across the load-tuned circuit becomes

$$V_{out} = \frac{V_c V_m}{4V_t} \frac{R_L}{R_E} f(t) \cos(\omega_c t) \tag{5.45}$$

which is apparently a *BPSK* signal. As such, in Figure 5.20, the baseband signal, which is equal to ± 1, is upconverted to RF frequency and goes through the air by the antenna. Then, the coupled signal with the BALUN (balanced to unbalanced) feeds two antenna branches.

5.7.6 Generating and Detecting the Quadrature Phase Shift Keying Signal

It is possible to have a structure to transmit the baseband in *QPSK* form. The required architecture is shown in Figure 5.21. Each point on the constellation can be conceived by a vector such that $\vec{V} = Ae^{j\varphi}$. As a result, Figure 5.21 makes it possible to generate each point of the constellation. For instance, for *QPSK* modulation, we can choose $I = \pm\sqrt{2}/2$ and $Q = \mp\sqrt{2}/2$ and the modulated signal would have the following form

$$v_c(t) = \frac{\sqrt{2}}{2} \left[a_i(t) \cos \omega_c t \quad a_q(t) \sin \omega_c t \right] \tag{5.46}$$

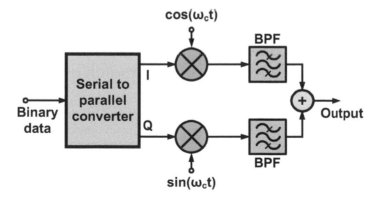

Figure 5.21: Implementation of a *QPSK* generator.

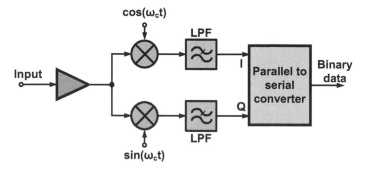

Figure 5.22: *QPSK* receiver architecture.

Likewise, an architecture as depicted in Figure 5.22 can be used for *QPSK* demodulation. The output voltages of the demodulator mixers can be expressed as

$$v_{o1}(t) = KV_c(t) \times \cos\omega_c t = K\frac{\sqrt{2}}{4}\left[a_i(t) + a_i(t)\cos(2\omega_c t) - a_q(t)\sin(2\omega_c t)\right]$$
$$(5.47a)$$

$$V_{o1,LP} = K\frac{\sqrt{2}}{4}a_i(t) \tag{5.47b}$$

$$v_{o2}(t) = KV_c(t) \times \sin\omega_c t = K\frac{\sqrt{2}}{4}\left[a_i(t)\sin(2\omega_c t) + a_q(t) - a_q(t)\cos(2\omega_c t)\right]$$
$$(5.47c)$$

$$V_{o2,LP} = K\frac{\sqrt{2}}{4}a_q(t) \tag{5.47d}$$

where K is the proportionality constant of the mixer.

In this architecture, the in-phase and the quadrature bit streams are synchronously detected and they are applied to the parallel to serial converter at the output of the low-pass filters.

5.8   Effect of Phase and Amplitude Mismatch on the Signal Constellation

In this section, we investigate the frequency response and the bandwidth efficiency of modern digital modulation schemes. Of the most important parameter of digital modulators is the trade-off between the bandwidth occupancy and the symbol rate. For instance, *QPSK* and 64-*QAM* modulations having the same bandwidth of 50 kHz to transmit the baseband data. The former would have a 100 kb/s bit rate and the latter would have a 300 kb/s bit rate. A transmitter shown in Figure 5.23 is called direct-conversion transmitter and it is used for modern digital modulation.

Figure 5.23 shows a quadrature transmitter. At the input, the serial-to-parallel converter maps the baseband bits two by two at the input of the mixers. Then, the upconverted signals are added at the output and drive the power amplifier to provide the

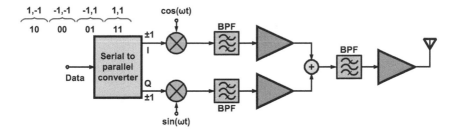

Figure 5.23: Direct-conversion transmitter architecture.

signal to be transmitted through the air. One of the main considerations in transmitter design is quadrature mismatches which can be frequency- and time-dependent. These errors cause the points on the constellation to change their phase/amplitude, and this results in difficult signal detection. Moreover, it also degrades the detection probability at the receiver. This phenomenon is shown in Figure 5.24. Here it is observed that if the SNR is diminished, each symbol in the constellation might interfere with its adjacent symbols, and therefore, introduce errors in the detection process.

Another assumption is the operation of mixers which act as ideal switches. However, in reality, this may not happen and RF harmonics might be troublesome. Figure 5.25 depicts the receiver for I/Q detection. Here it is assumed that the I and the Q channels introduce an amplitude error of $\Delta G/2$ and a phase error of $\Delta\varphi/2$ on each path.

Since the frequencies of transmission and reception in the direct-conversion system are the same, the oscillators and mixers can be used for both purposes. This architecture is called a coherent transceiver. Another drawback in this structure is frequency variations due to temperature. Suppose the outgoing signal is at 900 MHz which is generated by a temperature-compensated crystal oscillator (TCXO) which has a 3 ppm frequency variation. This means that a frequency error of 2.7 kHz may

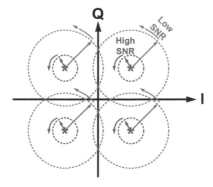

Figure 5.24: Phase and amplitude error effect in *QPSK* modulation due to noise, in two cases, low SNR and high SNR.

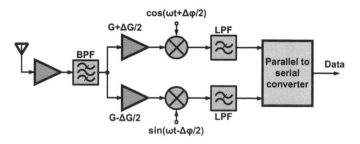

Figure 5.25: Quadrature receiver's architecture.

occur during the operation of the circuit. This frequency deviation will result in the rotation of the points on the constellation which in fact increases the probability of error in detection. To mitigate this issue, one of the solutions is using a PLL to lock the phase and the frequency of the LO to the received signal. Finally, we can summarize the signal distortions in the following

(1) **Gain mismatch results in rectangular constellation distortion (shown in Figure 5.26).**

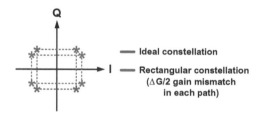

Figure 5.26: Rectangular constellation distortion due to $\Delta G/2$ gain mismatch in each path of the quadrature receiver.

(2) **Phase deviation results in parallelogram constellation (rotation) (shown in Figure 5.27).**

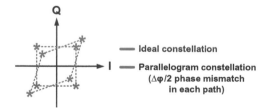

Figure 5.27: Parallelogram constellation distortion due to $\Delta \varphi/2$ phase mismatch in each path of quadrature receiver.

(3) **In** *QAM* **receivers, it is necessary to have a PLL to lock the LO to the input signal.**

5.8.1 Improvement of bandwidth efficiency

Consider Figure 5.28 which depicts signals in quadrature modulation.

By looking at the spectrum of Figure 5.28, a $sinc(u)$ function with side lobes will appear. However, if we assume that in *QPSK* modulation, a transition will happen that results in 180° phase shift between two signals, it makes stronger side lobes which is not desired. One of the methods to suppress this effect is to use offset QPSK (OQPSK), its concept is shown in Figure 5.28. If we insert a delay equal to half of the bit period, the mentioned transition will never happen and as a result, a lower side lobe may be achieved. Furthermore, the fast transition itself between data bits may result in more bandwidth occupancy. Nowadays, another efficient modulation is used which is called Gaussian minimum shift keying (GMSK). In this scheme, in addition to the above delay, we use a Gaussian low-pass filter to smooth the transition between data bits to lower the side lobes. The transmission standards limit the level of the side lobes, thus improving them is of great importance. The Gaussian filter has a low-pass

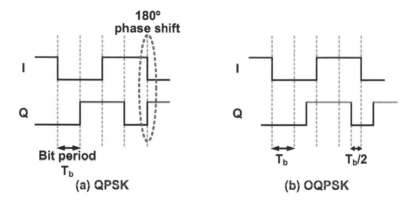

Figure 5.28: Signals in quadrature phase modulation, (a) a QPSK signal pair with a 180° phase change, and (b) an OQPSK signal pair where the 180° phase shift is avoided.

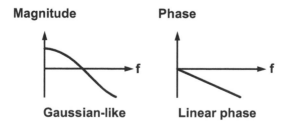

Figure 5.29: Frequency response of a Gaussian filter (magnitude and phase).

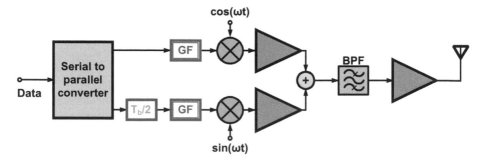

Figure 5.30: GMSK transmitter architecture.

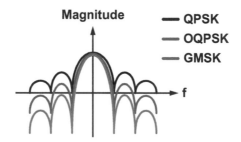

Figure 5.31: The comparison of the spectra of QPSK, OQPSK, and GMSK signals.

magnitude behavior and a linear phase response which is equivalent to a constant delay. Figure 5.29 shows the frequency response of a Gaussian filter (GF).

A typical GMSK transmitter block diagram is shown in Figure 5.30.

Figure 5.31 shows a comparison of the spectra of the three types of quadrature modulations. As it is obvious, the GMSK has the lowest side-lobes' level.

Although, nowadays, analog amplitude and FM for legacy radio and television broadcasting are used, there are modern digital receivers with compatible analog techniques for their detection, as well. The following example describes this issue to detect frequency-modulated signal with I/Q demodulator.

Example 5.1 If we consider FM-modulated signal as $X = A\cos\left(\omega_0 t + k\int V_{\mathrm{m}}\right.$ $\cos\left(\omega_{\mathrm{m}}t'\right)dt' + \phi\bigg)$, the instantaneous frequency will be $\omega_0 + kV_{\mathrm{m}}\cos(\omega_{\mathrm{m}}t)$. Suggest an structure to demodulate an FM signal with quadrature zero IF receiver (note that this can be done by two analog multipliers, two differentiators, and a voltage subtractor). In cell phones, FM signals are demodulated with this structure with digital signal processing right after the mixers.

Solution:
We can use the structure shown in Figure 5.32 for FM detection.

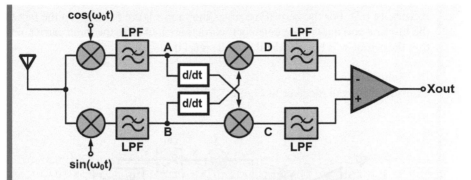

Figure 5.32: Quadrature FM demodulator.

For the signals at nodes A and B, with the assumption of filtering out other mixing products, we will have the low-pass components as

$$X_A = \frac{A}{2} \cos \left(k \int V_m \cos \left(\omega_m t' \right) dt' + \phi \right) \tag{5.48a}$$

$$X_B = -\frac{A}{2} \sin \left(k \int V_m \cos \left(\omega_m t' \right) dt' + \phi \right) \tag{5.48b}$$

By taking the derivatives of Equation 5.48, we then reach to

$$\frac{d}{dt} X_A = -\frac{A}{2} k V_m cos \left(\omega_m t \right) \sin \left(k \int V_m \cos \left(\omega_m t' \right) dt' + \phi \right) \tag{5.49a}$$

$$\frac{d}{dt} X_B = -\frac{A}{2} k V_m cos \left(\omega_m t \right) \cos \left(k \int V_m \cos \left(\omega_m t' \right) dt' + \phi \right) \tag{5.49h}$$

Finally, signals in nodes C and D with the assumption of low-pass filter at the output will be

$$X_C = \frac{A^2}{4} k V_m cos \left(\omega_m t \right) \tag{5.50a}$$

$$X_D = -\frac{A^2}{4} k V_m cos \left(\omega_m t \right) \tag{5.50b}$$

$$\Rightarrow X_{out} = X_C - X_D = \frac{A^2}{2} k V_m \cos \left(\omega_m t \right) \tag{5.50c}$$

Using digital signal processors, this process can be implemented in digital domain as well, as such a digital receiver could be compatible with an analog modulation technique. ∎

Example 5.2 For the zero IF receiver shown in Figure 5.33, given the fact that the in-phase and quadrature detectors' carriers are locked to the input carrier, prove that the output will be the AM detected signal if $m < 1$.

Solution:
Consider the given receiver in Figure 5.33.

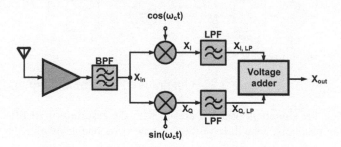

Figure 5.33: Zero IF quadrature receiver.

One may write the input signal as

$$X_{in} = A\cos\left(\omega_c t + \phi\right)\left(1 + m\cos\left(\omega_m t\right)\right) \tag{5.51}$$

Then, for the output signal, we will have

$$X_I = A\left(1 + m\cos\left(\omega_m t\right)\right)\cos\left(\omega_c t + \phi\right)\cos\left(\omega_c t\right) \tag{5.52a}$$
$$X_Q = A\left(1 + m\cos\left(\omega_m t\right)\right)\cos\left(\omega_c t + \phi\right)\sin\left(\omega_c t\right) \tag{5.52b}$$
$$X_{out} = X_{I,LP} + X_{Q,LP}$$

where LP stands for the low-pass component of the signal.

$$X_{out} = \frac{A}{2}\cos\left(\phi\right)\left(1 + m\cos\left(\omega_m t\right)\right) - \frac{A}{2}\sin\left(\phi\right)\left(1 + m\cos\left(\omega_m t\right)\right) \tag{5.53}$$
$$= \frac{A}{2}\left(1 + m\cos\left(\omega_m t\right)\right)\left(\cos\left(\phi\right) - \sin\left(\phi\right)\right)$$
$$= \frac{\sqrt{2}A}{2}\left(1 + m\cos\left(\omega_m t\right)\right)\left(\cos\left(\phi + \frac{\pi}{4}\right)\right)$$

The above equation suggests that the output signal is dependent on the phase of the input AM-modulated signal. For instance, if this phase is $45°$, the output will be zero. To solve this problem, indeed we should use a PLL to lock the LO signals to the transmitted carrier signal. In other words, the received signal is injected to a PLL, and then the phase values of the in-phase and the quadrature LO input of the mixers are chosen in order to make the maximum amplitude which is a phase difference of $-45°$ between the LO and the carrier signal. That is to say, the maximum value of $\frac{\sqrt{2}A}{2}(1+m\cos(\omega_m t))$ is achieved for $\phi = -45$.

In case of digital signal processor implementation of the detection process, one could implement the following relation

$$X_{\text{out}} = \sqrt{X_I^2 + X_Q^2} = \frac{A}{2}(1+m\cos(\omega_m t)) \tag{5.54}$$

Apparently here, the use of a PLL is not needed.   ∎

Example 5.3 For the given synchronous AM detector presented in Figure 5.34, the modulated signal has a form presented in Figure 5.35. Calculate the output detected signal amplitude and the unwanted second harmonic component at 2 MHz.

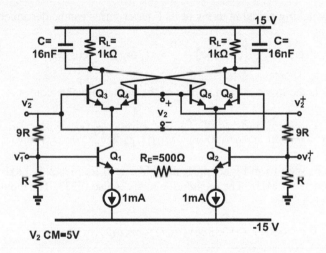

Figure 5.34: Zero IF synchronous detector.

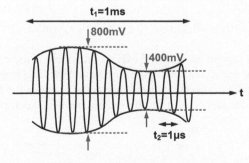

Figure 5.35: The input AM signal waveform (v_2).

Solution:
One can express the differential current of the Gilbert cell multiplier as (note that
here there is a linearizing resistance, $R_E = 500\,\Omega$, between the emitters of the lower
tree)

$$\Delta I_{EE} = \frac{I_{EE}}{1 + g_m \frac{R_E}{2}} \left(\frac{v_1}{2V_t} \right) \tanh \left(\frac{v_2}{2V_t} \right) \tag{5.55}$$

The large-signal input voltage of the upper tree switches the value of hyperbolic
tangent between $+1$ and -1 , so the differential output current can be expressed as

$$\Delta I_{EE} = \frac{I_{EE}}{1 + g_m \frac{R_E}{2}} \frac{v_1}{2V_T} S(\omega_0 t) \tag{5.56}$$

The switching signal in terms of its Fourier series can be described as

$$S(\omega_0 t) = \frac{4}{\pi} \left[\cos(\omega_0 t) - \frac{1}{3} \cos(3\omega_0 t) + \frac{1}{5} \cos(5\omega_0 t) - \cdots \right] \tag{5.57}$$

As such, the low-pass component of the output signal becomes

$$V_{out} = \frac{2}{\pi} \frac{V_1 (1 + m\cos(\omega_m t))}{10} \frac{g_m}{1 + g_m \frac{R_E}{2}} R_L \tag{5.58}$$

It can be seen from Figure 5.35, the AM signal frequency is 1 kHz and the carrier
frequency is 1 MHz. The modulation index is also 0.333. The input voltage has the
following form

$$V_{in} = 600 \left(1 + 0.333 \cos(\omega_m t) \right) \cos(\omega_0 t) \, (mV) \tag{5.59}$$

The output voltage can be calculated as

$$V_{out} = 600 \left(\frac{1 + 0.333 \cos(2\pi \times 1000 \times t)}{10} \right) \frac{2}{\pi} \frac{0.04}{1 + 0.04 \times 250} 1000 \tag{5.60}$$

$$= 139 \left(1 + 0.333 \cos(2\pi \times 1000 \times t) \right) (mV)$$

The unwanted second harmonic (2 MHz) component at the output can be calculated
by considering the second harmonic current and the load impedance at 2 MHz. The
second harmonic component amplitude is the same as the DC current amplitude
(why?). The load impedance for the second harmonic component becomes

$$Z_L(j\omega) = \frac{R_L}{1 + j \frac{\omega}{\omega_{cut-off}}} \tag{5.61}$$

where

$$\omega_{\text{cut-off}} = \frac{1}{R_{\text{L}}C} = 2\pi \times 10 \text{ kHz} \tag{5.62}$$

$$Z_{\text{L}}\left(j2\pi \times 2\,\text{MHz}\right) \approx -j\frac{R_{\text{L}}}{200} \tag{5.63}$$

The unwanted 2 MHz output voltage will take the following form

$$V_{\text{out}}\left(2\pi \times 2\,\text{MHz}\right) = 0.2\left(1 + 0.333\cos\left(2\pi \times 10^3 t\right)\right)\cos\left(4\pi \times 10^6 t - \frac{\pi}{2}\right)\text{(mV)} \tag{5.64}$$

∎

5.9 Conclusion

In this chapter, the AM modulation and demodulation techniques as well as double-sideband suppressed carrier AM and single-sideband suppressed carrier generation were studied. Different digital modulation techniques such as BPSK, QPSK, and QAM were presented as well. The quadrature digital modulator architecture as well as quadrature digital receiver were studied. The synchronous AM detection was presented and the quadrature demodulator was used for AM and FM detection as shown in the examples. Normally, the best modulation technique is the one that has the higher data rate within the specified bandwidth. The maximum achievable data rate is specified by the information theory formula $R_{\text{b}} = BW \times \log_2\left(1 + S/N\right)$. This means that the more we increase the discrete levels of digital modulation (in order to transmit more bits of data within a symbol), the higher S/N level we need to be able to distinguish between different symbol levels.

5.10 References and Further Reading

1. L.W. Couch, *Digital and Analog Communication systems*, eighth edition, New Jersey: Prentice-Hall, 2013.
2. D.H. Wolaver, *Phase-Locked Loop Circuit Design*, United Kingdom: Prentice Hall, 1991.
3. K.K. Clarke, D.T. Hess, *Communication Circuits, Analysis and Design*, United States: Krieger Publishing Company, 1994.
4. B. Razavi, *RF Microelectronics*, second edition, Castleton, NY: Prentice-Hall, 2011.
5. R.E. Ziemer, R.L. Peterson, *Digital Communication and Spread Spectrum Systems*, New York, NY: MacMillan, 1985.

6. D.O. Pederson, K. Marayam, *Analog Integrated Circuits for Communications*, Boston, MA: Kluwer Academic Publishers, 1990.

7. Robert Gallager, course materials for 6.450 *Principles of Digital Communications I*, Fall 2006. MIT OpenCourseWare (http://ocw.mit.edu/), Massachusetts Institute of Technology.

5.11 Problems

Problem 5.1 In the *QPSK* transmitter/receiver system shown in Figure 5.36 using MATLAB software,

1. Determine the transmitted spectrum with the random bit sequence generator at the input of the baseband data. Assume that the LO is locked by a PLL.

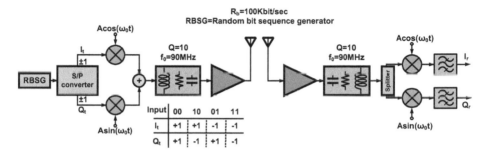

Figure 5.36: *QPSK* 90 MHz trnsmitter/receiver system.

Note that the band-pass filters' can be modeled as $H(j\omega) = \dfrac{1}{1+jQ\left(\frac{\omega}{\omega_0} - \frac{\omega_0}{\omega}\right)}$
where we have $\omega_0 = 2\pi(90\text{MHz})$ and $Q = 10$. Suppose the low-pass filters' 3 dB cut-off frequency is near 1 MHz and they are ideal.

2. For different values of I_t and Q_t, derive the constellation for the received signal.
3. If the gain in the I_t path is multiplied by 1.1, while the gain in Q_t path is unity, draw the constellation again.
4. If the phase of $A\sin(\omega t)$ at the transmitter varies by 5° in the transmitter, how the constellation changes and what happens?

Problem 5.2 An AM signal is applied between the bases of Q_1 and Q_2 in the analog multiplier as shown in Figure 5.37,

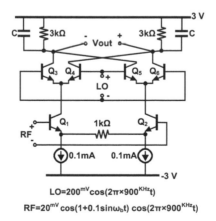

LO=200^{mV}cos(2π×900^{KHz}t)

RF=20^{mV} cos(1+0.1sinω_bt) cos(2π×900^{KHz}t)

Figure 5.37: Double-balanced synchronous detector.

1. Determine the capacitor C such that the second carrier harmonic at the output is 30 dB lower than the desired signal.
2. Determine the baseband component at the output. Assume that $\omega_b = 2\pi \times 1$ kHz.

Problem 5.3 In the 16-QAM modulator depicted in Figure 5.38, two double-balanced analog multipliers are employed like the one used in problem 2, terminated on 3 kΩ loads.

1. Find the necessary amplitude for the input I and Q channels to achieve a maximum output level of 100 mV. The LO signal has an amplitude of 200 mV at 1 MHz frequency. Furthermore, determine the output spectrum about 1 MHz for a bit rate of 4 kbit/s. (Note that in 16-QAM each symbol represents 4 bits of information).
2. If the quadrature signal has a finite phase error of $5°$, draw the generated constellation. Moreover, if the resistor on the Q input path has a 10% error (that is, it turns into 1.1 kΩ), how the constellation will be deformed.

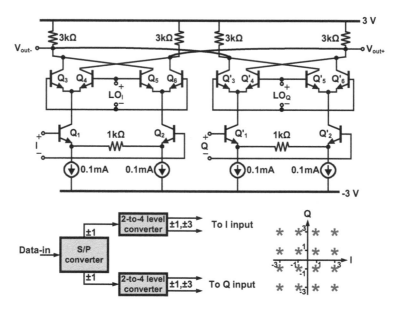

Figure 5.38: A typical 16QAM modulator using two balanced analog multipliers.

Problem 5.4 The circuit depicted in Figure 5.39 is a balanced AM modulator. Consider the MOS transistors have a threshold voltage V_{TH} and $K = \frac{1}{2}\mu_n C_{ox}\frac{W}{L}$, furthermore $v_S = V_S \cos(\omega_m t)$ and $v_{RF} = V_{RF}\cos(\omega_{RF} t)$.

1. Determine the modulation index of the current source, and then find an expression for the output voltage in terms of the input voltages, v_s and v_{RF}. Assume $V_{RF} \ll V_{GS} - V_{TH}$ and the DC voltage of the gate source junction is V_{GS0}.
2. If the RF voltage is sufficiently large to completely switch the differential pair, find an expression for the output in this case.

3. If the capacitor C is equal to 5 pF and the carrier frequency is 1 GHz, determine the required value of the inductors and the required value of the resistors in such a way that while the output circuits resonate at the RF frequency, the third output harmonic voltage will be 40 dB lower than the main harmonic.

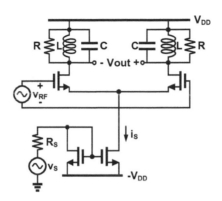

Figure 5.39: A balanced MOS differential pair AM modulator.

Problem 5.5 The circuit depicted in Figure 5.40 shows an envelope detector for AM signals. The input voltage is $v_{in} = 4^V (1 + 0.8\cos(2\pi f_m t))\cos(2\pi f_0 t)$ where $f_m = 10\,\text{kHz}$ and $f_0 = 1\,\text{MHz}$.

1. Find the spectrum of the output signal near 10 kHz, 1 MHz, 2 MHz, and 4 MHz through computer simulation.
2. What consideration must be taken into account for the values of R and C to have minimum distortion?

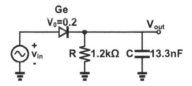

Figure 5.40: AM Ge envelope detector.

Problem 5.6 In the circuit depicted in Figure 5.41, which is a synchronous AM detector, the input is of the following form $V_{in} = \cos(\omega_c t)(1 + m\cos(\omega_m t))$.

1. Describe the operation of the circuit.
2. Determine the proper cut-off frequency for the load filters and the RC avalue.
3. Considering the value of the tail currents and the degeneration resistor, determine the maximum input voltage swing for the proper operation of the detector.
4. Find an expression for the first, the second, and the fourth parasitic RF harmonics at the output.

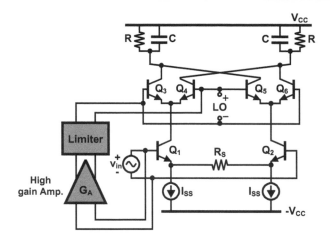

Figure 5.41: Self-carrier generating balanced synchronous AM detector.

Problem 5.7 Determine the output signal in the circuit of Figure 5.42 where the inputs are $V_1 = 20^{mv} f(t)$, and $V_2 = 700^{mv} \cos\left(2\pi \times 10^7 t\right)$ for two cases:

1. $f(t)$ is a normalized analog voice signal of 10 kHz bandwidth limited to ± 1. What kind of modulation is realized in this case?
2. $f(t)$ is a digital signal of 20 kbit/s rate, varying between ± 1. What kind of modulation is realized in this case?

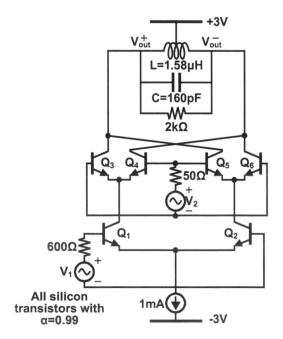

Figure 5.42: A double-balanced amplitude modulator.

Problem 5.8 In the amplitude detector depicted in Figure 5.43 for the given input voltage, determine the output voltage.

Hint: Note that the input amplifier stage (given the unbypassed emitter resistor R_E) operates in the linear region. Furthermore, the envelope detector loads the output-tuned circuit at the RF frequency by a value of $Z_L(j\omega_0) = R_0/2$.

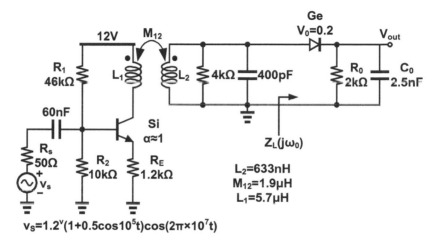

Figure 5.43: An envelope detector with the input RF amplifier.

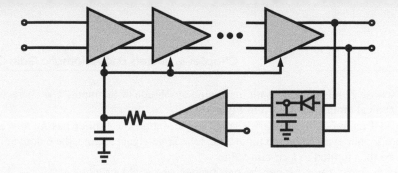

6. Limiters and Automatic Gain Control

Amplifiers and limiters are among the important building blocks of communication circuits. In this chapter, we deal with two different types of amplifiers which are limiting amplifiers and automatic gain controlled amplifiers. The former category has a high gain and we have introduced a few of its applications in Chapter 5; however, the latter is concerning methods to control and change the gain of amplifiers which is quite useful in radio receivers.

6.1 Limiting Versus Automatic Gain Control

The main amplifier goal is to amplify the received signal from a transimpedance amplifier. Amplification level must be enough to satisfy the required input signal for subsequent stages such as clock and data recovery circuits. Signal level for this goal is roughly a few hundred millivolts. The main amplifier is also called back-end amplifier because it is usually placed at the end of a receiver chain. Due to advantages of differential amplifiers such as higher signal swing and common-mode noise rejection, the main amplifiers are designed as fully differential at both the input and output. In different applications, signal distortion due to nonlinearity may be specified by a certain standard. However, automatic gain control (AGC) will adjust the circuit to mitigate the nonlinearity effect.

6.1.1 Limiting Circuits

When a small-signal is applied to an amplifier, we may assume a linear response without any distortion at the output. However, large-signal inputs might drive the amplifier into nonlinear operation and cause distortion at the output. In differential pair circuits which are of interest, part of the headroom voltage is dropped across the tail current source. At large input levels due to switching behavior of the pairs, the signal will be chopped. A limiter is a circuit which has a linear gain for small signals and whose gain is reduced with increasing amplitude of the input. The characteristics of a typical limiter are shown in Figure 6.1.

As Figure 6.1 suggests, for small-signal input, the limiter has merely a linear response; however, while the input enters the large-signal regime, the output amplitude will be then limited to a certain value.

As an example, for a differential pair limiter, one could write

$$v_O = (V_{CC} - R_L i_{c1}) - (V_{CC} - R_L i_{c2}) = R_L (i_{c1} - i_{c2}) \tag{6.1}$$

$$v_O = \frac{\alpha I_{EE} R_L}{2} \tanh\left(\frac{v_i}{2V_T}\right) \tag{6.2}$$

It is obvious here that for small values of v_i, the output would be linearly related to the input and for large values of v_i, the output will be saturated to a voltage of $\alpha R_L I_{EE}/2$.

6.1.2 AGC Amplifiers

AGC circuits are usually made of an amplifier in which its gain can be adjusted and a mechanism to control that gain for different input signals provides a desired output signal. Unlike the limiter which limits the large-signal input, AGC circuits decrease its own gain to suppress the effect of nonlinearity in circuits due to the large-signal input. Figure 6.2 depicts typical characteristics of an AGC.

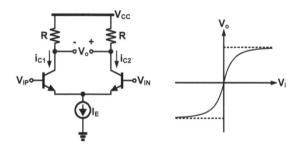

Figure 6.1: (a) A typical limiting differential amplifier. (b) Input–output characteristics of a limiting amplifier.

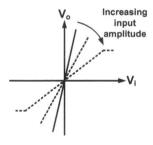

Figure 6.2: Input–output characteristics of an AGC.

6.2 Total Bandwidth with Multistage

One of the most important parameters in amplifiers is the gain bandwidth product or briefly GBW. To achieve high data rate, we may need an amplifier with very high GBW which may be greater than its unity-gain frequency. We can attain more GBW by cascading amplifiers. Unlike operational amplifiers which need stability check under feedback condition, the open-loop behavior of cascaded amplifiers doesn't have this restriction. The only feedback which is applied for these amplifiers is the offset cancellation feedback that has a low-frequency nature. We now focus on frequency response of cascaded amplifiers which have the same transfer function. Consider an amplifier with the DC gain of A and a dominant pole at f_0. Thus, cascading will lead to a DC gain equal to sum of the all gains (in decibels) and a cut-off frequency, f_0.

> **Example 6.1** Verify that a single-stage amplifier with an overall gain of 30 dB and 3 dB cut-off frequency of 3 GHz achieves a GBW product of 95 GHz and then determine the required GBW of each single stage, in a three-stage amplifier configuration, in order to achieve the same GBW of 95 GHz. Assume that all the amplifiers have a single dominant pole in their frequency response.
>
> **Solution:**
> Case I($n = 1$): In this case, the overall GBW will be equal to single-stage gain times its bandwidth, in other words
>
> $$GBW_S = GBW_{tot} = 10^{\frac{30}{20}} \times 3GHz = 95GHz \tag{6.3}$$
>
> Case II($n = 3$): The gain of an amplifier with a single dominant pole, in decibels, as a function of frequency can be expressed as
>
> $$A_{dB}(\omega) = 20\log \frac{A_0}{\sqrt{1 + \left(\frac{\omega}{\omega_C}\right)^2}} \tag{6.4}$$
>
> where ω_C is the 3 dB cut-off frequency of the amplifier and A_0 is the DC gain of the amplifier. Now if we consider to have three stages of amplifiers to have the same GBW, we should consider the 1 dB bandwidth of each stage (because the overall frequency response will be the product of the frequency responses of each stage). It is noteworthy that if one equates Equation 6.4 to $0.891A_0$, he/she would obtain the 1 dB cut-off frequency as $\omega_{C_{1dB}} = 0.5\omega_{C_{3dB}}$. Therefore, we would need three stages of amplifiers with a 3 dB bandwidth of 6 GHz and a DC gain of 10 dB each to achieve the same overall GBW. So, the GBW product of each stage is
>
> $$GBW_S = 10^{\frac{10}{20}} \times 6 \text{ GHz} = 19 \text{ GHz} \tag{6.5}$$
>
> In this case, the GBW of each stage is decreased approximately to one-fifth with respect to the first case, which is of great interest.
>
> Note that here the expression for the frequency response, in case II, would be of the following form

$$A_{dB}(\omega) = 20\log \frac{A_0'^3}{\sqrt{\left(1+\left(\frac{\omega}{\omega_C}\right)^2\right)^3}} \tag{6.6}$$

where A_0' is the DC gain of each stage (here, 10 dB). ∎

The question which may arise is that is it possible to increase the number of stages to achieve the same total GBW by lower GBW of each stage? This procedure may continue till each stage has a gain more than unity. But care must be taken that in the case of amplifiers, cascading the overall bandwidth would be reduced with respect to each stage's bandwidth as demonstrated in the previous example. In a real amplifier, the nature of frequency response due to finite resistance and capacitance of subsequent stage will be low-pass. Due to internal feedback at high frequencies and inductive loads, this behavior of trading gain for bandwidth may change. Thus, increasing the bandwidth may have an upper bound.

Example 6.2 Suppose an amplifier with a low-frequency gain of 12 dB and the given transfer function. Obtain an equation for frequency response for increasing number of stage for $f_0 = 1$ GHz and $H(j\omega) = 4/(1 + j\omega/\omega_0)$.

Solution:
General equation for the magnitude of n cascaded amplifier stages can be written as

$$|H_{\text{tot}}(j\omega)| = \left(\frac{4}{\sqrt{1+\left(\frac{\omega}{\omega_0}\right)^2}} \right)^n \tag{6.7}$$

Figure 6.3 illustrates the magnitude of frequency response of these cascaded amplifiers.

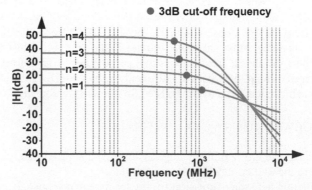

Figure 6.3: Magnitude of frequency response of the cascaded amplifiers.

> As Figure 6.3 suggests when the number of stages increases, their overall bandwidth decreases. Why all the curves pass through the same point at 0 dB gain? ∎

6.3 Offset Compensation Circuits

One of the most prominent effects which should be resolved in limiters is the offset problem. As a rule of thumb, we can state that this value must be roughly lower than $100\,\mu V$ and if this value increases, it has a destructive effect on the receiver's performance. A bipolar amplifier usually has a $3\,\sigma$ offset voltage of 1 to 2 millivolts, where σ is the standard deviation of the offset voltage. This value reaches about $10\,mV$ for high-speed MOSFET transistors. Both types of transistors have a very high offset voltage which obliges us to provide a method to cancel it out. Now, consider Figure 6.4.

Figure 6.4 shows a number of cascaded amplifiers. If we assume that due to inevitable process variations there is an offset voltage, the high gain value of A will saturate the final stages. Figure 6.5 illustrates an offset cancellation feedback loop which is also capable of setting the input impedance near to R_0 (note that feedback circuit capacitors are AC short-circuit).

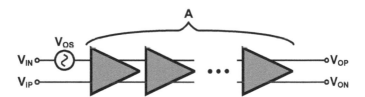

Figure 6.4: Offset voltage at the input of the limiting amplifiers.

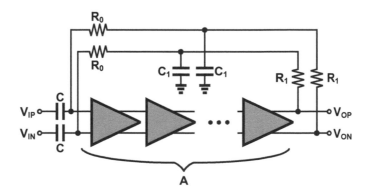

Figure 6.5: Offset cancellation loop with negative feedback, where $R_0 = 300\,\Omega$, $C_1 = 0.1\,\mu F$, and $R_1 = 20\,k\Omega$.

Limiting amplifiers are mostly used at the back-end of a receiver and thus are low-frequency (e.g., 455 kHz) building blocks in communication circuits. The input impedance in a limiter for a ceramic filter at 455 kHz might be near to $1.5\,\text{k}\Omega$ or for a ceramic filter at $10.7\,\text{MHz}$ might be $300\,\Omega$. As a result, the matching resistor in Figure 6.5 is shown to be $300\,\Omega$. Moreover, Figure 6.5 shows the offset cancellation loop which extracts the undesired DC component at the output and returns a fraction of it (ideally $1/A$ of it) with correct sign to the input. Another configuration for offset cancellation is depicted in Figure 6.6.

Input capacitances in Figure 6.6 are AC short and the $50\,\Omega$ impedance at the load of the error amplifier is for the matching purpose. Thus, the intrinsic output impedance of the error amplifier must be negligible. However, due to finite output resistance of the error amplifier, one may decrease $50\,\Omega$ resistance to attain a good matching. The loop mechanism is such that the low-pass filter at the end of the circuit extracts the offset error voltage and then the error amplifier amplifies the error voltage and returns it to the input with the opposite polarity. This feedback continues till the DC offset error reaches zero ideally. Another technique for offset cancellation is shown in Figure 6.7.

As Figure 6.7 shows, the input main amplifier has two differential inputs, the main input and the auxiliary feedback ones. The error amplifier (A_1) is placed for the sake of offset cancellation. Note that the matching criteria is satisfied by $50\,\Omega$ brute-force matching. There are two reasons that the structures shown in Figures 6.6 and 6.7 cannot completely remove the offset voltage. The first reason is the finite gain of the error amplifier and the second one is their own offset voltages V_{OS1} for amplifier A_1. We can easily develop a relation for the overall offset voltage in Figure 6.7 as

$$V'_{OS} = \frac{\sqrt{V_{OS}^2 + A_1^2 V_{OS_1}^2}}{AA_1 + 1} \approx \sqrt{\left(\frac{V_{OS}}{AA_1}\right)^2 + \left(\frac{V_{OS_1}}{A}\right)^2} \tag{6.8}$$

where in Equation 6.8 offset voltages are considered as random functions with Gaussian distribution and standard deviation of σ and those are also assumed to be independent.

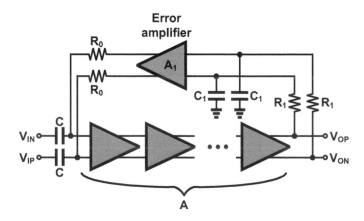

Figure 6.6: Offset cancellation loop with active negative feedback, where $R_0 = 50\,\Omega$.

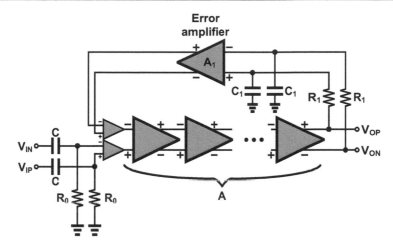

Figure 6.7: Offset cancellation loop with negative active feedback and two differential inputs, where $R_0 = 50\,\Omega$.

An approximation for Equation 6.8 is valid for the condition of $AA_1 \gg 1$. This is also valid for Figure 6.6, if the gain of the differential input main amplifier is the same for two paths. Note that the offset voltage of the main amplifier is decreased by the closed-loop gain; however, unfortunately the offset voltage of error amplifier is just divided by the main amplifier gain. These gains, however, are low-frequency gains. Thus, if the offset voltage of the error amplifier is not negligible (i.e., $V_{OS_1} \ll V_{OS}/A_1$ doesn't hold), this component will be dominant. While using high-speed error amplifiers are not necessary for offset cancellation, we can use larger devices with good matching to decrease their offset voltage. Finally, note that based on the degree of offset cancellation needed, we might employ either high-gain error amplifier $A_1 > 1$, or a buffer amplifier $A_1 = 1$, or a feedback loop without an amplifier.

6.3.1 Lower Cut-off Frequency of the Amplifier with Offset Compensation Loop

In Figures 6.6 and 6.7, the feedback loop of offset cancellation might eliminate the low-frequency components of the input signal. The transfer function for the negative feedback amplifier depicted in Figure 6.7 can be written as

$$H_1(s) = \frac{A(1+R_1C_1s)}{1+AA_1+R_1C_1s} \tag{6.9}$$

The lower cut-off frequency for this amplifier becomes

$$f_{co_1} = \frac{1+AA_1}{2\pi R_1 C_1} \tag{6.10}$$

The transfer function for the input AC coupling response can be written as

$$H_2(s) = \frac{R_0Cs}{1+R_0Cs} \tag{6.11}$$

The lower cut-off frequency for this transfer function becomes

$$f_{co_2} = \frac{1}{2\pi R_0 C} \tag{6.12}$$

Obviously, the overall lower cut-off frequency of this amplifier would be the maximum of f_{co_1} and f_{co_2}, or

$$f_{LF} = MAX \left\{ \frac{1+AA_1}{2\pi R_1 C_1}, \frac{1}{2\pi R_0 C} \right\} \tag{6.13}$$

Regarding the amplifier in Figure 6.6 as there is a voltage division at the output of the error amplifier, A_1 is replaced by $\frac{A_1}{2}$ in Equation 6.10, and therefore, the lower cut-off frequency in this amplifier becomes

$$f_{LF} = MAX \left\{ \frac{1+\frac{AA_1}{2}}{2\pi R_1 C_1}, \frac{1}{2\pi R_0 C} \right\} \tag{6.14}$$

The above results suggest that for proper operation of the circuit, we should choose $1/(2\pi R_1 C_1)$ very lower than the required f_{LF}. For instance, with a loop gain of 100, and if we need a lower cut-off frequency of 250 kHz, then the loop bandwidth should be lower than 2.5 kHz. It is possible to decrease the loop bandwidth by employing a Miller capacitance in the feedback to obtain a large capacitance, as $C_1 = (A_1 + 1)C_F$, at the input of the error amplifier (Figure 6.8). Like any feedback structure, we need to examine the stability condition for both the differential and the common-mode operation. Fortunately, regarding the open-loop gain, the dominant pole of the feedback circuit is sufficiently near the origin and it is far from the dominant poles of the main amplifier; as such, it doesn't pose any challenge to the stability condition.

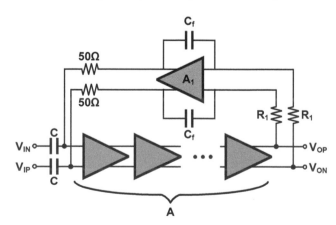

Figure 6.8: Offset cancellation loop with negative feedback and use of a Miller capacitance (instead of a large capacitance) in the feedback amplifier.

6.4 Automatic Gain Control

An amplifier which employs AGC is composed of a variable-gain amplifier alongside a DC feedback loop proportional to the output to control the amplifier gain, both are shown in Figure 6.9.

Similar to limiting amplifier circuits, AGCs are implemented as multistage amplifiers to achieve an optimum GBW product. Unlike limiters, here the gain of the circuit is controlled via V_{AGC}. The output signal amplitude is extracted by an amplitude detector and the corresponding voltage, compared to a V_{REF}, is applied to amplifiers to control their gain. The speed and the stability of this loop are dependent on the cut-off frequency of $R - C$ low-pass filter of the feedback loop. For the sake of simplicity, we assume V_{AGC} is applied to all the amplifiers and it is assumed to have the same value for all of them. However, in real applications, there might be some amplifiers to which the control signal is not applied. In the following sections, we discuss the gain stages and the detector circuit in detail.

6.4.1 Gain Control Methods

Single-stage gain can be controlled with different techniques. It is noteworthy that while changing the gain of a single stage, it should not affect drastically other parameters such as bandwidth, input dynamic range, noise figure, and common-mode rejection. Moreover, we expect the output to be linear (away from clipping) for the lowest gain and the maximum input, and for highest gain, we expect the noise figure to be minimum. In the subsequent sections, we introduce a number of popular techniques to control the gain of an amplifier stage.

Changing the transconductance of a transistor

It can be generally said that the gain of a single-stage amplifier is equal to $g_m R_L$, thus we can alter g_m to change the value of the gain. One of the methods to change g_m is changing the value of the tail current source which is shown alongside a differential pair in Figure 6.10.

It is possible to change the value of the tail current source to change the g_m of transistors. Unfortunately, changing the current source to a lower value results in the change of the voltage drop across the load resistors and increases the common-mode

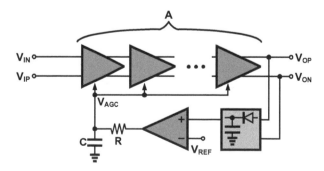

Figure 6.9: Automatic gain control structure.

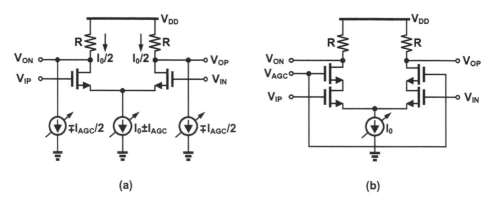

(a)   (b)

Figure 6.10: Variable-gain stage with g_m variation, (a) Through tail current variations, and (b) Through second gate voltage variations.

voltage. To solve this problem, we can add two current sources from the outputs to the ground. By this modification, the current reduced from the load is injected again to make the common-mode level constant. The constant current is selected to be $I_0/2$ for both loads. The main drawback of this circuit is the decrease in the output voltage swing of the circuit for lower gains. Moreover, since the gain is decreased, from input referred noise point of view, it may result in lower SNR. It can be asserted that with this technique, the amplifier bandwidth will be constant. Another method to decrease the g_m of the transistor is to push it in a triode region, which in fact lowers its output impedance and consequently its gain. This can be implemented by a cascode structure shown in Figure 6.10(b). The gate voltage of the cascode device is controlled by V_{AGC}. The gain and the cascode devices can be implemented via a single structure MOSFET as dual-gate. In this method where the current source is not changed, the common-mode level stays unchanged. However, the input dynamic range will be decreased. Moreover, this circuit can show an extreme nonlinear behavior for large-signal inputs.

Changing the load resistor

As stated earlier, the gain of an amplifier can be estimated as $g_m R_L$, thus another parameter that can be modified to change the gain is the load resistor. Although varying the load resistor itself alters the gain, this also changes the common-mode voltage which is not of interest. This issue can be resolved by inserting a resistor differentially. The resistor can be implemented by a MOS device which is biased in the triode region. The circuit is shown in Figure 6.11(a) which has a constant input dynamic range alongside proper noise behavior. The main drawback of this structure is the increase in the bandwidth due to the decreased gain.

Changing the amount of feedback

Figure 6.11(b) depicts a variable-gain stage with series feedback. As this figure shows, two current sources are tied to the sources of the transistors. Alternatively, we could tie equivalently the current source (I_0) to the middle of R_{AGC}. The main advantage of two current sources is eliminating the voltage drop on the emitter resistor. However,

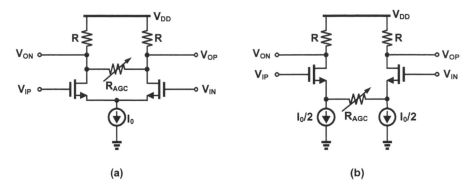

Figure 6.11: Variable-gain stage with (a) load resistor, and (b) series feedback.

one current source implementation has the merit of eliminating the common-mode noise whose elimination is of interest. The feedback resistor can be implemented by a MOSFET operation in the triode region. This structure has a loosely fixed bandwidth and a fixed common-mode voltage. Interestingly, decreasing the gain results in an increment in the input dynamic range. Moreover, the degeneration resistor (R_{AGC}) improves the linearity of the amplifier.

Switching between amplifiers

Figure 6.12 shows a structure that makes it possible to achieve different gains by turning on and off the corresponding switches. This can be achieved by a digital control circuit. This structure is indeed an amplifier bank whose gain is digitally controlled depending on the amplifiers which are switched in. The switches can be implemented using single-MOSFET structures.

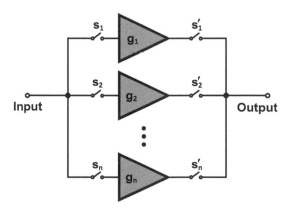

Figure 6.12: Implementation of automatic gain control circuit using multiple switches.

6.5 Amplitude Detectors

Figure 6.13 illustrates a peak detector which somehow rectifies the input signal.

When the input signal rises and reaches a value higher than the output (plus the cut-in voltage of the diode), it turns on and charges the capacitor to the peak value with a time constant equal to the diode on-resistance times the capacitance ($T_0 = r_D C$). The current source I discharges the capacitor slowly when the diode is off. The values for C and I must be chosen to provide proper system time constant and proper voltage drop across the load. This time constant can be evaluated as

$$\tau_0 = C \frac{V_{DC,out}}{I} \tag{6.15}$$

Note, we should have $\tau_0 \gg 1/\omega_0$. Otherwise, one can put a proper load resistance instead of the current source constituting a sufficiently large RC time constant with respect to the inverse of the carrier frequency ($RC \gg 1/\omega_0$). However, this structure can detect only the positive peaks of the signal and for detecting the negative peaks, we need a modified version. To have half a discharge time and consequently less ripple, one can consider a full-wave rectifier for this purpose. The full-wave rectifier is shown in Figure 6.13(b). Here, the in-phase and the out-of-phase components of the input signal are applied to the anodes of the two diodes. As such, despite the previous circuit, this circuit's operation is such that as if we use the absolute value of the signal at the input. This input signal can be implemented differentially and with the assumption of ideal diodes, the output voltage will be equal to the peak value of the input voltage. The large capacitance at the output roughly filters out the ripples of the output signal. One can observe that this circuit somehow solves the problem of output ripple. Note that this circuit for proper operation needs a relatively large input signal amplitude, that is, with an AM signal, one should have the following condition (to avoid distortion in the detector).

$$V_C \geq \frac{4V_D}{1-m} \tag{6.16}$$

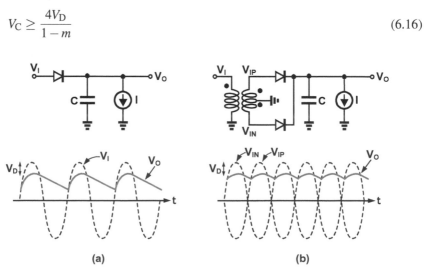

(a)   (b)

Figure 6.13: (a) Single-ended peak detector and (b) Full-wave peak detector.

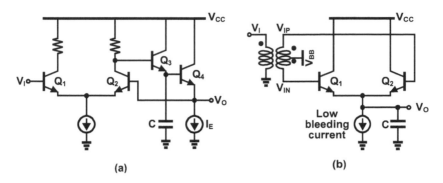

Figure 6.14: The implementation of a peak detector, (a) single-ended peak detector, and (b) full-wave peak detector.

where V_C is the carrier amplitude, V_D is the diode cut-in voltage, and m is the modulation index. Using differential pair transistors, one can detect AM signals with an amplitude less than V_D (why?). The implementation of these peak detectors is shown in Figure 6.14. In Figure 6.14(a), given the differential pair amplifier, one can detect small-signal AM inputs, while in Figure 6.14(b), full-wave detection is realized. In these two implementations, again a constant bleeding current is employed across the charging capacitor for AM detection. Care must be taken that in these two circuits a DC bias voltage is required at the input. Furthermore, in Figure 6.14(a), the input bias should have approximately the same value as the output DC voltage, and the DC bias of the base of the Q_3 should be approximately $2V_D + V_{DC,O}$. The discharge time constant of the AM detection in this circuit is approximately

$$\tau_0 = C\frac{\beta V_C}{I_E} \tag{6.17}$$

where β is the current gain of Q_4 and V_C is the carrier voltage amplitude.

In the AM detector circuit depicted in Figure 6.14(b), a bias voltage larger than V_D is required at the input and the discharging time constant at the output is approximately $\tau_0 = C\frac{V_C}{I}$, where V_C is the carrier voltage amplitude. As mentioned earlier, the advantage of these circuits with respect to the circuits of the Figure 6.13 is that both of them can operate with an AC input signal voltage amplitude of a fraction of V_D. Note that the input differential signals should have a low offset voltage to operate properly.

6.5.1 Logarithmic Signal Level Indicator

In this section, we introduce a structure to investigate the level of the signal passing through the receiver chain of amplifiers. This is a common circuit that records the signal strength, for example, in a common mobile phone receiver. Consider cascaded amplifiers along with the corresponding amplitude detectors shown in Figure 6.15. As the signal is amplified through the chain of cascaded amplifiers, the first detector detects the highest levels of the signal and the last detector indicates the lowest levels of the signal. The structure as a whole functions like a normal limiter.

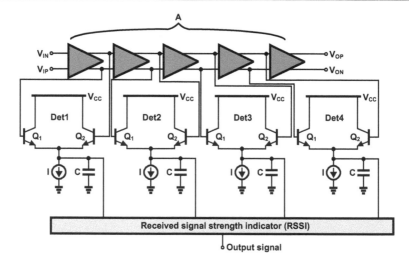

Figure 6.15: The configuration of a signal strength indicator.

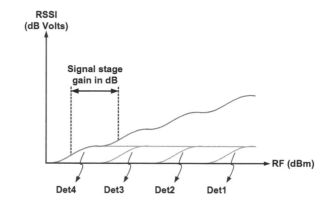

Figure 6.16: Typical indicator output as a function of the input RF power.

Figure 6.16 shows the typical output of the indicator versus the input RF power. As Figure 6.16 suggests, for low-power input signals, merely the final stages sense the power; however, when the input signal increases, the initial stages sense the power as well.

6.6 Amplifier Circuit with Gain Control Based on Analog Multipliers

Figures 6.17 and 6.18 show the structure of an amplifier with variable gain which is applicable in AGC circuit. In Figure 6.17, the signal is differentially applied to one port and the control signal is applied to the other port of the Gilbert cell. It is also possible to change the tail current source value to change the transconductance of the

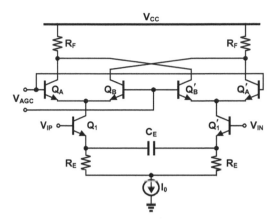

Figure 6.17: Amplifier implementation with gain control based on a multiplier.

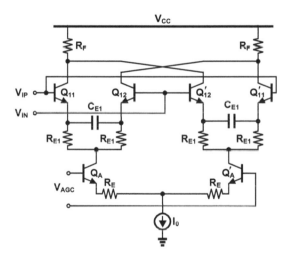

Figure 6.18: Modified amplifier with gain control based on a multiplier.

stage; interestingly due to crossing of the outputs in the Gilbert cell, the common mode is eliminated. If the control voltage V_{AGC} is high, the currents of transistors Q_1 and Q_1' flow through the transistors Q_A and Q_A' and a gain of $A = R_F/R_E$ is obtained. However, if the control voltage is small, the output current will tend to have small values and the Gilbert cell output will tend to be zero and a very small gain will be observed. In a special case where $V_{AGC} = 0$, the currents totally cancel out each other and a zero gain is obtained by the assumption of a complete match between the upper tree transistors. In this circuit, the degeneration resistors R_E are used to decrease the low-frequency gain and increase the dynamic range through the lower tree. The capacitor C_E is used to bypass the degeneration resistors at high frequency and consequently increase the gain at higher frequencies. The whole scenario is described by Equation 6.18.

$$\Delta I_{EE} = \begin{cases} \dfrac{v_{in}}{R_E} \tanh\left(\dfrac{v_{AGC}}{2V_T}\right) & f \ll f_{cutoff} \\[3mm] I_0 \tanh\left(\dfrac{v_{in}}{2V_T}\right) \tanh\left(\dfrac{v_{AGC}}{2V_T}\right) & f > f_{cutoff} \end{cases} \tag{6.18}$$

where $f_{cutoff} = 1/4\pi R_E C_E$. Note that $v_{out} = \Delta I_{EE} R_F$.

In Figure 6.18, another circuit is proposed for the variable-gain stage. In this structure, however, the control signal is applied to the lower tree and the differential signal is applied to the upper tree of the Gilbert cell. Again, in this structure, the degeneration resistors, R_{E1}, are bypassed by the capacitors, C_{E1}, so that the upper tree stages will have a high-pass behavior. When the control voltage is decreased, the current flows in both branches and it results in the lowering of the transconductance and correspondingly the gain of the circuit is reduced. The above discussion is well described in the following relations:

$$\Delta I_{EE} = \begin{cases} \dfrac{v_{AGC}}{R_E} \dfrac{v_{in}}{R_{E1}\frac{I_0}{2}} & f \ll f_{cutoff} \\[3mm] \dfrac{v_{AGC}}{R_E} \tanh\left(\dfrac{v_{in}}{2V_T}\right) & f > f_{cutoff} \end{cases} \tag{6.19}$$

where $f_{cutoff} = 1/4\pi R_{E1} C_{E1}$. Note that $v_{out} = \Delta I_{EE} R_F$.

Another implementation of the circuit is shown in Figure 6.19. In this implementation, no degeneration resistors are used and while the gain is high, the input dynamic range is limited. Because the upper tree will be easily saturated with large input signals. Furthermore, the gain control will be achieved by a larger gradient. Here, we have

$$\Delta I_{EE} = I_0 \tanh\left(\dfrac{v_{AGC}}{2V_T}\right) \tanh\left(\dfrac{v_{in}}{2V_T}\right) \tag{6.20}$$

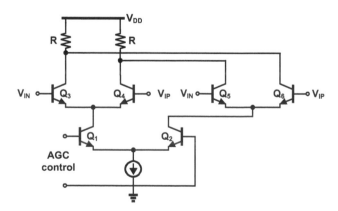

Figure 6.19: Automatic gain control circuit without degeneration resistors (without the high-pass response).

6.7 Increasing Bandwidth Methods

Although multistage amplifiers improve GBW product, we look for high-speed amplifiers or devices with large unity current gain frequency. In the following, we propose structures which can be used or combined to increase the bandwidth.

6.7.1 Employing High-Speed Transistors

To achieve high-speed stages, the optimum quiescent point and vital transistor topology should be considered. The parameter which shows the maximum operating frequency of a device which corresponds to unity current gain is defined by f_T. Another parameter which denotes for unity power gain is f_{max}. For a bipolar transistor, f_T can be computed as

$$f_T = \frac{1}{2\pi} \frac{g_m}{C_{be} + C_{bc}} \tag{6.21}$$

where in Equation 6.21, g_m is the device transconductance, C_{be} is the base–emitter capacitance, and C_{bc} is the base–collector capacitance. The maximum oscillation frequency (f_{max}) for a bipolar transistor then can be obtained as

$$f_{max} = \frac{1}{2} \sqrt{\frac{f_T}{2\pi R_b C_{bc}}} \tag{6.22}$$

where in Equation 6.22, R_b is the intrinsic base resistance. With respect to Equation 6.22, to achieve high-speed operation, we should decrease R_b and C_{bc}. Similarly, unity current gain frequency for a MOSFET device can be written as

$$f_T = \frac{1}{2\pi} \frac{g_m}{C_{gs} + C_{gd}} \approx \frac{3}{4\pi} \frac{\mu_n}{L^2} (V_{GS} - V_{TH}) \tag{6.23}$$

where in Equation 6.23, C_{gs} is the gate–source capacitance, C_{gd} is the gate–drain capacitance, μ_n is the electron mobility, and L is the gate length. For high-speed operation, NMOS transistors are preferred due to their better mobility. Moreover, the shorter the length of the device channel and the higher the overdrive voltage the higher unity current gain frequency, f_T, would be obtained. We can also write the maximum oscillation frequency of a MOS device as

$$f_{max} = \frac{1}{2} \sqrt{\frac{f_T}{2\pi R_g C_{gd}}} \tag{6.24}$$

where in Equation 6.24, R_g is the gate resistance. In this equation, we have neglected the effect of output resistance of the device due to channel length modulation. To increase the maximum frequency of the transistor, we should increase f_T and decrease R_g and C_{gd}.

6.7.2 Increasing Unity Current Gain Frequency

The unity current gain frequency (f_T) is inversely proportional to the carrier transit time in the channel, that is, the carrier transient time. For a bipolar transistor, the transit time with neglecting parasitic capacitances at the emitter and the collector is $\tau_F = 1/(2\pi f_T)$. The same phenomenon holds for a MOS device. Thus, to achieve high-speed operation, we need short gate length devices with small access resistances to the gate and to the source. In the following, we deal with the circuit-level methods to increase the unity current gain frequency. As stated before, one can realize that this parameter is loosely proportional to the device transconductance divided by its input capacitance. Therefore, if one decreases the input capacitance while maintaining the transconductance as unchanged, the higher-speed operation will be achieved. A f_T doubler circuit is shown in Figure 6.20.

In this circuit, the input signal is divided between transistors Q_1 and Q_3. Since half of the input voltage drops on the base–emitter junction of Q_1 and the other half drops on the base–emitter junction of Q_3, the collector current will not change and the transconductance stays the same (the collector currents of Q_1 and Q_2 are added with half-input voltages), while the input capacitance will be halved due to series connection of Q_1 and Q_3 and in fact the unity current gain frequency will be multiplied by 2. Figure 6.20 also shows a differential implementation of f_T doubler. In this topology, the input voltage is divided between nodes B and B' in a manner that the overall transconductance remains unchanged (because the collector currents are added). However, the input capacitance is halved due to series connection of the two differential stages. In practice, however, these circuits are not ideal due to parasitic capacitances which also impose a finite phase shift and result in imperfection. Moreover, the base–collector capacitance is not further reflected to the input through the Miller effect. Another drawback of this structure arises from the fact that the collector–substrate capacitance is doubled. This may adversely affect the improvement in f_T doubler. Moreover, it consumes twice the power of a single circuit.

6.7.3 Inductive Load (Shunt Peaking)

Consider Figure 6.21 which shows a common-source amplifier with inductive load alongside a resistor.

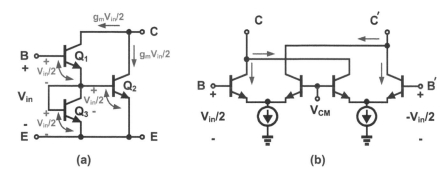

Figure 6.20: f_T doubler, (a) single-ended, and (b) differential implementation.

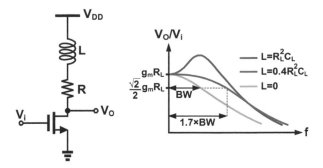

Figure 6.21: Inductive peaking in the common-source amplifier with the corresponding frequency response.

The inserted inductor is for canceling out the effect of the output capacitance. This technique is called inductive peaking because it results in a peaking in the frequency response of the output voltage. For instance, by choosing the value of the inductor as $L - 0.4R^2C_L$, the bandwidth will be increased by 70% with respect to the pure resistive load, i.e., $BW' = 1.7\,\mathrm{BW}$. The optimum value of the inductor from the bandwidth point of view doesn't lead to peaking indeed; however, inductive peaking alleviates the negative effect of multistage amplifiers in terms of bandwidth. It can be shown that by increasing the value of the inductor, the bandwidth will be increased and eventually reaches to its optimum point. If we continue to increase the inductor value, we will not attain the required bandwidth efficiency anymore and a peaking in the response will occur which makes the response nonflat. Regarding the inductor model itself and its self-resonance frequency, as a rule of thumb, the inductor should be chosen such that its self-resonance frequency is at least twice the cut-off frequency of the amplifier. Moreover, it is possible to replace the bulky spiral on-chip inductor with its active counterpart, however, at the cost of higher noise. Figure 6.22 shows the inductor model with its active counterpart.

Given the equivalent circuit model in Figure 6.22(b), the admittance of the active inductance circuit can be written as

$$I = V \left[\frac{g_m + jC_g\omega}{1 + jR_gC_g\omega} \right] \tag{6.25}$$

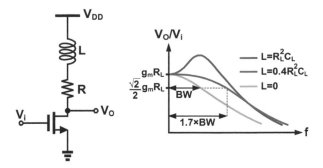

(a)   (b)

Figure 6.22: (a) Passive (spiral) inductor model. (b) The active MOS inductor topology and its equivalent circuit.

where C_g is the gate–source capacitance, and g_m is the MOSFET's transconductance. For

$$R_g \gg \frac{1}{C_g \omega} \tag{6.26}$$

or

$$R_g C_g \omega \gg 1 \tag{6.27}$$

The total admittance can be simplified to

$$Y = \frac{g_m}{\left(R_g C_g \omega\right)^2} - j\frac{g_m}{R_g C_g \omega} \tag{6.28}$$

and finally the total impedance can be simplified to

$$Z = \frac{1}{g_m} + j\frac{R_g C_g \omega}{g_m} = \frac{1}{g_m} + j\frac{R_g}{\omega_T}\omega \tag{6.29}$$

which is evidently the description of an inductance in series with a resistance as shown in Figure 6.22(b).

6.7.4 Decreasing Input Capacitance by Series Feedback

One of the main drawbacks in bipolar transistors is their speed issue which is due to their input pole made up of intrinsic base resistance and the base–emitter capacitance. For the sake of simplicity, we now neglect the effect of base–collector capacitance, i.e., assume $C_{bc} \approx 0$. We know that in bipolar transistors, the carrier injection in base is low, and thus this may result in higher intrinsic resistance and consequently lower input pole frequency. One may write the base–emitter capacitance from $C_{be} = C_{je} + (I_C/V_T) \cdot \tau_F$. Note that the base–emitter capacitance is proportional to the collector current and will be increased linearly with the collector current. As an example, suppose a bipolar transistor of $R_b = 120\,\Omega$, $C_{be} = 170\,\text{fF}$, $g_m = 40\,\text{mS}$, and $f_T = 30\,\text{GHz}$ which is operating at $I_C = 1\,\text{mA}$. In this case, the low-pass response of the base–emitter input will have a 3 dB cut-off frequency of 7.8 GHz which is much lower than the unity current gain frequency of the device. One of the well-known techniques to resolve this issue is to insert a degeneration resistor at the transistor's emitter as a series feedback which is shown in Figure 6.23.

Due to series feedback, the emitter will follow the base voltage, and therefore the equivalent capacitance seen from the base will be decreased. This phenomenon pushes the input pole farther from the origin. Here, in a similar way to the f_T doubler circuit, the Miller effect decreases the input capacitance; however, due to lowering the transconductance, no improvement in unity current gain frequency is achieved. It can be shown that with neglecting the base intrinsic resistance and the output conductance of the transistor, the gain from the base to the emitter can be written as $g_m R_E / (1 + g_m R_E)$. Therefore, regarding the Miller effect, the equivalent capacitance can be computed as $C_{eq} = C_{be}(1 - A)$, and we can write $C_{eq} = C_{be}/(1 + g_m R_E)$, which results in pushing the input pole farther by a $1 + g_m R_E$ factor. Interestingly, this will result in lowering the capacitance seen from the previous stage which finally results in increasing its

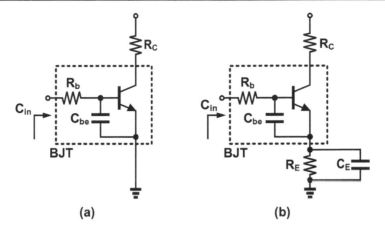

Figure 6.23: Decreasing device input capacitance via series feedback.

bandwidth. The side effect of series feedback is lowering the low-frequency gain of the amplifier by a $1 + g_m R_E$ factor. However, it is possible to increase the load resistance proportionally to maintain the gain of the amplifier as unchanged. However, this may result in lowering the output pole frequency due to larger resistance of the load along with parasitic capacitances. If this pole is not compensated, this would lead to decrease in the amplifier bandwidth. It is instructive to note that if we choose $C_E = 1/(2\pi f_T \cdot R_E)$, the emitter capacitance produces a zero in the frequency response which neutralizes the pole in the response and maintains the amplifier bandwidth. For example, by insertion of a resistor $R_E = 100\,\Omega$ and a transconductance of $40\,\text{mS}$, the input capacitance will be lowered by a factor of 5 and we obtain $C_{in} = 34\,\text{fF}$ which results in a cut-off frequency of 39 GHz for the input low-pass response which is higher than the unity current gain frequency. Now, to maintain the gain, we should multiply the output resistance by 5, and to neutralize the high-frequency pole of the emitter, we choose $C_E = 50\,\text{fF}$. It is possible to increase the emitter capacitance to achieve more bandwidth which is called emitter peaking. The degeneration resistor has a lot of advantages, in addition to lowering the input capacitance and increasing the circuit bandwidth by moving the input pole farther, which are (1) precise gain control with the ratio of resistors as $A = -(R_C/R_E)$ with the criteria of $R_E \gg 1/g_m$, (2) increasing the input resistance, (3) improving the circuit linearity for large-signal input, and (4) increasing the dynamic range for a differential stage. In the MOS design, this issue is not of great concern due to the low intrinsic gate resistance provided by a proper layout. However, in a similar way to bipolar devices, this resistor can be employed for the gain control, the amplifier linearity, decreasing the input capacitance, and increasing the bandwidth with peaking in the source.

6.8 Oscillation in Limiting Stages

In an amplifier, the amplified signal can leak from the supply voltage line, the ground, the substrate, and the radiation through the air. For instance, with an assumption of 80 dB loss in the leakage and a signal amplitude of 200 mv at the output, the returned

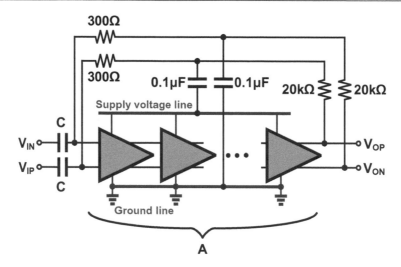

Figure 6.24: The possible feedback leakage path and the oscillation issue in multistage amplifiers.

signal to the input will have an amplitude of $20\,\mu v$. Thus, for the signals that have an amplitude greater than $20\,\mu v$, the input signal captures the limiting amplifier circuit; however, if this signal is less than the mentioned value, the returned signal captures the circuit itself. Most of designers assume this issue is due to poor noise figure; however, this may happen due to improper layout that worsens the feedback leakage phenomenon. Furthermore, if the total gain of the amplifiers chain is greater than 80 dB, there would be a possibility of oscillation (if the feedback phase is constructive). Figure 6.24 illustrates this issue. As it is seen here, there is a possibility of feedback leakage through either the supply voltage line or the ground line to the input. This regenerative feedback is one of the pitfalls of designing a limiting stage and thus proper layout and isolation must be taken into account.

6.9 Conclusion

In this chapter, we discussed different types of limiter circuits. The limiter circuit is a nonlinear circuit which has a high gain for a limited range of the input signal. When the signal starts to become large, this circuit limits the signal amplitude to an upper value. DC offset voltage is another intricate problem in an electronic circuit design. Since the limiters provide a high gain, offset cancellation techniques are required. We discussed different types of offset cancellation loops in this chapter. Limiter circuits are used in FM applications to remove the unwanted amplitude modulation. The AGC in an amplifier chain is another method to increase the dynamic range of the receiver. The AGC control loop mainly consists of an amplitude detector plus a negative feedback bias loop. Furthermore, analog multiplier circuits can be devised in such a way that it realizes the AGC. The problem of bandwidth enhancement in the RF/IF amplifiers was presented in this chapter as well. Different techniques were presented in this regard, including f_T doubling, inductive peaking, and input capacitance reduction. The problem of signal leakage and the oscillations due to the feedback path was discussed as well.

6.10 References and Further Reading

1. K.K. Clarke, D.T. Hess, *Communication Circuits, Analysis and Design*, United States: Krieger Publishing Company, 1994.
2. P.R. Gray, P.J. Hurst, S.H. Lewis, R.G. Meyer, *Analysis and Design of Analog Integrated Circuits*, fifth edition, Hoboken, NJ: J. Wiley & Sons, Inc., 2009.
3. J.P. Alegre Pérez, S. Celma Pueyo, D. Calvo López, *Automatic Gain Control: Techniques and Architectures for RF Receivers*, New York, NY: Springer, 2011.
4. R. Wu, J.H. Huijsing, K.A.A Makinwa, *Precision Instrumentation Amplifiers and Read-Out Integrated Circuits*, New York, NY: Springer, 2013.
5. J.R. Smith, *Modern Communication Circuits*, second edition, New York, NY: McGraw Hill, 1997.
6. B.Razavi, *Design of Integrated Circuit for Optical communications* , second edition, Hoboken, NJ: J. Wiley & Sons, Inc., 2012.
7. J.F. Witte, K.A.A. Makinwa, J.H. Huijsing, *Dynamic Offset Compensated CMOS Amplifiers*, New York, NY: Springer, 2009.
8. J. Dostal, *Operational Amplifiers*, second edition, Stoneham, MA: Butterworth-Heinemann, 1993.

Part 3

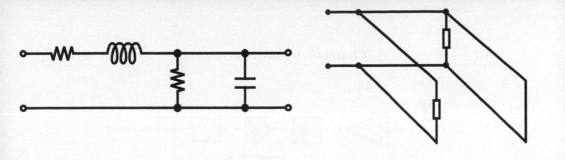

Transmission lines (T-lines) bridge the gap between the field and wave analysis, on the one hand, and circuit analysis on the other. This makes transmission line theory an integral part in understanding microwave and mm-Wave devices and circuits. As it will be seen throughout this chapter, wave propagation through the T-lines can be formulated by extending the circuit theory basics and making use of some specific solutions of the Maxwell equations. In this chapter, we provide a profound understanding of circuit equations governing T-lines using differential equations. The readers are encouraged to refer to Ref. [1] for a more comprehensive account of the microwave theory. T-lines play a very important role in modern wireless circuits and systems which find applications in antenna interfacing to TRX, impedance matching in mixers and amplifiers, resonator in oscillators and filters, etc.

7.1   An Introduction to Radio-Frequency Amplifiers in Receivers

Low-noise amplifiers (LNAs) are one of the most challenging constituents of a high-frequency receiver. Consider Figure 7.1, in which, the high-frequency signal is received by an LNA and passed on to the next building blocks. This suggests that the overall noise and sensitivity behavior of the receiver tightly depends on the LNA. In RF communication circuits, building blocks are designed to be matched to a $50\,\Omega$ impedance, both at the input and the output.

Given the fact that the impedances of RF devices, antennas, and passive and active components mostly vary between a value of few ohms to few hundreds, a commensurate value of $50\,\Omega$ has been chosen as the reference impedance for RF circuits. As such, matching RF devices to the value of $50\,\Omega$ would be feasible in most of the cases. Matching to $50\,\Omega$ is therefore necessary to obtain the maximum power transfer. We discuss later that this matching comes at certain circumstances at the price area occupation in RF integrated circuits. Another problematic issue in high-frequency amplifiers is oscillation. As the frequency increases, parasitic elements,

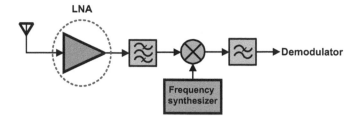

Figure 7.1: A generic RF front-end.

namely, capacitors and inductors, tend to have a larger effect, the likelihood of a positive feedback also increases. For example, consider Figure 7.2, where the model of a high-frequency amplifier is depicted. As it is seen in Figure 7.2, the matching circuit elements could be set such that the impedance seen from the antenna and that of the load are both equal to $50\,\Omega$. It is possible that the circuit could oscillate because of the existence of parasitic C_{gd} or L_s, both of which may cause unexpected feedback at higher frequencies.

7.1.1 Transmission Line

It is not hard to imagine that wires (interconnects) against the ground could be modeled as a sequential combination of inductive and capacitive components. On the other hand, there should also be a physical means of sending a signal to a transistor, and then extracting it out to the next stage. A lossless T-line, in principle, delivers the signal to the load unattenuated, while introducing an associated propagation delay. Besides the signal transmission, the main application of T-lines is in impedance matching. However, T-lines introduce both uncharacterized distortion and delay. Figure 7.3 depicts the phase and amplitude response of a typical T-line.

As it is obvious from Figure 7.3, the gain of an ideal T-line is equal to unity and it exhibits a linear phase behavior (constant delay behavior). The gain of a nonideal

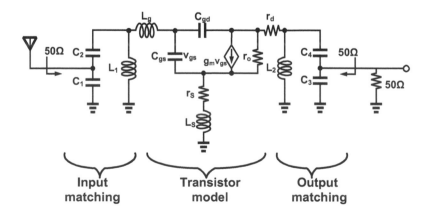

Figure 7.2: The model of a high-frequency amplifier.

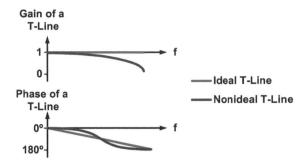

Figure 7.3. The phase and gain response of a lossless T-line.

T-line falls with frequency, furthermore its phase changes in a nonlinear manner with frequency, and therefore it has a variable delay with frequency which causes phase distortion (or dispersion). Furthermore, a characteristic impedance is defined for a T-line. As the operating frequency increases, the circuit dimensions become comparable to the carrier wavelength, the wave behavior of the electromagnetic waves should be taken into consideration rather than using the lumped element KVL and KCL relations. As suggested by the maximum power transfer theorem, in the case of matched terminations, the signal will be completely absorbed by the load. However, as we will see shortly, if the circuit suffers from a nonzero reflection coefficient, a portion of the signal, and hence the power, is reflected back.

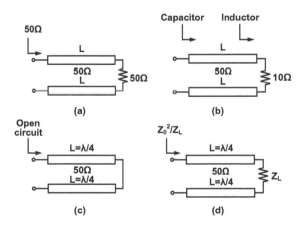

Figure 7.4: Impedance transformation property of the T-line, (a) The line terminated by $50\,\Omega$ exhibits an impedance of $50\,\Omega$ throughout the line, (b) The line terminated by an unmatched load can exhibit both an inductive or capacitive impedance seen through it depending on the position, (c) Transformation of a short circuit to an open circuit using a quarter wavelength T-line, and (d) Load impedance inversion using a quarter wavelength T-line.

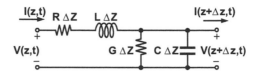

Figure 7.5: The lumped model of a differential transmission line for a differential length ΔZ.

Another important property of T-lines is impedance transformation. Figure 7.4 elaborates on this property. In case (a), the impedance is equal to $50\,\Omega$ regardless of the position throughout the line. For case (b), however, the impedance seen through the line can be either capacitive or inductive, depending on the position. Interestingly, in case (c), the T-line can transform a short circuit to an open circuit, for the quarter wavelength, and T-line in case (d) operates as an impedance transformer (impedance inverter).

The principal difference between the standard circuit theory and microwave circuit theory lies within the electrical size of the circuits and devices. Recall from circuit theory that a circuit can be viewed as a lumped one if the physical length is smaller than the wavelength of the operating signal. This allows us to conclude that the voltages and currents do not alter through a conductive wire according to the position. However, in microwave circuits, circuit size can be as large as the wavelength or even larger, which calls for a new perspective into design and analysis of such circuits. To begin with, consider the lumped model of a differential TEM T-line depicted in Figure 7.5.

7.2 Wave propagation Equations in Transmission Line for $R = 0$ and $G = 0$

As illustrated in Figure 7.5, a T-line is generally made up of two conductors. In electromagnetic wave theory, it is proved that this kind of structure can support (transmit) a TEM wave. A TEM wave is the one in which both the electric field and the magnetic field are perpendicular to the direction of the wave propagation. Most of the transmission lines used in the modern electronic circuits are of TEM type. This type of transmission line can be modeled with a distributed cells of series R and L alongside parallel G and C. We consider the transmission line model as depicted in Figure 7.5 consists of a series resistance per unit length R, a series inductance per unit length L, a parallel conductance per unit length G, and a parallel capacitance per unit length C. It can be observed that for a tiny fraction of the length, the circuit is lumped, and hence, KVL and KCL are still valid. It follows that the voltage and the current can be written as

$$v(z,t) = L\Delta z \frac{\partial i(z,t)}{\partial t} + v(z + \Delta z, t) \tag{7.1}$$

and

$$i(z,t) = i(z + \Delta z, t) + C\Delta z \frac{\partial v(z + \Delta z, t)}{\partial t} \tag{7.2}$$

respectively. Equations 7.1 and 7.2 can be rewritten as

$$\frac{v(z+\Delta z,t) - v(z,t)}{\Delta z} = -L\frac{\partial i(z,t)}{\partial t} \tag{7.3}$$

and

$$\frac{i(z+\Delta z,t) - i(z,t)}{\Delta z} = -C\frac{\partial v(z+\Delta z,t)}{\partial t} \tag{7.4}$$

respectively. The length of the line can approach to zero in order to comply with our lumped treatment of the circuit. Therefore, Equations 7.3 and 7.4 can be written in differential form as

$$\frac{\partial v}{\partial z} = -L\frac{\partial i}{\partial t} \tag{7.5}$$

and

$$\frac{\partial i}{\partial z} = -C\frac{\partial v}{\partial t} \tag{7.6}$$

respectively. These time-domain equations are famously known as the telegraphic equations. To arrive at a unified solution, partial derivative with respect to position is taken from Equation 7.5, which yields

$$\frac{\partial^2 v}{\partial z^2} = -L\frac{\partial}{\partial z}\left(\frac{\partial i}{\partial t}\right) \tag{7.7}$$

Changing the order of differentiation in Equation 7.7 and using Equation 7.6, we obtain

$$\frac{\partial^2 v}{\partial z^2} = LC\frac{\partial^2 v}{\partial t^2} \tag{7.8}$$

which corresponds to

$$\frac{\partial^2 v}{\partial z^2} - LC\left(\frac{\partial^2 v}{\partial t^2}\right) = 0 \tag{7.9}$$

This is a simple one-dimensional wave equation for the voltage on the line. It is left to the reader to find a similar equation for the current. Now, for the sake of simplicity, we assume that the input RF voltage is sinusoidal, having a phasor representation of $v(z,t) = \text{Re}\{V(z)e^{j\omega t}\}$. It follows that

$$\frac{d^2 V}{dz^2} - LC(-\omega^2)V = 0 \tag{7.10}$$

which can be rewritten as

$$\frac{d^2 V}{dz^2} = -\left(\omega^2 LC\right)V \tag{7.11}$$

Defining $Z = j\omega L$ and $Y = j\omega C$, Equation 7.11 can be rewritten as

$$\frac{d^2V}{dz^2} = (ZY)V \tag{7.12}$$

Let's define $\gamma^2 = ZY$, where γ is the propagation constant, and then find the solution to Equation 7.12 as

$$V(z) = Ae^{-\gamma z} + Be^{+\gamma z} \tag{7.13}$$

The constants in Equation 7.13 can be found using the initial conditions. The first term to the left represents the wave prorogating in the positive direction of the z-axis, and the second term to the left represents the wave prorogating in the negative direction of the z-axis. Therefore, a more meaningful representation of Equation 7.13 would look like

$$V(z) = V_0^+ e^{-\gamma z} + V_0^- Be^{+\gamma z} \tag{7.14}$$

where, the propagation constant, γ is given by

$$\gamma = j\beta = \sqrt{(j\omega L)(j\omega C)} = j\omega\sqrt{LC} \tag{7.15}$$

One can observe from Equation 7.15 that γ is frequency dependent. Now, to obtain the current wave, we substitute $V(z)$ from Equation 7.13 in Equation 7.5, and hence obtain

$$I(z) = \frac{j\beta}{j\omega L}\left(V_0^+ e^{-j\beta z} - V_0^- e^{+j\beta z}\right) = \frac{1}{Z_0}\left(V_0^+ e^{-j\beta z} - V_0^- e^{+j\beta z}\right) \tag{7.16}$$

where

$$Z_0 = \frac{\omega L}{\beta} = \sqrt{\frac{L}{C}} \tag{7.17}$$

Here, Z_0 is the transmission line's characteristic impedance. This impedance always shows the quotient of incident voltage to incident current or the quotient of reflected voltage to reflected current traveling along the transmission line. The voltage traveling wave equation can be easily obtained by assuming $V_0^+ = |V_0^+|\angle\phi$ and arriving at

$$v^+(z,t) = |V_0^+|\cos(\omega t - \beta z + \phi^+) \tag{7.18}$$

The same equation as that derived in Equation 7.18 can also be obtained for the reflected wave, by means of which the total voltage waveform is given by

$$v(z,t) = |V_0^+|\cos(\omega t - \beta z + \phi^+) + |V_0^-|\cos(\omega t + \beta z + \phi^-) \tag{7.19}$$

Now, we introduce two new quantities. The distance between two successive planes in z-direction having the same phase is defined as the wavelength, that is, $\beta(z_2 - z_1) = 2\pi$, where $z_2 = z_1 + \lambda$ and therefore

$$\lambda = \frac{2\pi}{\beta} \tag{7.20}$$

The wave's phase velocity is also defined assuming $\omega t - \beta z = cte$ as velocity with which a specific point on the wave front travels. That is

$$v_p = \frac{dz}{dt} = \frac{\omega}{\beta} = \frac{2\pi f}{\frac{2\pi}{\lambda}} = \lambda f \tag{7.21}$$

In other words, the phase velocity can be written as

$$v_p = \frac{\omega}{\beta} = \frac{\omega}{\omega\sqrt{LC}} = \frac{1}{\sqrt{LC}} \tag{7.22}$$

To grasp a better understanding of the meaning of the phase velocity, note that this quantity describes the speed with which the plane of constant phase travel in the space. This velocity in air-filled cables and transmission lines is the same as the velocity of light, while for those cables filled with other dielectric materials, it is the speed of light divided by the square root of the relative permittivity.

7.2.1 General Wave Propagation Relations in lossy Transmission Lines

For a lossy transmission line ($R \neq 0$ and $G \neq 0$) as depicted in Figure 7.6, a two-wire representation for a T-line can be thought of two conductors separated by a dielectric and excited by a source with a Thevenin voltage, V_{TH}, and the Thevenin impedance, Z_{TH}. As stated earlier, it can be shown that for a tiny fraction of the wire, the circuit can be assumed to be lumped, and therefore, KVL and KCL still hold. It therefore follows that

$$v(z,t) - R\Delta z \times i(z,t) + L\Delta z \frac{\partial i(z,t)}{\partial t} + v(z+\Delta z,t) \tag{7.23}$$

and

$$i(z,t) = i(z+\Delta z,t) + G\Delta z \times v(z+\Delta z,t) + C\Delta z \frac{\partial v(z+\Delta z,t)}{\partial t} \tag{7.24}$$

Equations 7.23 and 7.24 can be rewritten as

$$\frac{v(z+\Delta z,t) - v(z,t)}{\Delta z} = -Ri(z,t) - L\frac{\partial i(z,t)}{\partial t} \tag{7.25}$$

and

$$\frac{i(z+\Delta z,t) - i(z,t)}{\Delta z} = -Gv(z+\Delta z,t) - C\frac{\partial v(z+\Delta z,t)}{\partial t} \tag{7.26}$$

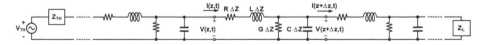

Figure 7.6: A transmission line divided into consecutive differential lumped sections.

As the length of the sample line approaches zero, differential forms for Equations 7.25 and 7.26 are obtained as

$$\frac{\partial v}{\partial z} = -Ri - L\frac{\partial i}{\partial t} \tag{7.27}$$

and

$$\frac{\partial i}{\partial z} = -Gv - C\frac{\partial v}{\partial t} \tag{7.28}$$

Now taking a derivative of Equation 7.27, we obtain

$$\frac{\partial^2 v}{\partial z^2} = -R\frac{\partial i}{\partial z} - L\frac{\partial}{\partial z}\left(\frac{\partial i}{\partial t}\right) \tag{7.29}$$

By changing the order of derivation and using Equation 7.28, we arrive at

$$\frac{\partial^2 v}{\partial z^2} = -R\left(-Gv - C\frac{\partial v}{\partial t}\right) - L\left(-G\frac{\partial v}{\partial t} - C\frac{\partial^2 v}{\partial t^2}\right) \tag{7.30}$$

The overall differential equation for the voltage is given by

$$\frac{\partial^2 v}{\partial z^2} - (RG)v - (RC + LG)\frac{\partial v}{\partial t} - LC\left(\frac{\partial^2 v}{\partial t^2}\right) = 0 \tag{7.31}$$

In sinusoidal regime with phasor representation of $v(z,t) = \mathrm{Re}\{V(z)e^{j\omega t}\}$, we arrive at

$$\frac{d^2 V}{dz^2} - (RG)V - (RC + LG)j\omega V - LC(-\omega^2)V = 0 \tag{7.32}$$

Equation 7.32 can be rewritten as

$$\frac{d^2 V}{dz^2} = (RG)V + j\omega(RC + LG)V - (\omega^2 LC)V \tag{7.33}$$

Defining $Z = R + j\omega L$ and $Y = G + j\omega C$, Equation 7.32 can be written as

$$\frac{d^2 V}{dz^2} = (ZY)V \tag{7.34}$$

The propagation constant, γ, is defined as $\gamma^2 = ZY$, and then Equation 7.34 can be recast as

$$V(z) = Ae^{-\gamma z} + Be^{+\gamma z} \tag{7.35}$$

The constants in Equation 7.35 can be computed using the initial conditions. The first term to the left represents the wave prorogating in the positive direction of the z-axis, and the second term represents the wave prorogating in the negative direction of the z-axis. Therefore, a more meaningful representation of Equation 7.35 would look like

$$V(z) = V_0^+ e^{-\gamma z} + V_0^- e^{+\gamma z} \tag{7.36}$$

It also follows that

$$\gamma = \alpha + j\beta = \sqrt{(R + j\omega L)(G + j\omega C)} \tag{7.37}$$

where the real and imaginary parts describe the attenuation and the phase constants, respectively. Here the propagation constant has a nonlinear frequency dependency, and therefore the transmission line is dispersive. The current waveform can be obtained by substituting Equation 7.36 in the phasor form of Equation 7.27 and we obtain

$$I(z) - \frac{\gamma}{R + j\omega L} \left(V_0^+ e^{-\gamma z} - V_0^- e^{+\gamma z} \right) \tag{7.38}$$

As before, the wave traveling in the positive z-direction is

$$v^+(z,t) = |V_0^+| e^{-\alpha z} \cos(\omega t - \beta z + \phi^+) \tag{7.39}$$

And hence the overall voltage waveform at the input is given by

$$v(z,t) = |V_0^+| e^{-\alpha z} \cos(\omega t - \beta z + \phi^+) + |V_0^-| e^{+\alpha z} \cos(\omega t + \beta z + \phi^-) \tag{7.40}$$

The phase velocity can be defined as before for the lossy case, here we have

$$v_p = \frac{\omega}{\operatorname{Im}\left\{ \sqrt{(R + j\omega L)(G + j\omega C)} \right\}} \tag{7.41}$$

7.3 Characteristic Impedance of a Line

The characteristic impedance of the line is defined as the ratio of the positive z-direction traveling voltage to the positive z-direction traveling current as

$$Z_0 = \frac{V_0^+(z)}{I_0^+(z)} \tag{7.42}$$

Writing Equation 7.27 in the phasor domain, we arrive at

$$\frac{dV}{dz} - -RI - j\omega LI - -ZI \tag{7.43}$$

Considering only the forward-propagating waves, V^+ and I^+, and using Equation 7.43 we can write

$$-\gamma V_0^+ e^{-\gamma z} = -Z I_0^+ e^{-\gamma z} \tag{7.44}$$

where in using Equation 7.42, the characteristic impedance can be found to be

$$Z_0 = \frac{V_0^+}{I_0^+} = \frac{Z}{\gamma} = \sqrt{\frac{Z}{Y}} = \sqrt{\frac{R + j\omega L}{G + j\omega C}} \tag{7.45}$$

As it can be observed from Equation 7.45, the characteristic impedance is a function of intrinsic properties of the line as well as the frequency. But once the line is lossless, the characteristic impedance becomes independent of frequency.

Example 7.1 Consider Equation 7.45 for a line with $L = 0.273 \frac{\text{nH}}{\text{mm}}$, $C = 93.5 \frac{\text{fF}}{\text{mm}}$, $R = 170 \frac{\text{m}\Omega}{\text{mm}}$, and $G = 60 \frac{\mu\text{S}}{\text{mm}}$. Derive an expression for the phase and amplitude of the line characteristic impedance from 0 to 1 GHz. $\left(\frac{\ell}{\lambda} = \frac{2n-2}{4} (n \geq 1) \right)$, and plot the real and the imaginary parts as a function of frequency.

Solution:
It follows from Equation 7.45 that

$$Z_0 = \sqrt{\frac{R + j\omega L}{G + j\omega C}} = \sqrt{\frac{0.17 + j\omega(0.273)10^{-9}}{60 \times 10^{-6} + j\omega(93.5)10^{-15}}} \qquad (7.46)$$

which is depicted in Figure 7.7.

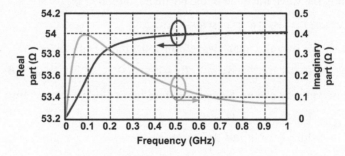

Figure 7.7: Characteristic impedance of a line.

As it can be seen in the above figure, Z_0 becomes almost pure real and independent of frequency at the higher portion of the spectrum. ∎

The current waveform as a function of the line characteristic impedance can also be written as

$$I(z) = \frac{1}{Z_0} \left(V_0^+ e^{-\gamma z} - V_0^- e^{+\gamma z} \right) \qquad (7.47)$$

7.3.1 Lossless Transmission Line

The equations derived earlier can be well extended to any line in TEM mode, and as it was shown, the propagation constant and line impedance in their most general forms are complex quantities. In many practical cases, however, losses are small enough to be neglected, that is the line resistance is sufficiently small compared to the line series reactance, and the line conductance is sufficiently small compared to the line parallel susceptance. Therefore, it follows from 7.37 that

$$\gamma = \alpha + j\beta = \sqrt{(0 + j\omega L)(0 + j\omega C)} = j\omega\sqrt{LC} \qquad (7.48)$$

This means that the attenuation constant is equal to zero and the phase constant $\beta = \omega\sqrt{LC}$. Hence, the voltage and the current waveforms traveling through this line will no longer experience any loss and will experience only a phase shift. The line characteristic impedance becomes in this case as

$$Z = \sqrt{\frac{0 + j\omega L}{0 + j\omega C}} = \sqrt{\frac{L}{C}} \tag{7.49}$$

As it is observed from Equation 7.49, the line characteristic impedance becomes purely real and it is also independent of frequency. The voltage and the current waveforms can also be derived in the lossless case as

$$V(z) = V_0^+ e^{-j\beta z} + V_0^- e^{+j\beta z} \tag{7.50}$$

and

$$I(z) = \frac{V_0^+}{Z_0} e^{-j\beta z} - \frac{V_0^-}{Z_0} e^{+j\beta z} \tag{7.51}$$

The phase velocity is given by

$$\upsilon_p = \frac{\omega}{\beta} = \frac{\omega}{\omega\sqrt{LC}} = \frac{1}{\sqrt{LC}} \tag{7.52}$$

As an example, for the values given in Example 7.1, the phase velocity amounts to 1.98×10^8 m/sec. If the medium in between the two conductors is homogeneous with the permittivity ε and the permeability μ, it can be shown that for the transmission line, one can write

$$LC = \mu\varepsilon \tag{7.53}$$

The wavelength can be readily derived from Equation 7.48 as

$$\lambda = \frac{2\pi}{\beta} = \frac{2\pi}{\omega\sqrt{LC}} \tag{7.54}$$

We also know that the speed of TEM electromagnetic propagation wave in a dielectric/magnetic medium is

$$\upsilon_g = \frac{1}{\sqrt{\mu\varepsilon}} = \frac{c}{\sqrt{\mu_r \varepsilon_r}} \tag{7.55}$$

where c is the speed of light in vacuum. Using Equation 7.53, we end up with $\upsilon_P = \upsilon_g$ which means the phase velocity in a TEM transmission line is the same as the speed of propagation in a free space medium.

7.4 Terminated Transmission Lines

In this section, we discuss the behavior of an arbitrarily terminated line, as depicted in Figure 7.8. For now, we assume that the origin of the z-axis is located at the load

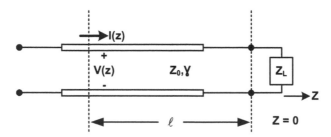

Figure 7.8: A terminated transmission line.

end. The voltage at the source end of the line is of interest, which can be written from Equation 7.36 as

$$V(-\ell) = V_0^+ e^{+\gamma\ell} + V_0^- e^{-\gamma\ell} \tag{7.56}$$

where ℓ is the distance from the load and the first term in the right-hand side is the traveling wave in the positive z-direction and the second term is the reflected wave in the negative z-direction. The current waveform can also be derived in a similar manner as

$$I(-\ell) = \frac{V_0^+}{Z_0} e^{\gamma\ell} - \frac{V_0^-}{Z_0} e^{-\gamma\ell} \tag{7.57}$$

Now, Equation 7.56 can be rewritten as

$$V(-\ell) = V_0^+ e^{\gamma\ell} + V_0^- e^{-\gamma\ell} = V_0^+ e^{\gamma\ell} \left(1 + \frac{V_0^-}{V_0^+} e^{-2\gamma\ell}\right) \tag{7.58}$$

where V_0^+ is the traveling wave vector toward the load at $Z = 0$ and V_0^- is the reflected wave from the load at $Z = 0$. The ratio of the reflected wave to the incident wave at the load is defined as reflection coefficient and can be computed as

$$\Gamma_L(-\ell) = \frac{V_0^- e^{-\gamma\ell}}{V_0^+ e^{+\gamma\ell}} = \Gamma_L(0) e^{-2\gamma\ell} \tag{7.59}$$

Using the above definition, the voltage and the current waveforms can be written using the reflection coefficient as

$$V(-\ell) = V_0^+ e^{\gamma\ell} (1 + \Gamma_L e^{-2\gamma\ell}) \tag{7.60}$$

and

$$I(-\ell) = \frac{V_0^+}{Z_0} e^{\gamma\ell} (1 - \Gamma_L e^{-2\gamma\ell}) \tag{7.61}$$

It is easy to compute the input impedance by finding the ratio of the voltage to the current phasors as

$$Z(-\ell) = \frac{V(-\ell)}{I(-\ell)} = Z_0 \left(\frac{1 + \Gamma_L e^{-2\gamma\ell}}{1 - \Gamma_L e^{-2\gamma\ell}} \right) \tag{7.62}$$

which gives the input impedance of an arbitrarily terminated T-line. If one chooses $\ell = 0$, the load impedance is given as

$$Z(0) = Z_0 \left(\frac{1 + \Gamma_L}{1 - \Gamma_L} \right) = Z_L \tag{7.63}$$

The reflection coefficient in terms of load impedance becomes

$$\Gamma_L = \frac{Z_L - Z_0}{Z_L + Z_0} \tag{7.64}$$

As suggested by Equation 7.64, if the load impedance is equal to the characteristic impedance, the reflection coefficient becomes zero, in which case the matching is achieved. Now, by substituting Equation 7.64 into Equation 7.62, an explicit expression for the input impedance can be obtained as

$$Z(-\ell) = Z_0 \left(\frac{1 + \left(\frac{Z_L - Z_0}{Z_L + Z_0} \right) e^{-2\gamma\ell}}{1 - \left(\frac{Z_L - Z_0}{Z_L + Z_0} \right) e^{-2\gamma\ell}} \right) \tag{7.65}$$

Now, if the explicit expression of the input impedance of a T-line is to be derived, Equation 7.65 can be rewritten as

$$Z(-\ell) = Z_0 \left(\frac{(Z_L + Z_0) + (Z_L - Z_0) e^{-2\gamma\ell}}{(Z_L + Z_0) - (Z_L - Z_0) e^{-2\gamma\ell}} \right) \tag{7.66}$$

Multiplying the numerator and the denominator by $e^{+\gamma\ell}$, it follows that

$$Z(-\ell) = Z_0 \left(\frac{(Z_L + Z_0) e^{+\gamma\ell} + (Z_L - Z_0) e^{-\gamma\ell}}{(Z_L + Z_0) e^{+\gamma\ell} - (Z_L - Z_0) e^{-\gamma\ell}} \right) \tag{7.67}$$

Factoring Z_0 and Z_L out, Equation 7.67 can be rewritten using hyperbolic functions as

$$Z(-\ell) = Z_0 \left(\frac{Z_L \cosh(\gamma\ell) + Z_0 \sinh(\gamma\ell)}{Z_0 \cosh(\gamma\ell) + Z_L \sinh(\gamma\ell)} \right) \tag{7.68}$$

By dividing the numerator and the denominator by $\cosh(\gamma\ell)$, we have

$$Z(-\ell) = Z_0 \left(\frac{Z_L + Z_0 \tanh(\gamma\ell)}{Z_0 + Z_L \tanh(\gamma\ell)} \right) \tag{7.69}$$

Now, we derive the above expressions for a lossless T-line. The voltage and the current waveforms are given by

$$V(-\ell) = V_0^+ e^{j\beta\ell}\left(1 + \Gamma_L e^{-2j\beta\ell}\right) \tag{7.70}$$

and

$$I(-\ell) = \frac{V_0^+}{Z_0} e^{j\beta\ell}\left(1 - \Gamma_L e^{-2j\beta\ell}\right) \tag{7.71}$$

Therefore, it follows that

$$Z(-\ell) = \frac{V(-\ell)}{I(-\ell)} = Z_0\left(\frac{1 + \Gamma_L e^{-2j\beta\ell}}{1 - \Gamma_L e^{-2j\beta\ell}}\right) = Z_0\left(\frac{Z_L + jZ_0\tan(\beta\ell)}{Z_0 + jZ_L\tan(\beta\ell)}\right) \tag{7.72}$$

Since in the lossless T-line $\tanh(\gamma\ell) = \tanh(j\beta\ell) = j\tan(\beta\ell)$, by replacing $\tanh(\gamma\ell)$ in Equation 7.69, we can obtain the same result as above. It follows from Equation 7.72, the input impedance has a periodic property as a function of distance, and we have

$$\beta\ell = n\pi \Rightarrow \frac{2\pi}{\lambda}\ell = n\pi \Rightarrow \ell = \frac{n\lambda}{2} \tag{7.73}$$

This means that the input impedance attains the same value for each $\frac{\lambda}{2}$ length. Equation 7.72 is used for a lossless transmission line while Equation 7.69 is used for a lossy transmission line. The calculation of these complex equations can be simplified through the use of the Smith chart which is provided in section 7.10.

> **Example 7.2** For the circuit in Figure 7.9, compute the input impedance and the reflection coefficient at the input and at the load side.
> (a) First, consider the following parameters for the line and find the results at 1 GHz.
> (b) Consider the line is lossless and repeat part (a).
> $L = 1.125\frac{\mu H}{m}, C = 450\frac{pF}{m}, R = 5\frac{\Omega}{m}, G = 0.01\frac{S}{m}$

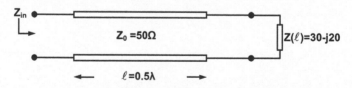

Figure 7.9: The terminated T-line.

Solution:
(a) We have

$$\gamma = \alpha + j\beta = \sqrt{(R + j\omega L)(G + j\omega C)} \tag{7.74}$$

$$= \sqrt{\left(5 + j(2\pi\times 10^9)\times 1.125\times 10^{-6}\right)\left(0.01 + j(2\pi\times 10^9)\times 450\times 10^{-12}\right)}$$

$$= 0.3 + j141.371 \ \frac{1}{m}$$

$$Z_0 = \sqrt{\frac{R+j\omega L}{G+j\omega C}} \approx 50\,\Omega \tag{7.75}$$

Then, using Equation 7.68, the input impedance is given by

$$Z(-\ell) = Z_0\left(\frac{Z_L + Z_0\tanh(\gamma\ell)}{Z_0 + Z_L\tanh(\gamma\ell)}\right) = 50\left(\frac{30 - j20 + 50\tanh(\gamma\ell)}{50 + (30 - j20)\tanh(\gamma\ell)}\right) \tag{7.76}$$

Since $\ell = \frac{\lambda}{2}$ and $\beta = \frac{2\pi}{\lambda}$, then from Equation 7.76, the line length becomes

$$\ell = \frac{\lambda}{2} = \frac{\frac{2\pi}{\beta}}{2} = \frac{\pi}{\beta} = 2.22\,\text{cm} \tag{7.77}$$

Therefore

$$Z(-\ell) = 50\left(\frac{30 - j20 + 50\tanh\left((\alpha + j\beta)\frac{\pi}{\beta}\right)}{50 + (30 - j20)\tanh\left((\alpha + j\beta)\frac{\pi}{\beta}\right)}\right) \tag{7.78}$$

From Equation 7.78, we have

$$Z(-\ell) = 30.2651 - j19.8399\,\Omega \tag{7.79}$$

First, the reflection coefficient at the load side can be computed as

$$\Gamma_L(0) = \frac{Z_L - Z_0}{Z_L + Z_0} = \frac{30 - j20 - 50}{30 - j20 + 50} = -0.1765 \quad j\,0.2941 \tag{7.80}$$

and for the source side, the reflection coefficient is given by

$$\Gamma_L(-\ell) = \Gamma_L(0)e^{-2\gamma\ell} = -0.1741 - j\,0.2902 \tag{7.81}$$

As it is seen here the input reflection coefficient is slightly different from the load reflection coefficient (because the line length is $\frac{\lambda}{2}$ and the line is lossy).
(b) For the input impedance, we have from Equation 7.72

$$Z(-\ell) = Z_0\left(\frac{Z_L + jZ_0\tan(\beta\ell)}{Z_0 + jZ_L\tan(\beta\ell)}\right) = 50\left(\frac{30 - j20 + j50\tan\left(\frac{2\pi}{\lambda}0.5\lambda\right)}{50 + j(30 - j20)\tan\left(\frac{2\pi}{\lambda}0.5\lambda\right)}\right)$$
$$= 30 - j20 \tag{7.82}$$

The reflection coefficient can be found as

$$\Gamma_L(-\ell) = \Gamma_L(0)e^{-2\beta\ell} = -0.1765 - j0.2941 \tag{7.83}$$

As it is seen here the reflection coefficient has exactly the same value as the input (because the line length is $\frac{\lambda}{2}$ and the line is considered as lossless). ∎

7.5 Special Cases of a Terminated Line

Some special cases of line loading are of specific interest and find their application in filter design. These cases include terminations to the intrinsic characteristic impedance of the T-line, short circuit, and open circuit.

7.5.1 Termination to the Line Characteristic Impedance

In this case, we assume that the load impedance is equal to Z_0, as depicted in Figure 7.10. Therefore, for the input impedance, we have from Equation 7.69

$$Z(-\ell) = Z_0 \left(\frac{Z_0 + Z_0 \tanh(\gamma \ell)}{Z_0 + Z_0 \tanh(\gamma \ell)} \right) = Z_0 \tag{7.84}$$

for the reflection coefficient at the load and the source, we have

$$\Gamma_L = \frac{Z_L - Z_0}{Z_L + Z_0} = \frac{Z_0 - Z_0}{Z_0 + Z_0} = 0 \tag{7.85}$$

Therefore, for this case, the input impedance is equal to the characteristic impedance and the reflection coefficient will be equal to zero throughout the line, irrespective of the location. No reflection will therefore occur on the line and perfect matching has been achieved, and the voltage and the current waveforms can be derived as

$$V(-\ell) = V_0^+ e^{+j\gamma \ell} \tag{7.86}$$

and

$$I(-\ell) = \frac{V_0^+}{Z_0} e^{+j\gamma \ell} \tag{7.87}$$

It implies that a voltage or a current wave can be moved from the point $-\ell$ to point zero with only a phase shift and then they will be absorbed by the load (with no reflection). It should be noted that the normal PCB circuit wires cannot transmit the high-frequency signals. Using transmission lines in PCBs, we can transfer very high frequency signals to any devised distance.

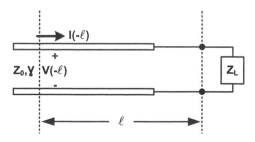

Figure 7.10: A terminated T-line to the intrinsic impedance.

7.5.2 Short-circuit load impedance

In this case, the load is terminated to a short circuit, i.e., $Z_L = 0$, which is shown in Figure 7.11.

It follows from Equation 7.69

$$Z(-\ell) = Z_0 \left(\frac{0 + Z_0 \tanh(\gamma\ell)}{Z_0 + 0 \times \tanh(\gamma\ell)} \right) = Z_0 \tanh(\gamma\ell) \qquad (7.88)$$

which in case of a lossless line can be written as

$$Z(-\ell) = Z_0 \tanh(j\beta\ell) = jZ_0 \tan(\beta\ell) \qquad (7.89)$$

As it is obvious from Equation 7.89, the input impedance in this case is always an imaginary quantity. This suggests that any arbitrary value of reactance can be obtained with a T-line terminated to a short circuit. The value of the input impedance in this case is depicted in Figure 7.12.

As it is obvious from Figure 7.12, the impedance assumes the values of zero and infinity at even multiples of $\frac{\lambda}{4}$ and at odd multiples of $\frac{\lambda}{4}$, respectively. This result shows

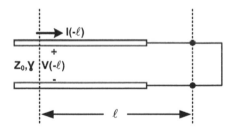

Figure 7.11: A short-terminated T-line.

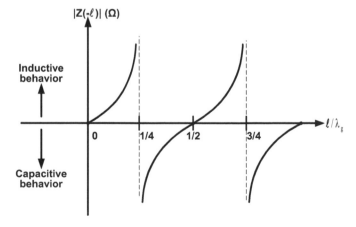

Figure 7.12: The input impedance of a short-terminated T-line.

that moving away from a short-ended transmission line first for less than quarter wave length distances, the input becomes inductive and for a length in excess of a quarter-wave length, it becomes capacitive. In practice, we use short-ended transmission lines shorter than a quarter of a wavelength to realize inductive impedances. The reflection coefficient at the load can be computed from Equation 7.64 as

$$\Gamma_L = \frac{Z_L - Z_0}{Z_L + Z_0} = \frac{0 - Z_0}{Z_0 + 0} = -1 \tag{7.90}$$

7.5.3 Open-circuit load

In this case, the load impedance is assumed to be infinite, that is, $Z_L \to \infty$, which is depicted in Figure 7.13. We first calculate the input impedance in this case as

$$Z(-\ell) = Z_0 \left(\frac{Z_L + Z_0 \tanh(\gamma\ell)}{Z_0 + Z_L \tanh(\gamma\ell)} \right)_{Z_L \to \infty} = Z_0 \coth(\gamma\ell) \tag{7.91}$$

which for the lossless case reduces to

$$Z(-\ell) = Z_0 \coth(j\beta\ell) = -jZ_0 \cot(\beta\ell) \tag{7.92}$$

As it is obvious from Equation 7.92, the input impedance in this case is always imaginary, meaning that any value of a reactive impedance can be realized by this line. The input impedance in this case is depicted in Figure 7.14.

As it is obvious from Figure 7.14, for odd multiples of $\frac{\lambda}{4}$ and even multiples of $\frac{\lambda}{4}$, the input impedance is that of short circuit and open circuit, respectively. This result shows that moving away from an open-ended transmission line first (for the lengths less than a quarter wave length), the input becomes capacitive and for a length in excess of a quarter wave length, it becomes inductive. In practice, we use open-ended transmission lines shorter than a quarter of a wavelength to realize capacitive impedances. The reflection coefficient can be computed from Equation 7.64 as

$$\Gamma_L = \left(\frac{Z_L - Z_0}{Z_0 + Z_L} \right)_{Z_L \to \infty} = 1 \tag{7.93}$$

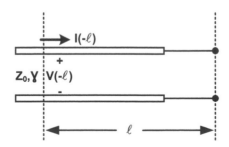

Figure 7.13: The input impedance of an open-terminated T-line.

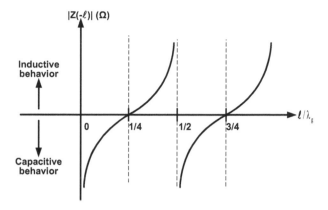

Figure 7.14: The input impedance of an open-terminated T-line.

7.6 Source and Load Mismatch in Lossless Lines (A Reflection Coefficient Perspective)

We have assumed thus far that the input source has an impedance equal to the characteristic impedance of the lines, resulting in no voltage or current reflection at the source. This might not be true for practical purposes and in real cases, a small fraction of incident wave is reflected off the line. An arbitrarily terminated line is depicted in Figure 7.15. As waves are reflected from both ends, there are countless waves propagating along the line, nevertheless, in the steady state, there is only one wave propagating toward the load and one wave toward the source. We shortly derive equations for the current and voltage in this case. Figure 7.15 can be redrawn as shown in Figure 7.16 by calculating the input impedance. From Figure 7.16, it follows using voltage division that

$$V(-d) = V_{\text{TH}} \left(\frac{Z_{\text{in}}}{Z_{\text{in}} + Z_{\text{TH}}} \right) \tag{7.94}$$

and the input impedance can be calculated from Equation 7.72 as

$$Z(-d) = Z_0 \left(\frac{Z_{\text{L}} + jZ_0 \tan(\beta d)}{Z_0 + jZ_{\text{L}} \tan(\beta d)} \right) \tag{7.95}$$

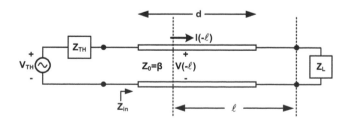

Figure 7.15: A T-line with its load and source impedances.

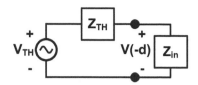

Figure 7.16: Equivalent model of Figure 7.15 for input impedance calculation.

The voltage wave at $Z = -\ell$ can be obtained from Equation 7.58 as

$$V(-\ell) = V_0^+ e^{j\beta\ell}\left(1 + \Gamma_L e^{-2j\beta\ell}\right) \tag{7.96}$$

Equation 7.96 can be written for the input, and using Equation 7.94, we arrive at

$$V(-d) = V_0^+ e^{j\beta d}\left(1 + \Gamma_L e^{-j2\beta d}\right) = V_{TH}\left(\frac{Z_{in}}{Z_{in} + Z_{TH}}\right) \tag{7.97}$$

The incident wave phasor can be obtained from Equation 7.96; it follows that

$$V_0^+ = V_{TH}\left(\frac{Z_{in}}{Z_{in} + Z_{TH}}\right) e^{-j\beta d}\left(\frac{1}{1 + \Gamma_L e^{-j2\beta d}}\right) \tag{7.98}$$

Now, by substituting Equation 7.98 in Equation 7.96, the voltage can be found at any point along the T-line as

$$V(-\ell) = V_{TH}\left(\frac{Z_{in}}{Z_{in} + Z_{TH}}\right) e^{-j\beta(d-\ell)}\left(\frac{1 + \Gamma_L e^{-j2\beta\ell}}{1 + \Gamma_L e^{-j2\beta d}}\right) \tag{7.99}$$

It follows from Equation 7.99 that the impedance at distance d from the load as

$$Z_{in} = Z_{in}(-d) = \frac{V(-d)}{I(-d)} = Z_0\left(\frac{1 + \Gamma_L e^{-j2\beta d}}{1 - \Gamma_L e^{-j2\beta d}}\right) \tag{7.100}$$

Then, an explicit expression of the ratio of the source impedance divided by the input impedance can be found as

$$\frac{Z_{in}(-d)}{Z_{in}(-d) + Z_{TH}} = \frac{Z_0\left(\frac{1+\Gamma_L e^{-j2\beta d}}{1-\Gamma_L e^{-j2\beta d}}\right)}{Z_0\left(\frac{1+\Gamma_L e^{-j2\beta d}}{1-\Gamma_L e^{-j2\beta d}}\right) + Z_{TH}} \tag{7.101}$$

which reduces to

$$\frac{Z_{in}(-d)}{Z_{in}(-d) + Z_{TH}} = \left(\frac{Z_0}{Z_{TH} + Z_0}\right)\frac{\left(1 + \Gamma_L e^{-j2\beta d}\right)}{1 - \Gamma_L e^{-j2\beta d}\left(\frac{Z_{TH} - Z_0}{Z_{TH} + Z_0}\right)} \tag{7.102}$$

Considering the input reflection coefficient, we have

$$\Gamma_S = \frac{Z_{TH} - Z_0}{Z_{TH} + Z_0} \tag{7.103}$$

If we substitute Equation 7.103 into Equation 7.102, we obtain

$$\frac{Z_{in}(-d)}{Z_{in}(-d) + Z_{TH}} = \left(\frac{Z_0}{Z_0 + Z_{TH}}\right) \left(\frac{1 + \Gamma_L e^{-j2\beta d}}{1 - \Gamma_S \Gamma_L e^{-j2\beta d}}\right) \tag{7.104}$$

Hence, substituting Equation 7.104 into Equation 7.99, we arrive at

$$V(-\ell) = V_{TH} \left(\frac{Z_0}{Z_0 + Z_{TH}}\right) e^{-j\beta(d-\ell)} \left(\frac{1 + \Gamma_L e^{-j2\beta\ell}}{1 - \Gamma_S \Gamma_L e^{-j2\beta d}}\right) \tag{7.105}$$

Finally, the voltage and current waveforms can be found at any point along the line using Equation 7.105.

Example 7.3 Consider the circuit shown in Figure 7.17(a). Find the voltage waveform at the input. (Hint: Derive the input such that it exhibits wave transmission back and forth toward the load and from the source. This behavior is depicted in Figure 7.17(b).)

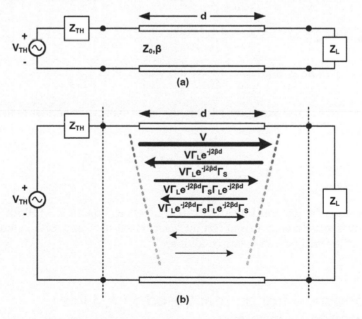

Figure 7.17: (a) Demonstration of a T-line with its load and source impedances, and (b) transmission of the wave back and forth on the line.

Solution:
First, the voltage is divided between the source impedance and the input impedance. Then, it moves along the line and a fraction of it is reflected, proportional to the reflection coefficient, Γ_L. This repeats while the wave reaches the sources and is reflected back proportional to the source reflection coefficient, Γ_S. In order to quantify this behavior, we have

$$V(-d) = V_{TH} \left(\frac{Z_0}{Z_0 + Z_{TH}} \right) \left(1 + \Gamma_L e^{-j2\beta d} + \left(\Gamma_L e^{-j2\beta d} \right) \Gamma_S + \cdots \right)$$

(7.106)

$$= V_{TH} \left(\frac{Z_0}{Z_0 + Z_{TH}} \right) \left(1 + (\Gamma_S + 1) \sum_{i=1}^{\infty} \left(\Gamma_L e^{-j2\beta d} \right)^i \Gamma_S^{i-1} \right)$$

Now, using the following expansion while $|z| < 1$,

$$\sum_{n=0}^{\infty} z^n = 1 + z + z^2 + \ldots = \frac{1}{1-z}, \quad \text{where } z = \Gamma_L \Gamma_S \, e^{-j2\beta d}$$

(7.107)

We can rewrite Equation 7.106 as

$$V(-d) = V_{TH} \left(\frac{Z_0}{Z_0 + Z_{TH}} \right) \left(1 + \frac{\Gamma_S + 1}{\Gamma_S} \sum_{i=1}^{\infty} \left(\Gamma_L \Gamma_S e^{-j2\beta d} \right)^i \right)$$

(7.108)

which results in

$$V(-d) = V_{TH} \left(\frac{Z_0}{Z_0 + Z_{TH}} \right) \left(1 + \frac{\Gamma_S + 1}{\Gamma_S} \left(\frac{1}{1 - \Gamma_L \Gamma_S e^{-j2\beta d}} - 1 \right) \right)$$

(7.109)

Now, we can use Equation 7.109 to derive an expression similar to that derived in Equation 7.105 for $d = \ell$

$$V(-d) = V_{TH} \left(\frac{Z_0}{Z_0 + Z_{TH}} \right) \left(\frac{1 + \Gamma_L e^{-j2\beta d}}{1 - \Gamma_L \Gamma_S e^{-j2\beta d}} \right)$$

(7.110)

As such, the general Equation 7.105 reduces to Equation 7.110 for $\ell = d$. This should make it clear that in fact the wave travels back and forth in multiple reflections to reach a steady-state condition.   ∎

7.7 Impedance Transformer Based on $\lambda/4$ line (Impedance Inverter)

One of the interesting properties of T-lines occurs when its length is equal to $\ell = \frac{\lambda}{4} + n\frac{\lambda}{2} (n \geq 1)$. It follows from Equation 7.58 that

$$Z(-d) = Z_0 \left(\frac{Z_L + jZ_0 \tan\left(\frac{\pi}{2}\right)}{Z_0 + jZ_L \tan\left(\frac{\pi}{2}\right)} \right) = \frac{Z_0^2}{Z_L} \tag{7.111}$$

The important property revealed by Equation 7.111 is that it can transform the load impedance to a value proportional to the inverse of load impedance (multiplied by the square of characteristic impedance of the line). This property finds an important application in impedance matching networks where one intends to change a real impedance value to another real impedance value.

7.7.1 Synthesis of an Inductor and a Capacitor with a Transmission Line

One of the most important applications of T-lines is in matching networks. As it was discussed earlier, the impedance seen through the line changes as a function of the load and we observed that inductive and capacitive impedances can be achieved when the line is terminated to an open or a short circuit. It therefore is tempting to replace the bulky matching elements like capacitors and inductors with T-lines terminated to a short or an open circuit, which are referred to as a short stub and an open stub, respectively. Therefore, an open or short-terminated line can be used to emulate a capacitor or an inductor, respectively, albeit for narrowband applications. This application is conceptually demonstrated in Figure 7.18

7.8 Voltage Standing Wave Ratio

From Equation 7.70, we can calculate the voltage waveform along the line; specifically, we can write

$$V(\ell) = V_0^+ e^{j\beta\ell} \left(1 + |\Gamma_L| e^{j\phi_L} e^{-2j\beta\ell} \right) \tag{7.112}$$

As it can be seen, the voltage along the line varies between a maximum and a minimum value depending on the phase of the second term in the parentheses, i.e., $\phi_L - 2\beta\ell$. Along the transmission line at the point (points) where this phase is equal to zero, there will be a maximum voltage and at the point (points) where this phase is equal to π, there is a minimum voltage. The ratio of these two quantities is called voltage standing wave ratio (VSWR).

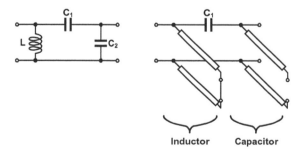

Figure 7.18: Implementation of a narrowband inductor and a narrowband capacitor using a short and an open-terminated line.

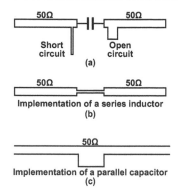

Figure 7.19: (a) Implementation of a parallel inductor and a parallel capacitor using short-circuit and open-circuit stubs, respectively, (b) Implementation of a series inductor using a short-length high-impedance transmission line, (c) Implementation of a parallel capacitor using a short-length low-impedance stub, all in microstrip technology.

$$VSWR = \frac{V_{\max}}{V_{\min}} = \frac{\left|V_0^+\right|(1+|\Gamma_L|)}{\left|V_0^+\right|(1-|\Gamma_L|)} = \frac{1+|\Gamma_L|}{1-|\Gamma_L|} \tag{7.113}$$

Equation 7.113 provides a measure of input impedance matching as well. As an VSWR equal to unity is only possible once $\Gamma_L = 0$. In the case of a mismatch, VSWR assumes a value greater than unity. A large value of VSWR is an indicator of a large value of the reflection coefficient as well as a greater value of the reflected power which is equal to $|\Gamma_L|^2 P_{\text{inc}}$ where P_{inc} is the incident power.

Normally, the RF equipment designers give the input VSWR of their device (instead of Γ_{in}) as a measure of their device matching. The nearer its value is to unity, the better the circuit is matched and the larger the value of VSWR indicates poorer matching. Nowadays, with the emergence of advanced network analyzers the input and output reflection coefficients of any device can be readily measured and transformed into impedance or admittance values. The generalization of input and output reflection coefficients concept can be interpreted as S-parameters that will be studied in Chapter 8.

7.9   Impedance Matching: The L-Section Approach

The basic principle behind impedance matching was first introduced for connecting a line with a certain characteristic impedance to a specific load. With matching, reflections from the load on the line can be avoided, and therefore the traveling wave is received at the other end with a sufficient power level. Impedance matching is of utmost importance in wireless systems and circuits, by means of which the maximum power transmission, the maximum SNR, and hence a higher data rate in the receivers, the minimum required power in the transmitters, a higher lifetime for the power supply, and a lower risk of undesired radiation can be achieved. Impedance matching also finds another application in amplifiers, where in order to achieve the maximum power transmission in both input and the output, matching must be present between the source and the input on the one hand, and the load and the output on the other. There are quite a few points which merit specific attention in matching networks:

(1) Complexity: designing the most simple matching network possible is crucial. After all, a smaller matching network is less expensive, more reliable and suffers from a smaller attenuation with respect to its more complex counterparts;

(2) Bandwidth: every single matching network can only provide matching in a specific bandwidth given by its specifications. This is, however, inadequate for many applications where a wideband matching is desired. There exists a myriad of techniques for bandwidth extension, which as expected, come at the price of complexity;

(3) Implementation: the designer may prefer one matching structure to another based on the type of the T-line, lumped elements, or the waveguide. For example, tuning screws in waveguide arms provide better flexibility than the quarter wave length T-line, and

(4) Tunability: In some applications, it is desirable to be able to fine-tune the matching network so as to achieve the maximum power transmission to the load. Some architectures are better fit for this property.

Consider the circuit depicted in Figure 7.20. We will call circuits of this kind which have a frequency-selective behavior a resonant or a tank circuit thereafter. Before dealing with the circuit details, let's first discuss a parameter which provides a measure of loss in energy storage element. Referred to as quality factor, Q, it is defined as

$$Q = \frac{f_0}{\Delta f} \tag{7.114}$$

wherein f_0 is the resonant frequency of the circuit and Δf is the corresponding 3 dB bandwidth. Nevertheless, the presence of load and source impedances degrades the overall quality factor of the circuit. We call the matching in this case a loaded one. The effect of the source impedance on the frequency response of the resonant network is depicted in Figure 7.21. As it is evident in Figure 7.21, when the input impedance is equal to $50\,\Omega$, the frequency response is wider, while for an input impedance of $1000\,\Omega$, a much narrower response is observed. The inductance of the inductor is also important in the overall quality factor of the circuit. As depicted in Figure 7.22, with a higher inductance, a lower Q is obtained mainly due to ohmic losses. In this circuit, $Q = \frac{R_S}{L\omega_0} = R_S C \omega_0$.

As stated in the beginning of this chapter, one of the properties of matching network is maximum power transmission. As you can recall from the basic circuit theory, in a DC circuit, maximum power transmission between the load and the source

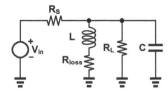

Figure 7.20: A loaded resonant circuit.

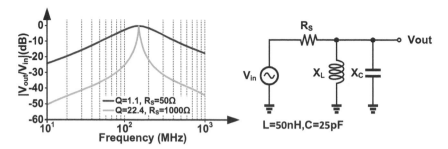

Figure 7.21: The effect of source impedance on Q.

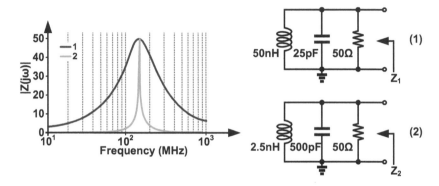

Figure 7.22: The effect of inductance on Q.

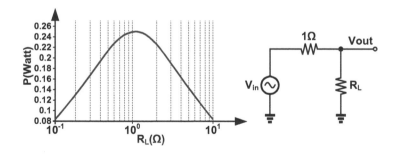

Figure 7.23: Output power characteristic for a DC circuit.

impedances occurs when the two have equal values. Depicted in Figure 7.23 is the power that is delivered to a variable load for a 1 V source.

Now, let's consider the more general case depicted in Figure 7.24, which includes a source impedance having both resistive and inductive components, and a load consisting of a capacitive and a resistive component. If the capacitor and the inductor resonate at some frequency, the circuit then reduces to that shown in Figure 7.22, and maximum power transmission is achieved by choosing equal load and source impedances. That is the total load impedance should be the complex conjugate of the

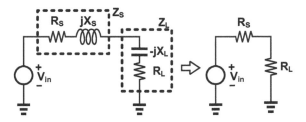

Figure 7.24: The general case of maximum power transmission to the load.

source impedance ($Z_L = Z_S^*$). In order to achieve this condition, we now proceed to introduce some popular matching structures, known as L-sections. Different variants of an L-section are depicted in Figure 7.25.

We analyze, in the final section of this chapter, all the structures depicted in Figure 7.25 using a powerful tool called Smith chart. Now, we consider the basic impedance matching methods. See Figure 7.27.

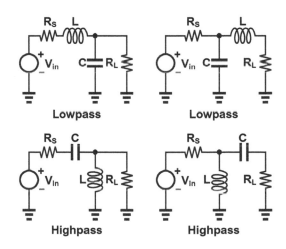

Figure 7.25: Four possible variants of an L-section matching network.

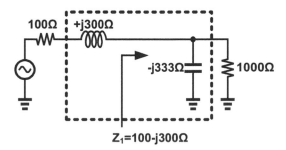

Figure 7.26: An example of an L-section matching network.

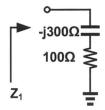

Figure 7.27: The equivalent circuit of the network shown in Figure 7.26 at the operating frequency.

Firstly assume that the source is terminated to an $R_L = 100\,\Omega$ load and we calculate the power delivered to this load. Secondly assume that in Figure 7.26, the LC matching network (L-section) is omitted and the source is terminated to an $R_L = 1000\,\Omega$ load and we recalculate the power delivered to this load. The ratio of this latter power to the former power is defined as the mismatch loss.

$$L = -10\log\frac{P_{\text{mismatch}}}{P_{\text{match}}} = -10\log\frac{\frac{v_{\text{mismatch}}^2}{R_{\text{mismatch}}}}{\frac{v_{\text{match}}^2}{R_{\text{match}}}} = -10\log\left(\frac{v_{\text{mismatch}}}{v_{\text{match}}}\right)^2\frac{R_{\text{match}}}{R_{\text{mismatch}}} = \tag{7.115}$$

$$-10\log\left(\frac{0.909v_{\text{in}}}{0.5v_{\text{in}}}\right)^2\frac{100}{1000} = 4.8\,\text{dB}$$

As demonstrated by Equation 7.115, the loss is equal to 4.8 dB. Now, if we employ the circuit shown in Figure 7.26, the impedance seen at the load, Z_1, will amount to $100 - j300\,\Omega$. Let's calculate the equivalent impedance of the RC section as

$$Z = \frac{jX_C R_L}{jX_C + R_L} = \frac{-j333(1000)}{-j333 + 1000} = 100 - j300 \tag{7.116}$$

This is equivalent to the circuit shown in Figure 7.27 at the operating frequency. Now, if an inductor of $300\,j\Omega$ is added in series to this network, then the load impedance will be purely resistive and equal to $100\,\Omega$. This is conceptually depicted in Figure 7.28. What we have learned from the concept of impedance matching thus far suggests that a parallel element can lower the real part of the impedance level of the load, which for the

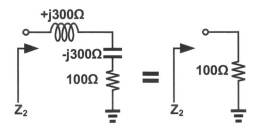

Figure 7.28: Impedance matching by adding an inductor.

case studied here, a $1000\,\Omega$ load was lowered to $100\,\Omega$. The series element can resonate with the other energy-storing element, resulting in a purely resistive impedance equal to the source impedance. This, however, points to a possible limitation of series or shunt resonant matching network, which is their narrowband response due to single frequency matching or resonance.

7.9.1 A New Definition of the Quality Factor

As we discussed in the previous section, Q factor plays a crucial role in matching networks of all kind. We consider the Q factor of a reactance (either inductive or capacitive) either in series or in parallel definition, as follows

$$Q_S = \frac{X_S(\omega_0)}{R_S}, \quad Q_P = \frac{R_L}{X_P(\omega_0)} \tag{7.117}$$

where Q_s and Q_p are the series and parallel quality factors, and X_s and X_p are the series and the parallel reactances, respectively. R_s and R_p are the equivalent series and parallel resistances of the considered reactance, respectively.

As shown in Figure 7.29, equating the series and the parallel impedances, we arrive at

$$R_s + jX_s = \frac{R_p(jX_p)}{R_p + jX_p} \tag{7.118}$$

Equating the real and the imaginary parts of Equation 7.118 to each other, we have

$$R_s = \frac{R_p X_p^{\,2}}{R_p^{\,2} + X_p^{\,2}} \tag{7.119}$$

and

$$X_s = \frac{R_p^{\,2} X_p}{R_p^{\,2} + X_p^{\,2}} \tag{7.120}$$

which represent the equivalent series and parallel impedances, respectively. Given the fact that $Q_P = Q_S$, it follows from Equation 7.119 that

$$\frac{R_p}{R_s} = 1 + Q_p^{\,2} = 1 + Q_s^{\,2} \tag{7.121}$$

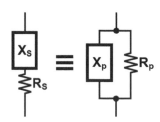

Figure 7.29: The equivalent circuit of a reactance with a limited quality factor.

Therefore, the quality factor is given by

$$Q_s = Q_p = \sqrt{\frac{R_p}{R_s} - 1} \tag{7.122}$$

Now, for an L-section matching, using two different reactances, one can use the source resistance, the load resistance, and Equation 7.122, as the starting point and then one chooses two reactances of the opposite signs to realize the conjugate matching condition. It is obvious that this matching procedure is applicable once $R_p > R_s$. This principal matching procedure is demonstrated in Figure 7.30.

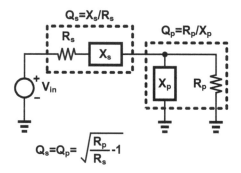

Figure 7.30: The definition of Q in a matching network.

Example 7.4 Consider the circuit shown in Figure 7.31. Design a matching network at 100 MHz which matches a source impedance of 100 Ω to the load impedance of 1000 Ω. Also, you may assume that the circuit is DC coupled.

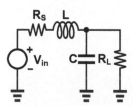

Figure 7.31: An L-section matching network.

Solution:
We first compute the quality factor from Equation 7.122 as

$$Q_s = Q_p = \sqrt{\frac{R_p}{R_s} - 1} = \sqrt{\frac{1000}{100} - 1} = 3 \tag{7.123}$$

The series and parallel reactances are also given by

$$X_s = Q_s R_s = 3 \times 100 = 300 \, \Omega \tag{7.124}$$

and

$$X_p = \frac{-R_p}{Q_p} = \frac{-1000}{3} = -333.3 \, \Omega \tag{7.125}$$

respectively. The values of the inductor and the capacitor can both be easily calculated as

$$L = \frac{X_s}{\omega} = \frac{300}{2\pi(100 \times 10^6)} = 477 \, \text{nH} \tag{7.126}$$

and

$$C = -\frac{1}{\omega X_p} = \frac{1}{2\pi(100 \times 10^6)(333.3)} = 4.8 \, \text{pF} \tag{7.127}$$

respectively. ∎

Now that we have learned the impedance matching concept, we turn to complex load and source impedances. This can be the case while circuits are being interfaced to real-world impedances like antennas, mixers, T-lines, transistors, and other components, where their input impedance is both complex and frequency-dependent. A possible solution can be absorption of impedances within the matching network. This can be done by absorbing the stray capacitances into parallel matching capacitors, and by absorbing the stray inductances into series matching inductors. We elaborate on this point in Example 7.5 in which series resonance occurs at the desired frequency.

Example 7.5 Consider the circuit depicted in Figure 7.32. Using impedance absorption technique, match the source and load impedances at 100 MHz.

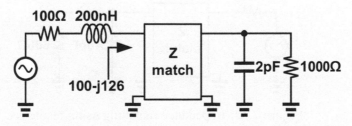

Figure 7.32: Impedance matching at the source and the load.

Solution:
The first step is to consider only the real part of the source impedance which is $100\,\Omega$ at $100\,$MHz and the real part of the load impedance which is $1000\,\Omega$ at $100\,$MHz. Using the numerical results of Example 7.4, the matching network will look like that depicted in Figure 7.33. The matching of the $1000\,\Omega$ to the $100\,\Omega$ source would need a $477\,$nH series inductor and a parallel $4.8\,$pF shunt capacitor as demonstrated in Example 7.4. By subtracting the existing $200\,$nH inductance from the $477\,$nH needed one and subtracting the $2\,$pF capacitance from the $4.8\,$pF needed one, we obtain the resultant values shown in the dashed recangle in Figure 7.33.

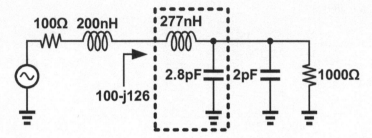

Figure 7.33: The proper matching network for complex source and complex load matching.

Note that here we had positive values for the series inductance and the parallel capacitance, given the fact that the total needed inductance and the total needed capacitance in the matching network were larger than the stray inductance and the stray capacitance, respectively. ∎

In Example 7.6, we discuss impedance matching using resonating load.

Example 7.6 Design a matching network at $75\,$MHz for the circuit shown in Figure 7.34. Employ DC blocking.

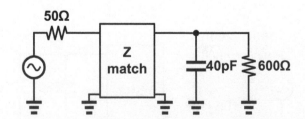

Figure 7.34: Impedance matching using resonance.

Solution:
The desired matching network which employs DC blocking is illustrated in Figure 7.35.

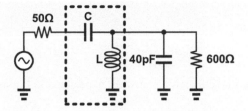

Figure 7.35: Impedance matching network using a resonating load.

We first place an inductor in parallel with the 40 pF stray capacitance so that they resonate at 75 MHz. It then follows that

$$L = \frac{1}{\omega^2 C} = \frac{1}{\left(2\pi(75 \times 10^6)\right)^2 \times 40 \times 10^{-12}} = 112.6\,\text{nH} \tag{7.128}$$

A part of the matching network is illustrated in Figure 7.36.

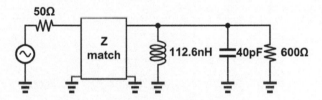

Figure 7.36: A part of the matching network.

Now that the stray capacitance has been tuned out, we should match the 50 Ω source to 600 Ω load resistor. Hence,

$$Q_s = Q_p = \sqrt{\frac{R_p}{R_s} - 1} = \sqrt{\frac{60}{5} - 1} = 3.32 \tag{7.129}$$

It then follows from Equation 7.117 that

$$X_s = -Q_s R_s = -3.32 \times 50 = -166\,\Omega \tag{7.130}$$

and

$$X_p = \frac{R_p}{Q_p} = \frac{600}{3.32} = 181\,\Omega \tag{7.131}$$

which corresponds to

$$L = \frac{X_p}{\omega} = \frac{181}{2\pi(75 \times 10^6)} = 384\,\text{nH} \tag{7.132}$$

and

$$C = -\frac{1}{\omega X_s} = \frac{1}{2\pi(75 \times 10^6)(166)} = 12.7\,\text{pF} \tag{7.133}$$

respectively, and the matching network is depicted in Figure 7.37.

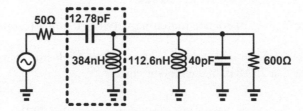

Figure 7.37: The designed matching network.

Now, if we replace two parallel inductors with one, the final matching network will be as depicted in Figure 7.38.

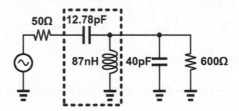

Figure 7.38: The final matching network in Example 7.11.

■

Another method which is used for impedance matching is known as the π and T method, which is illustrated in Figure 7.39.

A π network can be formed by cascading two L-sections with a virtual resistor in between, as shown in Figure 7.40.

The negative signs in the parallel reactances are just for representation purposes; however, the important point is that they differ with their series counterparts in sign. Therefore, if X_{p1} is a capacitor, $X_{s,1}$ must be an inductor and *vice versa*. Similarly, if X_{p2}

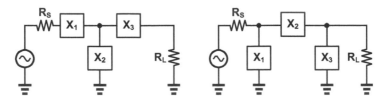

Figure 7.39: π and T matching networks.

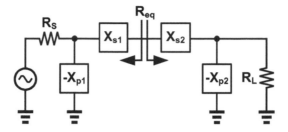

Figure 7.40: π matching network.

is a capacitor, then X_{s2} must be an inductor and *vice versa*. Now, from Equation 7.122, it follows that

$$Q = \sqrt{\frac{R_H}{R_{eq}} - 1} \tag{7.134}$$

where R_{II} is the larger resistor on the source and the load side, and R_{eq} is the virtual resistor between the two networks. Evidently, R_H should be larger than R_{eq}.

Example 7.7 Consider the circuit shown in Figure 7.40. Design a π matching network such that the input impedance of $100\,\Omega$ is matched to a $1000\,\Omega$ load impedance. The loaded Q of each L-section is equal to 15.

Solution:
We first consider Q from Equation 7.134, and calculate the virtual resistor as

$$R_{eq} = \frac{R_H}{Q_L^2 + 1} = \frac{1000}{226} = 4.42\,\Omega \tag{7.135}$$

We can then use Equation 7.117, the series and parallel reactances, as depicted in Figure 7.41, can be calculated as

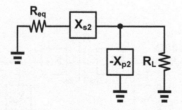

Figure 7.41: The right-hand equivalent half-circuit of the π matching network.

$$X_{s2} = Q_{s2}R_{eq} = 15 \times 4.42 = 66.3\,\Omega \tag{7.136}$$

Note that $Q_{s2} = Q_{p2}$ in this section. Therefore

$$X_{p2} = \frac{R_L}{Q_{p2}} = \frac{1000}{15} = 66.7\,\Omega \qquad (7.137)$$

Now, we analyze the left-hand L-section. The quality factor for this circuit, as depicted in Figure 7.42, can be found as

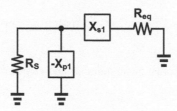

Figure 7.42: The left-hand equivalent half-circuit of the π matching network.

$$Q_{s1} = Q_{p1} = \sqrt{\frac{R_S}{R_{eq}} - 1} = \sqrt{\frac{100}{4.42} - 1} = 4.6 \qquad (7.138)$$

and using Equation 7.117, the parallel and series reactances can be calculated as

$$X_{s1} = Q_{s1}R_{eq} = 4.6 \times 4.42 = 20.51\,\Omega \qquad (7.139)$$

and

$$X_{p1} = \frac{R_S}{Q_{p1}} = \frac{100}{4.6} = 21.7\,\Omega \qquad (7.140)$$

Now that we have the values of all the reactances, the overall matching network can be chosen as in Figure 7.43 in a different version depending on the chosen sign of the reactances.

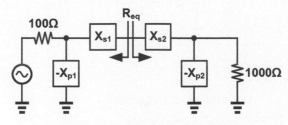

Figure 7.43: π matching network.

Figure 7.44 shows a DC-coupled (low-pass) version of the π matching network.

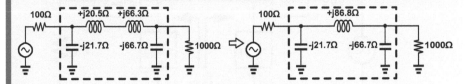

Figure 7.44: Low-pass version of the π matching network.

The main point to remember here is that the reactances in each branch of L-sections have a different sign. Therefore, the other structures can also be used for matching, all of which are depicted in Figure 7.45. In total, four combinations of π section matching network (depending on low-pass/DC coupled or high-pass/AC coupled L-section being chosen) are possible to realize.

It could be verified through circuit simulations that the overall quality factor and therefore the corresponding bandwidth of this matching circuit could be found through the following relation

$$Q_{\text{total}} = \frac{Q_{s1} + Q_{s2}}{2} = \frac{Q_{p1} + Q_{p2}}{2} \tag{7.141}$$

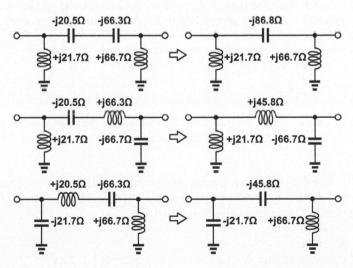

Figure 7.45: Variants of the matching network with inductors and capacitors in a π structure.

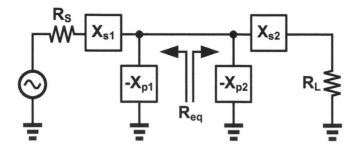

Figure 7.46: T matching network.

The reader may wonder that which one of the parameters should be chosen for the matching network. In order to pick one of them, it should be noted that the following parameters might be taken into account: DC coupling or AC coupling, the values of stray capacitances or stray inductances to be tuned out (to be deduced from the matching network), harmonic voltages or harmonic currents to be eliminated, the frequency response of the matching network, and finally realizable values for the matching elements.

We now proceed by discussing the T matching network. Matching with a T network is very much similar to its π counterpart and it is done by two L-sections as well. With the exception that both of the L-sections are matched to a larger value of virtual resistance and the horizontal part of the L-section, that is, the series reactance comes in series with the load or the source. As a result, the parallel arms of the two L-sections will eventually appear in parallel. The T matching network is depicted in Figure 7.46.

Let's now return to our definition of Q as

$$Q = \sqrt{\frac{R_{eq}}{R_1} - 1} \tag{7.142}$$

where R_{eq} is the virtual resistance to be matched to the load and the source, and R_1 is the minimum of the load and the source resistances. Example 7.8 clarifies this point.

Example 7.8 Consider the circuit illustrated in Figure 7.46. Design four different matching networks which match a $10\,\Omega$ source impedance to a $50\,\Omega$ load impedance. The loaded Q is chosen to be 10.

Solution:
The virtual resistance can be found from Equation 7.142 as

$$R_{eq} = R_1(Q_L^2 + 1) = 10(101) = 1010\,\Omega \tag{7.143}$$

and the values of the series and the parallel reactances, as depicted in Figure 7.47, can be calculated using Equation 7.117 as

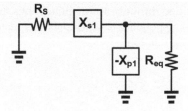

Figure 7.47: The left-hand equivalent half-circuit of the T matching network.

$$X_{s1} = Q_{s1}R_S = 10 \times 10 = 100\,\Omega \qquad (7.144)$$
$$X_{p1} = \frac{R_{eq}}{Q_{p1}} = \frac{1010}{10} = 101\,\Omega \qquad (7.145)$$

Now, it follows from Equation 7.142 that the quality factor of the right-hand L-section, as depicted in Figure 7.48, is

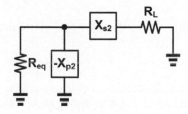

Figure 7.48: The right-hand equivalent half-circuit of the T matching network.

$$Q_{s2} = Q_{p2} = \sqrt{\frac{R_{eq}}{R_L} - 1} = \sqrt{\frac{1010}{50} - 1} = 4.4 \qquad (7.146)$$

and using Equation 7.142, we have

$$X_{s2} = Q_{s2}R_L = 4.4 \times 50 = 220\,\Omega \qquad (7.147)$$

and

$$X_{p2} = \frac{R_{eq}}{Q_{p2}} = \frac{1010}{4.4} = 230\,\Omega \qquad (7.148)$$

Finally, the four possible T matching networks are depicted in Figure 7.49. Note that the final parallel reactance is the resultant parallel combination of the two middle reactances of each L-section.

It could be verified through circuit simulations that the overall quality factor and therefore the corresponding bandwidth of this matching circuit could be found through the following relation

$$Q_{\text{total}} = \frac{Q_{s1} + Q_{s2}}{2} = \frac{Q_{p1} + Q_{p2}}{2} \qquad (7.149)$$

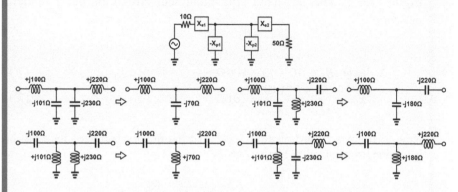

Figure 7.49: Variants of T matching network.

As in the previous example, here each one of the four variants could be chosen according to the following criteria: DC coupling or AC coupling, the required frequency response, low-pass or high-pass response, the requirement to deduce the stray capacitances or inductances from the matching network, harmonic voltages or harmonic currents to be eliminated, and finally realizable values for the matching elements. Thus far, we have learned to match any passive load to any passive source impedance using L, π, or T networks. We should have realized in this section, that matching networks are indispensable parts of any radio-frequency circuit. In the next section, we introduce a powerful tool which could be used for matching networks calculations, that is the Smith chart application.

7.10 Smith Chart Mapping

First developed in 1930 at Bell Labs by Philip Smith, the Smith chart was first thought of as a simple tool which would circumvent lengthy impedance matching calculations in RF circuits. Before dealing with the equations, we should note that Smith chart is nothing but a mapping of impedances to the reflection coefficient plane. We now proceed by carrying out the required derivations to reach to this chart. One may recall from section 7.2 that the reflection coefficient can be written as

$$\Gamma_L = \frac{Z_L - Z_0}{Z_L + Z_0} = \frac{\frac{Z_L}{Z_0} - 1}{\frac{Z_L}{Z_0} + 1} \qquad (7.150)$$

Now, if we define Z_L/Z_0 as the normalized impedance and writing Γ_L in terms of its real and imaginary parts, it follows that

$$\Gamma_r + j\Gamma_i = \frac{Z_n - 1}{Z_n + 1} = \frac{R + jX - 1}{R + jX + 1} \tag{7.151}$$

In order to find an explicit expression for the Smith chart contours, we rewrite Equation 7.151 as

$$Z_n = \frac{1 + \Gamma_L}{1 - \Gamma_L} \tag{7.152}$$

We proceed by replacing the real and the imaginary parts of the reflection coefficient and the normalized impedance in Equation 7.152, hence,

$$R + jX = \frac{1 + \Gamma_r + j\Gamma_i}{1 - \Gamma_r - j\Gamma_i} \tag{7.153}$$

By multiplying the right-hand side of Equation 7.153 by the complex conjugate of the denominator, we arrive at

$$R + jX = \frac{1 + \Gamma_r + j\Gamma_i}{1 - \Gamma_r - j\Gamma_i} \times \frac{1 - \Gamma_r + j\Gamma_i}{1 - \Gamma_r + j\Gamma_i} = \frac{1 - \Gamma_r^2 - \Gamma_i^2 + 2j\Gamma_i}{(1 - \Gamma_r)^2 + \Gamma_i^2} \tag{7.154}$$

Equating the real and the imaginary parts of Equation 7.154, we have

$$R = \frac{1 - \Gamma_r^2 - \Gamma_i^2}{(1 - \Gamma_r)^2 + \Gamma_i^2} \tag{7.155}$$

and

$$X = \frac{2\Gamma_i}{(1 - \Gamma_r)^2 + \Gamma_i^2} \tag{7.156}$$

Equations 7.155 and 7.156 play a crucial role in finding an expression for the Smith chart contours. Rewriting Equations 7.155 and 7.156 in two complete square forms in terms of Γ_r and Γ_i, we arrive at

$$\left(\Gamma_r - \frac{R}{1 + R}\right)^2 + \Gamma_i^2 = \left(\frac{1}{1 + R}\right)^2 \tag{7.157}$$

and

$$(\Gamma_r - 1)^2 + \left(\Gamma_i - \frac{1}{X}\right)^2 = \left(\frac{1}{X}\right)^2 \tag{7.158}$$

Equations 7.157 and 7.158 represent constant resistance and constant reactance contours in the Γ_r-Γ_i plane. Each point in this plane corresponds to a unique reflection

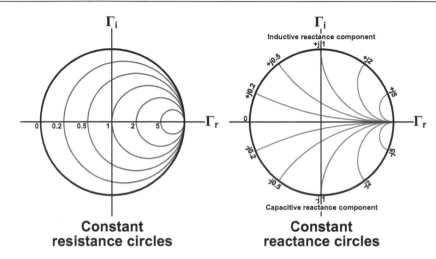

Figure 7.50: Constant-resistance and constant-reactance circles within $|\Gamma| \le 1$ plane.

coefficient and therefore a unique impedance in that plane. Equation 7.157 represents a group of circles centered at $(x,y) = (R/(1+R),0)$ and with a radius of $r = 1/(1+R)$. Equation 7.158 represents a group of circles centered at $(x,y) = (1,1/X)$ and with a radius of $r = 1/X$. Drawing the two groups of circles for loads having a positive real part (within a circle centered at the origin and with the unity radius) results in the Smith chart. The intersection of the reactance and the resistance circles results in a specific impedance in the Γ_r-Γ_i plane. Constant-resistance and constant-reactance circles are shown in Figure 7.50.

Now, we mention a few points regarding the Smith chart. (1) The intersections of constant-resistance circles with the Γ_r axis (x-axis) are purely real impedance $(R + j0)$ points. (2) The intersections of constant-reactance circles with the $|\Gamma| = 1$ circle are the purely imaginary impedance $(0 + jX)$ points. (3) The center of the chart corresponds to $R = 1$ point. (4) The perimeter of the chart corresponds to $R = 0$ points. (5) The extreme right-hand point on the x-axis that is $\Gamma = 1$ corresponds to an open circuit. (6) The extreme left-hand point on the x-axis that is $\Gamma = -1$ corresponds to a short circuit. (7) Load impedances having a negative real part would be projected out of the chart, that is, out of the $|\Gamma| = 1$ circle. Evidently for these impedances the corresponding reflection coefficient would have an amplitude greater than unity.

Example 7.9 Show $Z_1 = 50 + j50$ and $Z_2 = 50 - j50$ on the Smith chart. Assume the line impedance is equal to $50\,\Omega$.

Solution:
We begin by calculating the normalized values of the impedances,

$$Z_{1n} = \frac{Z_1}{50} = 1 + j \qquad (7.159)$$

and

$$Z_{2n} = \frac{Z_2}{50} = 1 - j \qquad (7.160)$$

which can be found using the Smith chart by finding the intersections of $R = 1$ and $X = 1$, and $R = 1$ and $X = -1$ circles, respectively. This procedure is shown in Figure 7.51.

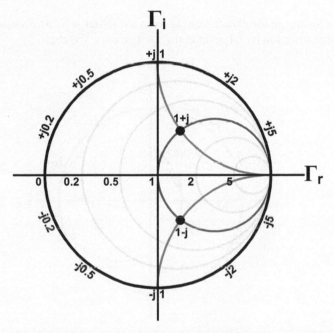

Figure 7.51: Representation of the two impedances in the Smith chart.

■

Recalling Equation 7.72, one can write the impedance value at a distance d from the load as follows

$$Z(-d) = Z_0 \left(\frac{1 + \Gamma(-d)}{1 - \Gamma(-d)} \right) = Z_0 \left(\frac{1 + \Gamma_L e^{-2j\beta\ell}}{1 - \Gamma_L e^{-2j\beta\ell}} \right) \qquad (7.161)$$

As it is evident from Equation 7.161, if one is moving toward the generator, by rotating an amount of $2\beta\ell$ clockwise from the initial impedance point, he/she would arrive at the new impedance point on the chart. Otherwise, if one is moving from the initial point toward the load, by rotating an amount of $2\beta\ell$ counterclockwise, he/she would arrive at the new impedance point on the chart.

Example 7.10 Consider the normalized impedance $Z = 0.5 + j0.7$ and assume that a $-1j$ reactance is added in series to this impedance. Sketch the initial and the resulting impedances on the Smith chart.

Solution:
The initial and the final points are shown on the Smith chart in Figure 7.52.

$$Z' = Z - j1 = 0.5 - j0.3 \tag{7.162}$$

Now moving on the $R = 0.5$ circle from the initial $X = j0.7$ toward $X = -j0.3$, we come from the initial point to the final point on the chart.

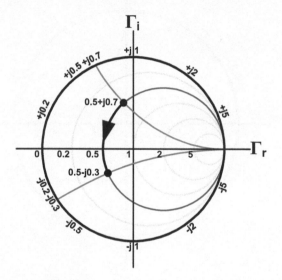

Figure 7.52: Representation of the effect of adding a series capacitive reactance in the Smith chart. ∎

We now turn our attention to Admittance Smith chart, that is, constant-conductance and constant-susceptance circles. Recall that

$$\Gamma = \frac{\frac{1}{Y_n} - 1}{\frac{1}{Y_n} + 1} = \frac{1 - Y_n}{1 + Y_n} \tag{7.163}$$

If we imagine having a load whose normalized impedance is inverse of the given normalized admittance ($Z_n = \frac{1}{Y_n}$), the Admittance Smith chart could be obtained by the same procedure as the Impedance Smith charts just by replacing Γ by $-\Gamma$ as shown in Equation 7.164. In other words, if we rotate the impedance Smith chart by 180°, the same chart would represent the Admittance Smith chart.

$$\Gamma_Z = \frac{\frac{1}{Y_n} - 1}{\frac{1}{Y_n} + 1} = \frac{1 - Y_n}{1 + Y_n} = -\Gamma \tag{7.164}$$

As such, the conductance and the susceptance circles can be derived from the resistance and the reactance circles, respectively. This procedure is useful in the sense that once we desire to add any reactance or resistance in series, we use the Impedance Smith chart and once we desire to add any susceptance or conductance in parallel, we use the Admittance Smith chart. A complete Smith chart is shown in Figure 7.53, in which the blue and the red circles denote the impedance and admittance circles, respectively.

Care should be taken by the reader while using the Smith chart. Notice that wherever on the chart the reactance is positive (upper half of the chart), then the susceptance is negative and wherever on the chart the reactance is negative (lower half of the chart), then the susceptance is positive. Furthermore, wherever on the chart the normalized resistance is greater than unity, then the normalized conductance is less than unity and wherever the normalized conductance is greater than unity, then the normalized resistance is less than unity.

7.10.1 Some simple application rules while using the Smith chart

Rule 1: For matching, we generally use circular paths on the chart.

Rule 2: Rotating clockwise on a constant resistance circle is equivalent to adding a series inductance.

Rule 3: Rotating counterclockwise on a constant resistance circle is equivalent to adding a series capacitance.

Rule 4: Rotating clockwise on a constant conductance circle is equivalent to adding a shunt capacitance.

Rule 5: Rotating counterclockwise on a constant conductance circle is equivalent to adding a shunt inductance.

Rule 6: Rotating clockwise on a constant radius circle about the chart center is equivalent to approaching the generator on the transmission line.

Rule 7: Rotating counterclockwise on a constant radius circle about the chart center is equivalent to approaching the load on the transmission line.

Rule 8: Jumping from a constant resistance circle to another constant resistance circle (on the real axis) is equivalent to putting a transformer on the transmission line, meaning that increasing or decreasing the input resistance (depending on the sense of the jump).

Rule 9: Jumping from a constant conductance circle to another constant conductance circle (on the real axis) is equivalent to putting a transformer on the transmission line, meaning that increasing or decreasing the input conductance (depending on the sense of the jump).

Rule 10: Just rotating by $180°$, one can read the admittance value on the same chart (once he/she has recorded the impedance value on the impedance chart). Otherwise, just rotating by $180°$, one can read the impedance value on the same chart (once he/she has recorded the admittance value on the admittance chart).

Rule 11: Using quarter wave transformers, one can travel clockwise on a half-circle from the starting point at the load impedance to the desired input impedance at the end point. As seen in Figure 7.54, this way one can transform a high impedance (point A) to a low impedance (point B), or one can transform a low impedance (point C) to a high impedance (point D).

The application of these rules is illustrated in Figure 7.54. We now revisit example 7.11 and repeat it using the Smith chart.

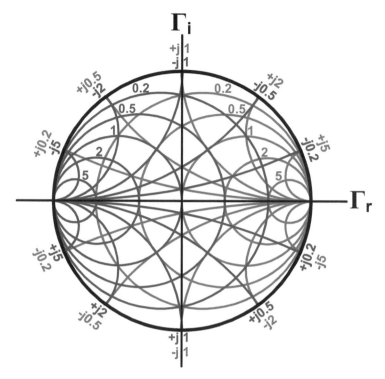

Figure 7.53: The complete Smith chart.

Example 7.11 Consider the circuit depicted in Figure 7.34. Use the Smith chart to determine a matching network at 75 MHz which incorporates DC current blocking.

Solution:

The load impedance consisting of a parallel resistor and a capacitor is depicted on the admittance chart as $Y_L = 1.7 + j18.9$ mS or in normalized form $\bar{Y}_L = 0.085 + j0.945$ or its normalized impedance is depicted as $\bar{Z}_L = 0.095 - j1.05$ at point 1. Now, we move on the constant normalized conductance circle of 0.085, on the admittance chart such that we come across the $R = 1$ circle at point 2. We now record the difference in normalized susceptances of point 2 and point 1 as

$$jB_2 - jB_1 = j0.945 - (-j0.265) = j1.21 \tag{7.165}$$

The inductive susceptance to be added in parallel with the load would become then

$$B_t = 1.21 \times 20\,\text{mS} = 24.2\,\text{mS} \tag{7.166}$$

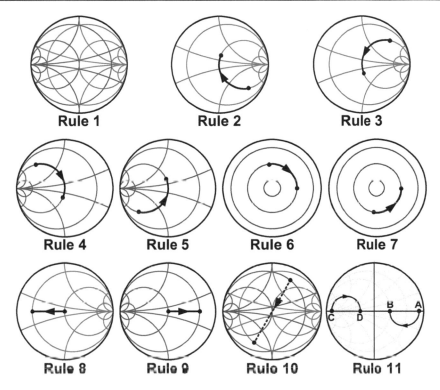

Figure 7.54: The Smith chart rules and applications.

$$L = \frac{1}{0.0242\omega} = \frac{1}{0.0242 \times 2\pi \left(75 \times 10^6\right)} = 87.7\,\text{nH} \qquad (7.167)$$

At point 2, we record the reactance $\bar{X}_2 = j3.4$ or $X_2 = 170\,\Omega$. We should add a capacitive reactance of $-170\,\Omega$ to tune out this inductance as follows

$$C = \frac{1}{170\omega} = \frac{1}{170 \times 2\pi \left(75 \times 10^6\right)} = 12.5\,\text{pF} \qquad (7.168)$$

With this series capacitance, we come to the center of the chart at point 3 as depicted in Figure 7.55.

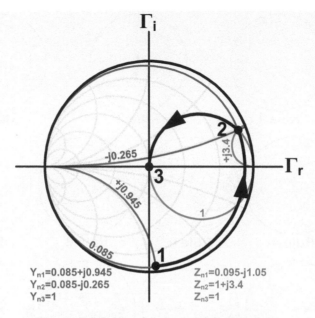

$Y_{n1}=0.085+j0.945$
$Y_{n2}=0.085-j0.265$
$Y_{n3}=1$

$Z_{n1}=0.095-j1.05$
$Z_{n2}=1+j3.4$
$Z_{n3}=1$

Figure 7.55: Matching on the Smith chart.

■

Example 7.12 Find a matching network to match the load impedance illustrated in Figure 7.56 to $50\,\Omega$ at $2.5\,\text{GHz}$. The reference impedance is $50\,\Omega$.

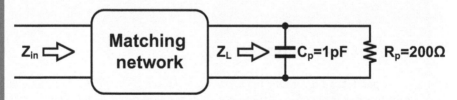

Figure 7.56: The load impedance of Example 7.12.

Solution:
We first compute the normalized admittance at $2.5\,\text{GHz}$

$$Y_{n1} = \frac{Y_L}{Y_0} = \frac{\left(1/200 + j2\pi(2.5\times10^9)10\times10^{-12}\right)}{0.02} = 0.25 + j0.7854 \quad (7.169)$$

This admittance is shown in Figure 7.57 on the Smith chart.

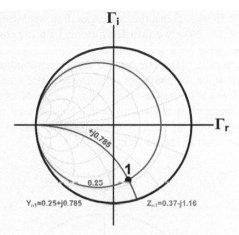

Figure 7.57: Representation of the load admittance on the Smith chart.

As it was discussed earlier, matching can be achieved using an L-section, as depicted in Figure 7.58.

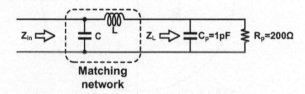

Figure 7.58: Matching based on an L-section.

Figure 7.59 explains how the proper value of the inductor can be found using the Smith chart.

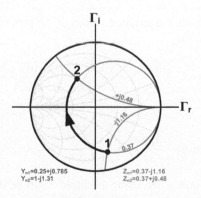

Figure 7.59: Adding an inductor to the load.

In Figure 7.59, we record the corresponding load impedance on the chart as $Z_{n1} = 0.37 + j1.16$. Now we turn from point 1 on the constant resistance circle ($R = 0.37$) clockwise until we intersect the constant conductance circle ($G = 1$) at $1 - j1.31$, point 2. Here we record the corresponding impedance value of $Z_{n2} = 0.37 + j0.48$ on the chart. As such, the required series reactance to be added to reach point 2 would be $j0.48 - (-j1.16) = j1.64$. Now to compute the required parallel susceptance, we turn on the constant conductance circle ($G = 1$) clockwise from point 2 to point 3, by adding a susceptance value of $j1.31$, where we reach the center of the chart as shown in Figure 7.60.

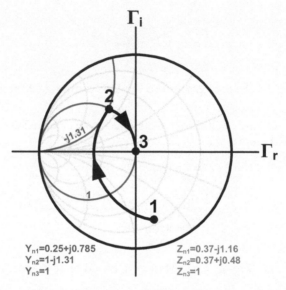

$Y_{n1}=0.25+j0.785$ $Z_{n1}=0.37-j1.16$
$Y_{n2}=1-j1.31$ $Z_{n2}=0.37+j0.48$
$Y_{n3}=1$ $Z_{n3}=1$

Figure 7.60: Adding a capacitor to the load to reach the point 3 from point 2.

Now, we compute the required values of the series inductance and the parallel capacitance as follows

$$\bar{X}_L = \frac{L\omega}{Z_0} \Rightarrow L = \frac{\bar{X}_L Z_0}{\omega} = \frac{(0.48 - (-1.16)) \times 50}{2\pi \times 2.5 \times 10^9} = 5.2\,\text{nH} \qquad (7.170)$$

and

$$\bar{Y}_C = \frac{C\omega}{Y_0} \Rightarrow C = \frac{\bar{Y}_C Y_0}{\omega} = \frac{1.31(0.02)}{2\pi \times 2.5 \times 10^9} = 1.67\,\text{pF} \qquad (7.171)$$

The reader might notice that using the Smith chart he/she would need only simple arithmetic calculations to design the matching circuit.   ∎

Example 7.13 The structure depicted in Figure 7.61 can be shown to match $1\,k\Omega$ to $50\,\Omega$ at $2\,\text{GHz}$. The relative permittivity is considered as unity. Verify this fact using the Smith chart.

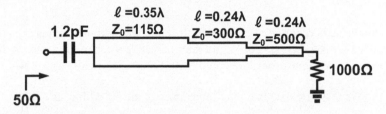

Figure 7.61: The matching network in Example 7.13.

Solution:
We first record the load impedance on the Smith chart for the first section of the transmission line:

$$Z_{n0} = 1000/500 = 2 \tag{7.172}$$

From point 0 on the Smith chart, we turn clockwise, for 0.24λ (or $173°$) to reach the point 1 and we record the input impedance as $0.501 - j0.047$, at point 1. For moving along the $300\,\Omega$ line, we normalized this impedance by the new reference impedance of $300\,\Omega$ as

$$Z_{n2} = (0.501 - j0.047) \times \frac{500}{300} = 0.835 - j0.0783 \tag{7.173}$$

We report this value on the chart as point 2. Now we turn clockwise about the center of the chart by 0.24λ (or $173°$) to reach the point 3. We record the value of the impedance at this point as $1.167 + j0.135$. For moving along the $115\,\Omega$ line, we normalized this impedance by the new reference impedance of $115\,\Omega$ as

$$Z_{n4} = (1.167 + j0.135) \times \frac{300}{115} = 3.047 + j0.351 \tag{7.174}$$

We report this value on the chart as point 4. Now we turn clockwise about the center of the chart by 0.35λ (or $252°$) to reach the point 5 on the chart. We record the value of the impedance at this point as $0.445 + j0.569$. We denormalize the value of the impedance here to achieve

$$Z_5 = (0.445 + j0.569) \times 115 = 51.26 + j65.4\,\Omega \tag{7.175}$$

We normalize the above value of impedance by $50\,\Omega$ to achieve a new value of normalized impedance at point 6:

$$Z_{n6} = \frac{(51.26 + j65.4)}{50} = 1.024 + j1.308 \tag{7.176}$$

This value is reported as point 6 on the chart. By adding a series capacitive reactance of $-j65.4$ ($\frac{1}{C_1\omega} = 65.4$), we approximately reach to the center of the chart at point 7. The same procedure could be followed analytically as follows:

We first calculate the input impedance of the $500\,\Omega$ line as

$$\begin{aligned} Z_1 &= Z_0\left(\frac{Z_L + jZ_0\tan(\beta\ell)}{Z_0 + jZ_L\tan(\beta\ell)}\right) = 500\left(\frac{1000 + j500\tan\left(\frac{2\pi}{\lambda}0.24\lambda\right)}{500 + j1000\tan\left(\frac{2\pi}{\lambda}0.24\lambda\right)}\right) \\ &= 250.74 - j23.57 \end{aligned} \tag{7.177}$$

and for the second section, we have

$$\begin{aligned} Z_3 &= Z_0\left(\frac{Z_1 + jZ_0\tan(\beta\ell)}{Z_0 + jZ_1\tan(\beta\ell)}\right) \\ &= 300\left(\frac{(250.74 - j23.57) + j300\tan\left(\frac{2\pi}{\lambda}0.24\lambda\right)}{300 + j(250.74 - j23.57)\tan\left(\frac{2\pi}{\lambda}0.24\lambda\right)}\right) \\ &= 350.32 + 40.427j \end{aligned} \tag{7.178}$$

and finally, the input impedance of the $115\,\Omega$ section is equal to

$$\begin{aligned} Z_5 &= Z_0\left(\frac{Z_3 + jZ_0\tan(\beta\ell)}{Z_0 + jZ_3\tan(\beta\ell)}\right) \\ &= 115\left(\frac{(350.32 + 40.427j) + j115\tan\left(\frac{2\pi}{\lambda}0.35\lambda\right)}{115 + j(350.32 + 40.427j)\tan\left(\frac{2\pi}{\lambda}0.35\lambda\right)}\right) \\ &= 51.26 + j65.4 \end{aligned} \tag{7.179}$$

By adding the series $1.21\,\text{pF}$ capacitor, the final input impedance becomes

$$Z_7 = Z_5 - j65.4 = 51.26\,\Omega \tag{7.180}$$

which is a good match for a $50\,\Omega$ source.

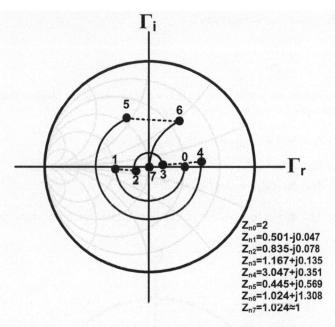

$Z_{n0}=2$
$Z_{n1}=0.501\text{-}j0.047$
$Z_{n2}=0.835\text{-}j0.078$
$Z_{n3}=1.167+j0.135$
$Z_{n4}=3.047+j0.351$
$Z_{n5}=0.445+j0.569$
$Z_{n6}=1.024+j1.308$
$Z_{n7}=1.024\approx1$

Figure 7.62: The steps of matching in the circuit shown in Figure 7.61.

Example 7.14 Determine the matching of a 1 kΩ load to a 50 Ω source using three section $\frac{\lambda}{4}$ transformers, as an example similar to the previous one.

Solution:
To transform a 1000 Ω load to a 50 Ω input impedance, we can use three transmission lines with characteristic impedances $Z_{0,1}$ and $Z_{0,2}$, and $Z_{0,3}$. If we want to reduce the input impedance level by a fraction of β at each stage, we should have

$$Z_1 = \frac{1000}{\beta} \quad Z_2 = \frac{\frac{1000}{\beta}}{\beta} \quad Z_3 = \frac{\frac{1000}{\beta^2}}{\beta} = 50\,\Omega \tag{7.181}$$

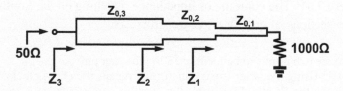

Figure 7.63: Impedance matching using three sections of quarter wave transformers.

Therefore, we would have

$$\beta = \sqrt[3]{\frac{1000}{20}} = 2.71 \tag{7.182}$$

Consequently, we would have the following values for the characteristic impedances and the input impedance of each section:

$$Z_1 = 368.4 \quad Z_{01} = \sqrt{368.4 \times 1000} = 607 \tag{7.183}$$

$$Z_2 = 135.7 \quad Z_{02} = \sqrt{135.7 \times 368.4} = 223 \tag{7.184}$$

$$Z_3 = 50 \quad Z_{03} = \sqrt{50 \times 135.72} = 82.4 \tag{7.185}$$

Figure 7.64 shows the contour of this impedance matching on the Smith chart.

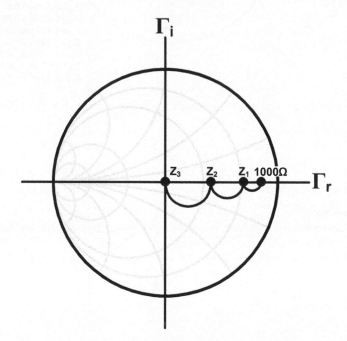

Figure 7.64: The contours of impedance matching on the Smith chart using three sections of quarter wave transformers.

There are two points to be mentioned in this example:
Point 1: Using a quarter wave transformer means travelling clockwise on a half-circle, on the Smith chart, from the starting point impedance to the end-point impedance.

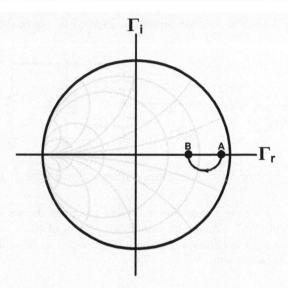

Figure 7.65: Half-circle contour showing the quarter wave impedance trans-formation on the Smith chart.

Point 2: Whenever the impedance transformation contours are held within the constant-Q locus, here $Q \leq \frac{\sqrt{2}}{2}$, the matching circuit would be wideband.

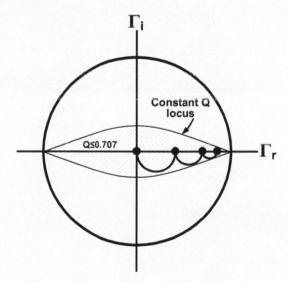

Figure 7.66: The impedance transformation contours and $Q = \frac{\sqrt{2}}{2}$ locus on the Smith chart.

Example 7.15 The structure depicted in Figure 7.67 can be shown to match $1\,\text{k}\Omega$ to $50\,\Omega$.

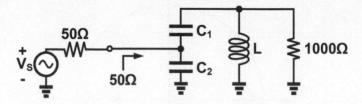

Figure 7.67: The lumped matching network at $1\,\text{GHz}$.

(a) Assuming we have matching at $1\,\text{GHz}$, calculate the values of C_1, C_2, and L, using Figure 7.67. Choose the input quality factor of 10.
(b) If Q for the capacitors and inductors are equal to 80 and 60, respectively, calculate the bandwidth.

Solution:
(a) Given an input quality factor of 10, let $(C_1 + C_2)\,\omega = 10\,G_S = 200\,\text{m}\mho$, so

$$C_1 + C_2 = 31.83\,\text{pF} \tag{7.186}$$

On the other hand

$$\left(\frac{C_1}{C_1 + C_2} \right)^2 = \frac{50}{1000} = \frac{1}{20} \tag{7.187}$$

Then

$$\frac{C_2}{C_1} = 3.47 \tag{7.188}$$

and

$$C_1 = 7.12\,\text{pF}, C_2 = 24.7\,\text{pF} \tag{7.189}$$

We calculate the value of the required inductance as

$$L = \frac{C_1 + C_2}{C_1 C_2 \omega^2} = 4.58\,\text{nH} \tag{7.190}$$

(b) Given the quality factors of the inductor and the capacitors, the parallel equivalent resistance is equal to (verification of the following is left to the reader):

$$R_P = R_L \parallel (Q_L L \omega) \parallel \left[\frac{Q_C}{C_1 \omega} \left(1 + \frac{C_1}{C_2} \right)^2 \right] \parallel \left[\left(R_S \parallel \frac{Q_C}{C_2 \omega} \right) \left(1 + \frac{C_2}{C_1} \right)^2 \right] = 331.9 \, \Omega$$

(7.191)

The bandwidth can be then calculated as

$$Q = \frac{\omega_0}{2\pi BW} \Rightarrow BW = \frac{\omega_0}{2\pi Q} = \frac{L \omega_0^2}{2\pi R_P} = 86.7 \, \text{MHz}$$

(7.192)

The following figure shows the frequency response and the bandwidth of this matching circuit through ADS simulation.

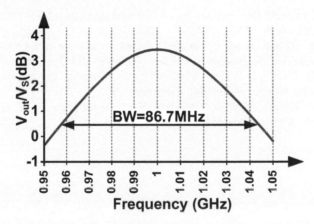

Figure 7.68: The frequency response and the bandwidth of the matching network (ADS simulation result).

■

7.11 Conclusion

Impedance matching is crucial in RF circuits, to transfer the maximum power from the source to the load. The impedance matching can be achieved through the use of reactive elements such as inductors, capacitors, or transformers. Transmission lines are widely used in RF circuits as well. Using the transmission line equations or the Smith chart, we can use them for impedance matching of the RF circuits as well. Note that while using the Smith chart in the design of the matching networks, generally inductors and capacitors are considered to be lossless to have a quick insight and a quick design procedure. Care should be taken while using the Smith chart not to approach the open-circuit vicinity on the impedance chart and not to approach the short-circuit vicinity on the admittance chart, in order to avoid significant errors in the matching network designs or evaluations. Open-circuit stubs or short-circuit stubs can be used as the reactive elements in the matching circuits. Quarter-wave transmission lines can be used as impedance transformers in the matching circuits as well. Furthermore, the combination of the lumped elements and transmission lines could be used in the matching circuit networks in any case.

7.12   References and Further Reading

1. D. Pozar, *Microwave Engineering*, fourth edition, Hoboken, NJ: J. Wiley & Sons, Inc., 2012.
2. S. Ramo, J.R. Whinnery, T. Van Duzer, *Fields and Waves in Communication Electronics*, New York: J. Wiley & Sons, 1994.
3. R. Chi-Hsi Li, *RF Circuit Design*, Hoboken, NJ: J. Wiley & Sons, Inc., 2009.
4. B. Razavi, *RF Microelectronics*, second edition, Castleton, NY: Prentice-Hall, 2011.
5. F. Farzaneh, *RF Communication Circuits* (in Persian), Tehran: Sharif University Press, 2005.
6. U.L. Rohde, A.M. Pavio, G.D. Vendelin, *Microwave Circuit Design Using Linear and Nonlinear Techniques*, Hoboken, NJ: J. Wiley & Sons, Inc., 2005.
7. K.K. Clarke, D.T. Hess, *Communication Circuits, Analysis and Design*, United States: Krieger Publishing Company, 1994.
8. H.L. Krauss, C.W.Bostian, F.H. Raab, *Solid State Radio Engineering*, New York, NY: J. Wiley & Sons, Inc., 1980.

7.13 Problems

Problem 7.1 Assume that a transmission line with a characteristic impedance of Z_0 is terminated to $75\,\Omega$. If $SWR = 1.5$, determine the characteristic impedance, Z_0.

Problem 7.2 Consider the circuit shown in Figure 7.69. Find the reflection coefficient, the input impedance $Z_{in}(\lambda/8)$, and SWR. Also, find the voltage and the current phasors at the source and the load sides. What is the delivered power to the load?

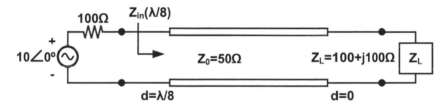

Figure 7.69: A transmission line with the source and the load terminations.

Problem 7.3 Consider the amplifier shown in Fig. 7.70 with the input and output matching networks. If the input impedance can be modeled as $15\,K\Omega \parallel 15\,pF$ and the output equivalent circuit can be modeled as $410\,\Omega \parallel 5\,pF$ and the operation frequency is 50 MHz, calculate the bandwidth at the input and at the output. Also calculate the input matching circuit loss in dB at the center frequency. All capacitors have a quality of factor of 100 at 50 MHz. Hint: The power delivered to the amplifier is the same power delivered to the $15\,k\Omega$.
$C_1 = 13\,pF, C_2 = 34\,pF, C_3 = C_4 = 28\,pF$
$L_1 = 436\,nH(Q = 11), L_2 = 336\,nH(Q = 32)$

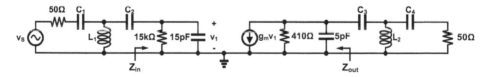

Figure 7.70: The circuit model for the amplifier.

Problem 7.4 Consider the load impedance at the point $Z = (30 + j40)\,\Omega$ on the Smith chart and also the two matching networks shown in Figure 7.71. Plot the matching circle paths on the Smith chart, and estimate the input impedance, and then compare it to your analytical computations. The operating frequency is 1 GHz. Is there any deficiency in these matching networks?

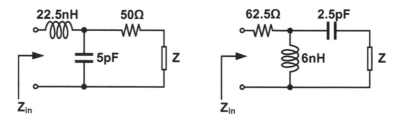

Figure 7.71: The circuits for input impedance evaluation.

Problem 7.5 We desire to match a load impedance $Z = (200 + j50)\,\Omega$ to $50\,\Omega$. In either of the matching networks depicted in Figure 7.72, determine the proper values of matching components at 900 MHz. Use C_2 to tune out the load's inductive reactance. Note that for the circuit (b) there isn't a unique solution (several values of L and C can satisfy the matching condition).

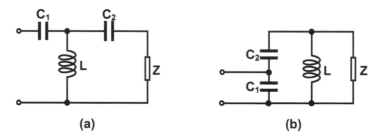

Figure 7.72: Two possible matching circuits for the specified load.

Problem 7.6 In the circuit shown in Figure 7.73, we desire to perform the matching to the complex source impedance. Calculate the length of the line, ℓ_1, the $\lambda/8$ stub characteristic impedance, $Z_{0,2}$, and the type of the stub (open-circuit stub or short circuit stub) for this purpose. The operation frequency is 1 GHz, and the effective permittivity of the line is 3.

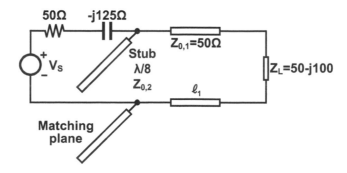

Figure 7.73: The $\lambda/8$ single stub matching for a complex source impedance.

Problem 7.7 A $50\,\Omega$ load is connected to a quarter-wavelength T-line, cascaded by a $75\,\Omega$ T-line of unknown length. Determine the length of the second T-line and the characteristic impedance of the first T-line such that the input impedance at 1 GHz is equal to $112.5\,\Omega$. Neglect T-line losses (Note that there are two possible solutions).

Problem 7.8 Sketch the frequency response of the circuit of the problem 7.7 for 300 MHz–3 GHz and determine whether the input impedance is capacitive or inductive at each frequency. Determine the input impedance at $f = 0.3, 0.5, 1, 1.5, 2.5, 3$ GHz.

Problem 7.9 In the circuit of problem 7.7, assume that the input source has an impedance of $112.5\,\Omega$ and a source voltage of 2 V and is operating at 3 GHz. Determine output voltage on the $50\,\Omega$ load. Note that the transmission line length is one-fourth of the wavelength at 1 GHz.

Problem 7.10 The output impedance of a ceramic filter is $50\,\Omega$ and the input impedance of an LNA can be represented by $900\,\Omega \parallel 0.5\,\mathrm{pF}$ at 900 MHz.
  1. Design a matching network consisting of two capacitors and one inductor. The acceptable range for the inductor is $1 - 8\,\mathrm{nH}$ and note that only capacitors larger than 500 fF are allowed. You might choose an inductor of 6 nH with a typical quality factor of 60 which is a relatively high achievable value in the range of $1 - 8\,\mathrm{nH}$. Noting that the Q of the capacitors is typically larger than 200 and quite larger than that of the inductor, find the loss of the matching network.
  2. In practice, if implemented on a chip, the inductor will have a Q as small as 5. Find the loss in this case as well.
  3. What is the difference between the bandwidths of the circuit in part 1 and part 2?

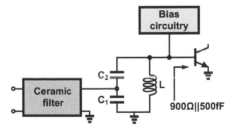

Figure 7.74: The matching circuit for the ceramic filter.

Problem 7.11 The VSWR in a lossless T-line terminated by an unknown load is 3. The distance between the two successive voltage minima is 20 cm, with the first minimum occurring at the distance of 5 cm from the load. The characteristic impedance of the line is equal to $50\,\Omega$.
  1. Find the reflection coefficient and the load impedance, Z_L.
  2. What a pure resistive load could be put at the T-line termination through which and using a minimum length T-line, one can reach to the same load impedance, Z_L? (Specify the amount of rotation required on the Smith chart.)

Problem 7.12 Determine the distance d_0, and the short-circuit stub length ℓ such that the input impedance in Figure 7.75 becomes $50\,\Omega$? Both the T-line and the short-circuit stub have a characteristic impedance of $50\,\Omega$. Express the lengths as fractions of the wavelength.

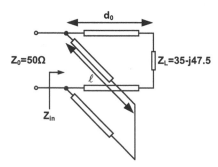

Figure 7.75: The single stub matching circuit.

Problem 7.13 Determine the values of ℓ_A and ℓ_B in Figure 7.76 such that the input is matched to $50\,\Omega$. Consider all the T-lines are of $50\,\Omega$ characteristic impedance. Note that X could be either a short circuit or an open circuit.

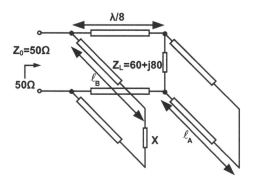

Figure 7.76: The double stub matching circuit.

Problem 7.14 Assuming a lossless $50\,\Omega$ line and the load impedance in the circuit depicted in Figure 7.77, calculate the power delivered to the load. Hint: calculate first the available power of the source and then the input reflection coefficient.

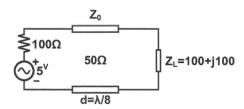

Figure 7.77: Determination of the delivered power to the load through a T-line.

Problem 7.15 A lossless line is terminated to a resistive $100\,\Omega$ load. If the reflected voltage at the load side is $V^-(z=0) = j$ Volts and the total current passing through the load is equal to $I = j(0.04)$ Amperes, determine the characteristic line impedance Z_0 (Fig. 7.78).

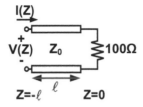

Figure 7.78: The characteristic impedance determination through the current and the voltage phasors.

Problem 7.16 In a lossless $50\,\Omega$ T-line, the voltage phasor at a distance of $\lambda/4$ from the load is equal to $(25 - j25)\,$mV and the current at this point is equal to $(0.5 - j1.5)\,$mA.

1. Determine the reflection coefficient at this point.
2. Calculate VSWR on this line.
3. What is the load impedance at the line termination?

Problem 7.17 In the lossless T-line depicted in Figure 7.79 which is terminated to $150\,\Omega$, the voltage phasor at endpoint of the line is $V(z=0) = j3$ V.

1. Determine the complex functions $\Gamma(z), V^+(z), V^-(z)$ along the line.
2. Determine the complex functions $Z(z), V(z), I(z)$ along the line.
3. Determine the values of the functions found in parts (1) and (2) at $z = -1$ m.

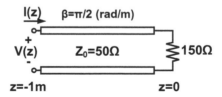

Figure 7.79: The circuit for determination of the incident and the reflected voltage phasors.

Problem 7.18 In the circuit depicted in Figure 7.80, compute $Z(-\ell)$ as a function of the T-line's length and plot the results in two separate curves expressing the real and the imaginary parts. Also, calculate the reflection coefficient at the points shown in the figure (at the input, this value can be expressed in terms of ℓ, where $0 < \ell < \frac{\lambda}{2}$).

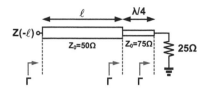

Figure 7.80: The circuit for reflection coefficient computation.

Problem 7.19 In the circuit depicted in Figure 7.81, we wish to perform the impedance matching between a $100\,\Omega$ source and a $1000\,\Omega$ load using an ideal capacitor and an ideal inductor at $1\,\text{GHz}$. Determine the required values of L, C, and the corresponding Q of the circuit. Sketch the frequency response ($|\frac{V_{out}}{V_{in}}|$) of the circuit as well.

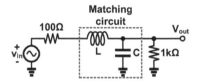

Figure 7.81: The circuit for L-section impedance matching.

Problem 7.20 Calculate the values of inductors and capacitors to achieve impedance matching between $100\,\Omega$ source and $1000\,\Omega$ load considering a load's L-section Q factor of 10 at $1\,\text{GHz}$.

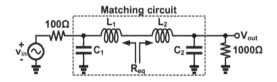

Figure 7.82: The π matching network.

Problem 7.21 Consider a DC source is connected to a lossless air line of $75\,\Omega$ characteristic impedance as in Figure 7.83. Draw the transient traveling waves along the line as a function of time.

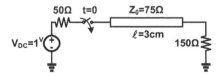

Figure 7.83: The circuit for transient voltage determination.

Problem 7.22 An air line is depicted in Figure 7.84 which is loaded by an *RLC* circuit. Draw the input reflection coefficient about 5 GHz (e.g., from 4.95 GHz to 5.05 GHz) on the Smith chart. What would be the equivalent circuit of this network then?

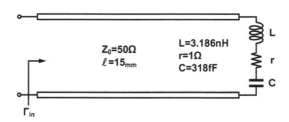

Figure 7.84: The circuit for impedance inversion.

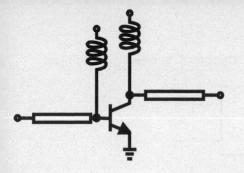

8. Scattering Parameters

8.1 Representation of Two-Port Networks

In order to characterize a linear two-port network, the measured parameters of the transfer functions are required. Two-ports networks are characterized by such parameters as Z, Y, G, H, and T which provide a basic understanding of a network function. These parameters are mostly used at lower frequencies; furthermore, they entail some difficulties in their measurement at higher frequencies. The most significant drawback of these parameters is that they need short- and open-terminations measurements which are not easily realizable at microwave and millimeter-wave frequencies. That's to say a short length of any wire would be equivalent to a small inductor and any open circuit at the end of a transmission line would represent a small capacitance. S-parameters have already been introduced in microwave circuit measurement and design as well as millimeter-wave application. Being based on the electromagnetic wave propagation model, S-parameters provide a comprehensive account of the circuit properties. In the next few sections, we first review common circuit parameters, and then introduce S-parameters.

8.1.1 Common Circuit Parameters of Two-Port Networks

A generic two-port network is depicted in Figure 8.1. The matrix representation of this two-port network parameters can be written as one of the following

$$\begin{bmatrix} v_1 \\ v_2 \end{bmatrix} = \begin{bmatrix} z_{11} & z_{12} \\ z_{21} & z_{22} \end{bmatrix} \begin{bmatrix} i_1 \\ i_2 \end{bmatrix} \tag{8.1a}$$

$$\begin{bmatrix} i_1 \\ i_2 \end{bmatrix} = \begin{bmatrix} y_{11} & y_{12} \\ y_{21} & y_{22} \end{bmatrix} \begin{bmatrix} v_1 \\ v_2 \end{bmatrix} \tag{8.1b}$$

$$\begin{bmatrix} v_1 \\ i_2 \end{bmatrix} = \begin{bmatrix} h_{11} & h_{12} \\ h_{21} & h_{22} \end{bmatrix} \begin{bmatrix} i_1 \\ v_2 \end{bmatrix} \tag{8.1c}$$

Figure 8.1: A generic two-port network.

$$\begin{bmatrix} i_1 \\ v_2 \end{bmatrix} = \begin{bmatrix} g_{11} & g_{12} \\ g_{21} & g_{22} \end{bmatrix} \begin{bmatrix} v_1 \\ i_2 \end{bmatrix} \tag{8.1d}$$

$$\begin{bmatrix} v_1 \\ i_1 \end{bmatrix} = \begin{bmatrix} A & B \\ C & D \end{bmatrix} \begin{bmatrix} v_2 \\ -i_2 \end{bmatrix} \tag{8.1e}$$

The five above parameters can be easily measured at low frequencies by open- or short-circuiting the network. Example 8.1 provides a better insight into this issue.

Example 8.1 Depicted in Figure 8.2 is a resistive two-port network. Derive the Z-parameters according to Equation 8.1a.

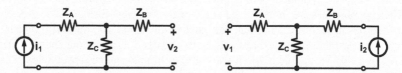

Figure 8.2: A resistive two-port and Z-parameters measurement.

Solution:
It follows from Equation 8.1a that

$$z_{11} = \frac{v_1}{i_1}\Big|_{i_2=0} = Z_A + Z_C \tag{8.2a}$$

$$z_{12} = \frac{v_1}{i_2}\Big|_{i_1=0} = Z_C \tag{8.2b}$$

$$z_{21} = \frac{v_2}{i_1}\Big|_{i_2=0} = Z_C \tag{8.2c}$$

$$z_{22} = \frac{v_2}{i_2}\Big|_{i_1=0} = Z_B + Z_C \tag{8.2d}$$

∎

Figure 8.3: Two port S-parameter matrix concept.

8.1.2 Scattering Parameters

Scattering parameters, or simply S-parameters, are defined according to the wave behavior of the signal and unlike low-frequency parameters, do not require open-circuit or short-circuit termination in their measurement. S-parameters can be obtained by terminating the circuit with the reference impedance (e.g., $50\,\Omega$). A generic microwave two-port is shown in Figure 8.3. As depicted in Figure 8.3, the incident waves at ports 1 and 2 are denoted by V_{i1} and V_{i2}, respectively. The reflected waves at ports 1 and 2 are denoted by V_{r1} and V_{r2}, respectively. For a linear two-port, there is always a linear relation between these four parameters as

$$V_{r1} = \rho_{11}V_{i1} + \tau_{12}V_{i2} \tag{8.3a}$$

$$V_{r2} = \tau_{21}V_{i1} + \rho_{22}V_{i2} \tag{8.3b}$$

Here, ρ_{11} is the reflection coefficient at port 1 once $V_{i2} = 0$ (port 2 is terminated by a matched load).
τ_{21} is the forward transmission coefficient once $V_{i2} = 0$.
ρ_{22} is the reflection coefficient at port 2 once $V_{i1} = 0$ (port 1 is terminated by a matched load).
τ_{12} is the reverse transmission coefficient once $V_{i1} = 0$.
Hence, the matrix representation of the network, relating the incident and the reflected waves at the output and the input ports, can be denoted in a matrix form by

$$\begin{bmatrix} V_{r1} \\ V_{r2} \end{bmatrix} = \begin{bmatrix} \rho_{11} & \tau_{12} \\ \tau_{21} & \rho_{22} \end{bmatrix} \begin{bmatrix} V_{i1} \\ V_{i2} \end{bmatrix} \tag{8.4}$$

The normalized incident and reflected waves can be defined in the following manner

$$a_1 = \frac{V_{i1}}{\sqrt{Z_0}}, \quad a_2 = \frac{V_{i2}}{\sqrt{Z_0}} \tag{8.5a}$$

$$b_1 = \frac{V_{r1}}{\sqrt{Z_0}}, \quad b_2 = \frac{V_{r2}}{\sqrt{Z_0}} \tag{8.5b}$$

where Z_0 is the reference (the matched load) impedance. Consequently, a and b will have a dimension of square root of Watts. Furthermore, the incident power and the reflected power at each port will be proportional to

$$P_i = \frac{1}{2}a_i a_i^* = \frac{1}{2}|a_i|^2 \tag{8.6a}$$

$$P_r = \frac{1}{2}b_i b_i^* = \frac{1}{2}|b_i|^2 \tag{8.6b}$$

The absorbed power of the two-port would be

$$P_{abs} = P_i - P_r = P_i \left(1 - |\Gamma_{in}|^2\right) \tag{8.7}$$

where Γ_{in} is the input reflection coefficient of the two-port. The standard representation of Equation 8.4 is given by

$$\begin{bmatrix} b_1 \\ b_2 \end{bmatrix} = \begin{bmatrix} S_{11} & S_{12} \\ S_{21} & S_{22} \end{bmatrix} \begin{bmatrix} a_1 \\ a_2 \end{bmatrix} \tag{8.8}$$

which is widely referred to as the S-parameters matrix. As with the other matrix representations of circuits, each entry carries an interpretation of its own. S_{11} is the input reflection coefficient once the output is terminated by the reference impedance, S_{22} is the output reflection coefficient once the input is terminated by the reference impedance, S_{21} is the forward transmission coefficient once the output is terminated by the reference impedance, and S_{12} is the reverse transmission coefficient once the input is terminated by the reference impedance. Each of the entries of the [S] matrix can be derived individually as

$$S_{11} = \left.\frac{V_{r1}}{V_{i1}}\right|_{V_{i2}=0} = \left.\frac{b_1}{a_1}\right|_{a_2=0} \tag{8.9a}$$

$$S_{12} = \left.\frac{V_{r1}}{V_{i2}}\right|_{V_{i1}=0} = \left.\frac{b_1}{a_2}\right|_{a_1=0} \tag{8.9b}$$

$$S_{21} = \left.\frac{V_{r2}}{V_{i1}}\right|_{V_{i2}=0} = \left.\frac{b_2}{a_1}\right|_{a_2=0} \tag{8.9c}$$

$$S_{22} = \left.\frac{V_{r2}}{V_{i2}}\right|_{V_{i1}=0} = \left.\frac{b_2}{a_2}\right|_{a_1=0} \tag{8.9d}$$

Equations 8.9a–8.9d also explicitly indicate how each one of the parameters can be measured. To measure S_{11}, for instance, one needs to measure the ratio of the reflected wave of port 1 to the incident wave of port 1 while the other ports are terminated by the reference impedance. A 5-port network is depicted in Figure 8.4. The scattering parameters for this network can be written as

$$\begin{bmatrix} b_1 \\ b_2 \\ b_3 \\ b_4 \\ b_5 \end{bmatrix} = \begin{bmatrix} S_{11} & S_{12} & S_{13} & S_{14} & S_{15} \\ S_{21} & S_{22} & S_{23} & S_{24} & S_{25} \\ S_{31} & S_{32} & S_{33} & S_{34} & S_{35} \\ S_{41} & S_{42} & S_{43} & S_{44} & S_{45} \\ S_{51} & S_{52} & S_{53} & S_{54} & S_{55} \end{bmatrix} \begin{bmatrix} a_1 \\ a_2 \\ a_3 \\ a_4 \\ a_5 \end{bmatrix} \tag{8.10}$$

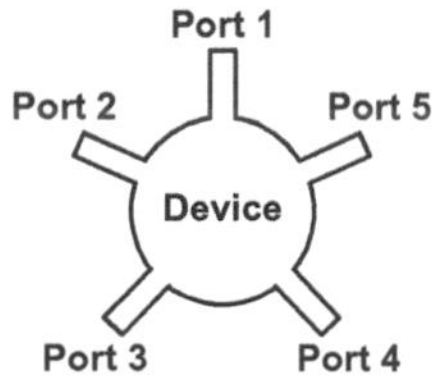

Figure 8.4: Illustration of a 5-port network.

Example 8.2 A two-port network is depicted in Figure 8.5. Determine the S-parameters.

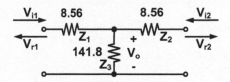

Figure 8.5: A resistive two-port network.

Solution:
Using Equation 8.9a–8.9d, it follows that

$$S_{11} = \left.\frac{V_{r1}}{V_{i1}}\right|_{V_{r2}=0} = \rho = \frac{Z_{in} - Z_0}{Z_{in} + Z_0} \tag{8.11}$$

Terminating the output by the reference impedance (50 Ω), the input impedance is given by

$$Z_{in} = Z_1 + [Z_3 \parallel (Z_2 + Z_0)] \tag{8.12}$$

$$= 8.56 + [141.8(8.56 + 50)/(141.8 + 8.56 + 50)] = 50 \ \Omega$$

Hence,

$$S_{11} = \frac{50 - 50}{50 + 50} = 0 \tag{8.13}$$

suggesting that we have matching at the input. Due to symmetry, S_{22} will also assume a value of zero. Given the matching at the output (note that the reflected wave at the output is zero), it follows that

$$V_{t2} = V_2 = V_1 \left(\frac{(Z_2 + Z_0) \parallel Z_3}{(Z_2 + Z_0) \parallel Z_3 + Z_1} \right) \left(\frac{Z_0}{Z_2 + Z_0} \right) = V_o \left(\frac{Z_0}{Z_2 + Z_0} \right)$$

$$= V_1 \left(\frac{41.44}{41.44 + 8.56} \right) \left(\frac{50}{50 + 8.56} \right) = 0.707 \, V_1$$

$$\tag{8.14}$$

Using the symmetry of the circuit, $S_{12} = S_{21} = 0.707$. Thus, the [S] matrix is given by

$$[S] = \begin{bmatrix} 0 & 0.707 \\ 0.707 & 0 \end{bmatrix} \tag{8.15}$$

It can be proved that, in general, Z-parameters can be converted to S-parameters, using the following transforms

$$S_{11} = \frac{(Z_{11} - Z_0)(Z_{22} + Z_0) - Z_{12}Z_{21}}{\Delta} \tag{8.16a}$$

$$S_{12} = \frac{2Z_0Z_{12}}{\Delta} \tag{8.16b}$$

$$S_{21} = \frac{2Z_0Z_{21}}{\Delta} \tag{8.16c}$$

$$S_{22} = \frac{(Z_{11} + Z_0)(Z_{22} - Z_0) - Z_{12}Z_{21}}{\Delta} \tag{8.16d}$$

$$\Delta = (Z_{11} + Z_0)(Z_{22} + Z_0) - Z_{12}Z_{21}$$

∎

Example 8.3 Find the S-parameters for the circuit depicted in Figure 8.6 at 1 GHz.

Solution:
It follows from Figure 8.6 that

$$Z_{in} = \left(R || \frac{1}{Cs} \right) + 50 = 75 - j25 \tag{8.17}$$

hence,

$$S_{11} = \frac{Z_{in} - Z_0}{Z_{in} + Z_0} = 0.277\angle - 33.7° \tag{8.18}$$

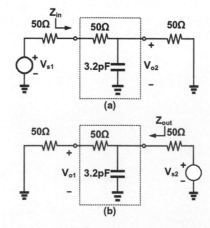

Figure 8.6: A low-pass filter for S-parameter calculation.

The output impedance can also be calculated as

$$Z_{\text{out}} = \left(2R\|\frac{1}{Cs}\right) = 20 - j40 \tag{8.19}$$

yielding

$$S_{22} = \frac{Z_{\text{out}} - Z_0}{Z_{\text{out}} + Z_0} = 0.624\angle - 97° \tag{8.20}$$

The forward gain and the reverse gain of the circuit can also be computed in a similar manner as

$$S_{21} = \frac{V_{O2}}{V_{S1}/2} = \frac{2\left(\frac{1}{Cs}\|R\right)}{\left(\frac{1}{Cs}\|R\right) + 2R} = 0.554\angle - 33.7° \tag{8.21}$$

and

$$S_{12} = \frac{V_{O1}}{V_{S2}/2} = \frac{2\left(\frac{1}{Cs}\|2R\right)}{\left(\frac{1}{Cs}\|2R\right) + R}\frac{R}{R + R} = 0.554\angle - 33.7° \tag{8.22}$$

∎

Example 8.4 Assume that in the oscillator depicted in Figure 8.7, an inductor is tied to the base of the bipolar transistor, Q. The S-parameters of the two-port at 1 GHz have been measured as $S_{11} = 1.8\angle 100°$, $S_{21} = 2.2\angle - 140°$, $S_{12} = 0.7\angle 140°$, and $S_{22} = 1.1\angle - 100°$.
(a) Determine the maximum value for R_s so that the oscillation occurs.
(b) If L is changed in such a way that the input reflection coefficient becomes $\Gamma_{\text{in}} = -2$, recalculate the maximum value of R_s.

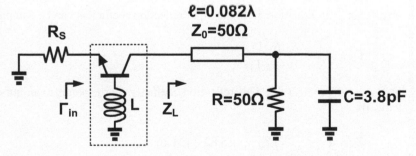

Figure 8.7: An oscillator with source impedance for S-parameter calculation.

Solution:

(a) Using Equation 7.72, the load impedance can be calculated as

$$Z_L = Z_0 \left(\frac{\left(R || \frac{1}{jC\omega}\right) + jZ_0 \tan(\beta\ell)}{Z_0 + j\left(R || \frac{1}{jC\omega}\right)\tan(\beta\ell)} \right) = 16.11 - j0.0496 \tag{8.23}$$

The load reflection coefficient becomes

$$\Gamma_L = \frac{Z_L - Z_0}{Z_L + Z_0} = -0.513 - j0.0011 \tag{8.24}$$

As we had from Equation 8.8

$$b_1 = S_{11}a_1 + S_{12}a_2 \tag{8.25}$$

or

$$\Gamma_{in} = S_{11} + S_{12}\frac{a_2}{a_1} \tag{8.26}$$

and as we had from Equation 8.8

$$b_2 = S_{21}a_1 + S_{22}a_2 \tag{8.27}$$

Given $b_2 = \frac{a_2}{\Gamma_L}$, one can write

$$\frac{a_2}{\Gamma_L} = S_{21}a_1 + S_{22}a_2 \tag{8.28}$$

Therefore,

$$\frac{a_2}{a_1} = \frac{S_{21}}{\frac{1}{\Gamma_L} - S_{22}} \tag{8.29}$$

Replacing $\frac{a_2}{a_1}$ in Equation 8.26, the input reflection coefficient can be computed as

$$\Gamma_{in} = S_{11} + \frac{S_{21}S_{12}\Gamma_L}{1 - S_{22}\Gamma_L} \tag{8.30}$$

or $\Gamma_{in} = -0.946 + j1.38$. This reflection coefficient corresponds to an impedance value of

$$Z_{in} = \left(\frac{1 + \Gamma_{in}}{1 - \Gamma_{in}}\right) Z_0 = -15.83 + j24.26 \tag{8.31}$$

We can then deduce that R_s should be less than $15.83\,\Omega$ for the oscillation to occur.

(b) The corresponding input impedance should first be calculated from the given reflection coefficient as

$$\frac{Z_{eq}}{Z_0} = \frac{1+\Gamma}{1-\Gamma} = \frac{1-2}{1+2} = -\frac{1}{3} \tag{8.32}$$

For the oscillation to start, R_s should be less than the absolute value of the negative resistance, i.e.,

$$R_S < \frac{1}{3}Z_0 = 16.67\,\Omega \tag{8.33}$$

$\blacksquare$

8.2 Measuring S-Parameters Using a Network Analyzer (For Advanced Readers)

In this section, we turn our attention to the important and practical issue of measuring S-parameters using a network analyzer. As you may already know, the numerous sources of error, namely, measurement device error and nonidealities, which specifically emerge at higher frequencies need some extent of calibration. We first focus on how a network analyzer operates and later will introduce a calibration technique based on electrical delay lines. Finally, a relatively more accurate technique will be introduced in order to measure the two-port S-parameters.

8.2.1 Operation of a Network Analyzer

Consider Figure 8.8, where the conceptual block diagram of a network analyzer is illustrated. First, assume we are to measure S-parameters from the port 1 (S_{11} and S_{21}). The device will then switch automatically so that the signal source is connected to port 1. The three resistors in Figure 8.8 simply serve as a power splitter, after which the divided signals are identical in phase and in amplitude. The amplitude and the phase of the direct signal are recorded using a detector at the output A_1, providing the phase and the amplitude of the signal source as the reference. The other part of the signal is fed through the port 1 to the DUT. Part of this signal is reflected back by the DUT at the port 1 and the other part of it passes through the DUT to the port 2. The reflected signal is measured using a directional coupler and its amplitude and phase are recorded at the output B_1. The signal passing through the DUT is measured at the ouput B_2.

The ratio of the reflected signal from the DUT to the incident reference signal at the DUT gives in S_{11} and the ratio of the output signal at B_2 of the DUT to the incident signal gives in S_{21}. Now to measure S_{22} and S_{12}, the signal source is switched to the port 2 and the port 1 is loaded by the reference impedance. As such, by the same procedure, the output reflection coefficient (S_{22}) and the reverse gain of the device (S_{12}) are measured.

The network analyzer system takes into account automatically the losses in the path of the signal, the coupling coefficient of the directional couplers, and the phase changes due to the transmission line lengths.

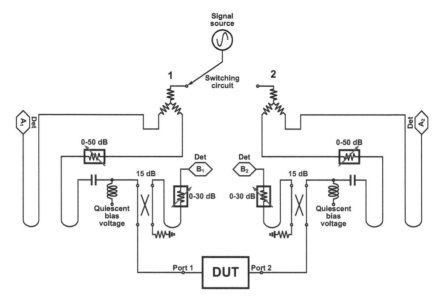

Figure 8.8: Conceptual block diagram of a network analyzer.

Therefore, all the scattering parameters can be evaluated using a network analyzer. The measurements, however, are subject to errors which may stem from numerous sources, as discussed earlier. In the next section, we discuss techniques to mitigate these errors.

8.2.2 Calibration Using Electrical Delay

One of the sources of error is the cables connecting the DUT and the network analyzer. In order to obtain the S-parameters accurately, one should compensate for the effect of delay and attenuation of these cables. To better understand this effect, assume that the length of the cable used for the reference signal amplitude measurement (A_1) might be smaller than that of the transmission line used for transmitting the signal to the device. This difference introduces a phase error in the measured S-parameters. Therefore, in order to make a correct comparison, an additional delay equal to the delay which the signal experiences from the source to the DUT should be added to the source signal path. This will provide a correct and free of error measurement.

8.2.3 Quiescent Point bias Circuit

Another valuable feature of a network analyzer is that it enables us to provide the quiescent bias point directly to the circuit. As depicted in Figure 8.8, this is achieved using an inductor along with a decoupling capacitor which might be mounted inside a network analyzer. Interestingly, as the bias is provided by the inductor or RF choke (RFC), it does not affect the S-parameter readings at high frequencies (as the RFC is open circuit at higher frequencies).

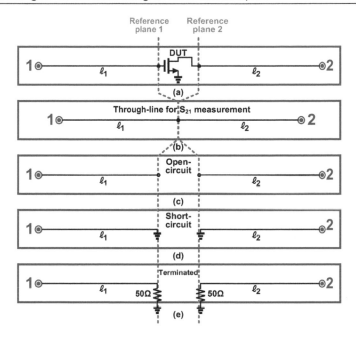

Figure 8.9: Calibration circuits needed in the calibration process for a network analyzer to measure the two-port S-parameters of the DUT.

8.2.4 One-Port and Two-Port Calibration for Short Circuit, Open Circuit, and the Characteristic Impedance

Calibration is of great importance if a network analyzer is to accurately measure an unknown impedance. This can be achieved by using components which simulate the behavior of an open circuit, a short circuit, or reference impedance. After calibration, only a slight difference can be observed between the real and the measured parameters. If carried out for only one port, the above procedure can only provide the input or the output matching accurately. To measure the device's gain or the isolation, however, the procedure should be performed for both ports. The calibration procedure is illustrated in Figure 8.9.

Consider we intend to measure the S-parameters of a transistor as depicted in Figure 8.9(a) where the transistor is accessed through the ports of the network analyzer with two transmission lines with the lengths ℓ_1 and ℓ_2, respectively.

As in Figure 8.9(b) a through-line with a length of $\ell_1 + \ell_2$ is realized, and it is referenced as $S_{21} = 1\angle 0$, in order to measure the parameters S_{12} and S_{21} with a correct phase.

As in Figure 8.9(c), two open circuits are realized at the ports 1 and 2 of the network analyzer through two transmission lines with the lengths ℓ_1 and ℓ_2, respectively. As such, here, at port 1, the reflection coefficient is referenced as $S_{11} = 1\angle 0$, and at port 2, the reflection coefficient is referenced as $S_{22} = 1\angle 0$.

As in Figure 8.9(d), two short circuits are realized at ports 1 and 2 of the network analyzer through two transmission lines with the lengths ℓ_1 and ℓ_2, respectively. As such, here, at port 1, the reflection coefficient is referenced as $S_{11} = 1\angle 180°$, and at port 2, the reflection coefficient is referenced as $S_{22} = 1\angle 180°$.

As in Figure 8.9(e), two matched loads are realized at ports 1 and 2 of the network analyzer through two transmission lines with the lengths ℓ_1 and ℓ_2, respectively. As such, here, at port 1, the reflection coefficient is referenced as $S_{11} = 0$, and at port 2, the reflection coefficient is referenced as $S_{22} = 0$.

Note that in all the five prototype circuits, depicted in Figure 8.9, used for calibration, the connectors and the printed circuits should be of the same type and the same fabrication. Through this calibration procedure, the length of the device is taken into account and any error associated with connector mismatches is deduced.

> **Example 8.5** In S_{11} measurements, the network analyzer is programmed to use the reference plane C_1, and the Smith chart given in Figure 8.10 (left) is resulted for open-circuit measurement. When the circuit is connected to the network analyzer, the plot given in Figure 8.10 (right) is obtained.
>
> (a) If the S_{11} of the transistor itself is to be measured, explain the errors in this measurement.
>
> (b) What can be done to obtain the correct plot of S_{11}?
>
> **Solution:**
>
> (a) The error is equal to the delay generated by a reference transmission line with the length of $\lambda/8$, that is about 90°, in reflection (assuming a lossless T-line).
>
> (b) For proper measurement, the delay resulting from the transmission line should be deduced from the phase of S_{11}. In other words
>
> $$S'_{11} = S_{11}e^{-j2\beta l} \tag{8.34}$$
>
>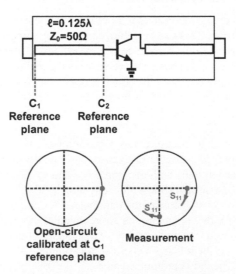

Figure 8.10: S-parameter measurement.

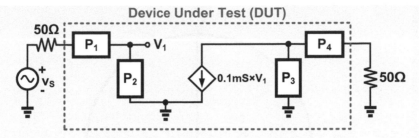

Figure 8.11: The proposed model for the Device Under Test.

where S'_{11} is the measured reflection coefficient at C_1 reference plane, and S_{11} is the measured reflection coefficient at C_2 reference plane.

$$S_{11} = S'_{11} e^{j2\beta l} \tag{8.35}$$

In other words, the measured plots of S_{11} at C_1 plane should be rotated 90° counter clockwise to obtain the value of S_{11} at C_2 reference plane.   ∎

Example 8.6 The S-parameters (S_{11} and S_{22}) for the circuit shown in Figure 8.11 in the frequency range of 500 kHz–3 GHz have been measured as

$$S_{11,A} = -0.111 \quad \text{at } 500\,\text{kHz} \tag{8.36}$$

$$S_{11,B} = 0.441\angle 158° \quad \text{at } 3\,\text{GHz} \tag{8.37}$$

$$S_{22,C} = 0.818 \quad \text{at } 500\,\text{kHz} \tag{8.38}$$

$$S_{22,D} = 0.807\angle -11.5° \quad \text{at } 3\,\text{GHz} \tag{8.39}$$

Sketch a reasonable and proper equivalent circuit for this network. The boxes contain passive devices.

Solution:
For an RF transistor, the input and the output impedances at 500 kHz are almost the same as the DC impedances. Given the fact that S_{11} is real and is equal to -0.111, then it could be deduced that a real impedance of the following value is seen at the input at low frequencies:

$$Z_{\text{in}} = Z_0 \frac{1 + S_{11}}{1 - S_{11}} = 40\,\Omega \tag{8.40}$$

As the input impedance on the chart progressively becomes capacitive and finally becomes inductive, we suggest the input equivalent circuit as depicted in Figure 8.13. This means $r_\pi = 40\,\Omega$.

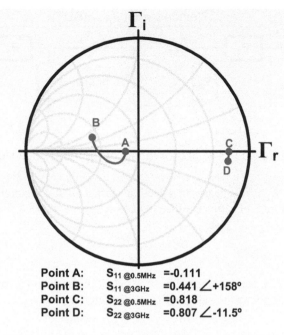

Point A: $S_{11\ @0.5MHz}$ =-0.111
Point B: $S_{11\ @3GHz}$ =0.441 $\angle$+158°
Point C: $S_{22\ @0.5MHz}$ =0.818
Point D: $S_{22\ @3GHz}$ =0.807 $\angle$-11.5°

Figure 8.12: Impedance matching on Smith chart.

The input impedance at 3 GHz becomes

$$Z_{in} = Z_0 \frac{1 + S_{11}}{1 - S_{11}} \approx 20 + j8.27 \tag{8.41}$$

Given the equivalent circuit, one can write

$$r_{in} = \frac{r_\pi}{1 + (\omega r_\pi C_\pi)^2} = 20 \tag{8.42}$$

Then

$$C_\pi \approx 1.33\,\text{pF} \tag{8.43}$$

Now given the equivalent circuit, one can write

$$X_{in} = jL_b\omega - \frac{jr_\pi^2 C_\pi \omega}{1 + (\omega r_\pi C_\pi)^2} \approx j8.27 \tag{8.44}$$

Then

$$L_b \approx 1.5\,\text{nH} \tag{8.45}$$

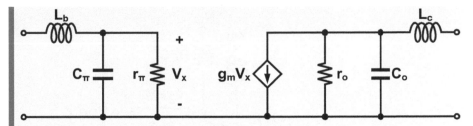

Figure 8.13: The proposed equivalent circuit for the measured S-parameters of the transistor.

By the fact that S_{22} (the output impedance) is real at 500 kHz, again one could say that the output impedance at DC is almost

$$Z_{out} = Z_0 \frac{1 + S_{22}}{1 - S_{22}} = 500\,\Omega \tag{8.46}$$

It is seen that S_{22} becomes gradually capacitive and its amplitude is slightly reduced, so we suggest the output equivalent circuit as depicted in Figure 8.13. This means $r_0 = 500\,\Omega$.

At 3 GHz, the output impedance becomes

$$Z_{out} = Z_0 \frac{1 + S_{22}}{1 - S_{22}} \approx 250 - j231.2 \tag{8.47}$$

Given the equivalent circuit, one can write

$$\frac{r_0}{1 + (\omega r_0 C_0)^2} = 250 \tag{8.48}$$

Then

$$C_0 \approx 0.106\,\text{pF} \tag{8.49}$$

Now given the equivalent circuit, one can write

$$jL_c\omega - \frac{jr_0^2 C_0 \omega}{1 + (\omega r_0 C_0)^2} \approx -j231.2 \tag{8.50}$$

Then

$$L_c \approx 1\,\text{nH} \tag{8.51}$$

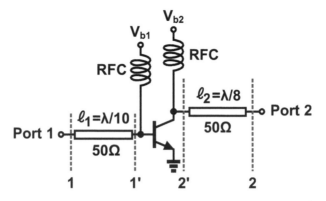

Figure 8.14: The effect of transmission line on the S-parameters.

Example 8.7 Derive the S-parameters for the transistor shown in Figure 8.14. The S-parameters measured at ports one and two are given at 1 GHz as $S_{11} = 0.3\angle -20°, S_{21} = 2\angle 20°, S_{12} = 0.02\angle -20°$, and $S_{22} = 0.15\angle -62°$. The T-lines are lossless.

Solution:
The new S_{11} is given by

$$S'_{11} = S_{11}e^{j2\beta l_1} = S_{11}e^{j\frac{4\pi}{\lambda}\frac{\lambda}{10}} = 0.3\angle(-20° +72°) = 0.3\angle 52° \tag{8.52}$$

Similarly, S_{22} can be computed as

$$S'_{22} = S_{22}e^{j2\beta l_2} = S_{22}e^{j\frac{4\pi}{\lambda}\frac{\lambda}{8}} = 0.15\angle(-62° +90°) = 0.15\angle 28° \tag{8.53}$$

It should be noted that forward gain and reverse isolation will also experience a 81° phase shift, yielding

$$S'_{12} = S_{12}e^{j\beta(l_1+l_2)} = S_{12}e^{j\frac{2\pi}{\lambda}\left(\frac{\lambda}{10}+\frac{\lambda}{8}\right)} = 0.02\angle(-20° +36° +45°)$$
$$= 0.02\angle 61° \tag{8.54}$$

$$S'_{21} = S_{21}e^{j\beta(l_1+l_2)} = S_{21}e^{j\frac{2\pi}{\lambda}\left(\frac{\lambda}{10}+\frac{\lambda}{8}\right)} = 2\angle(20° +36° +45°) = 2\angle 101° \tag{8.55}$$

■

Example 8.8 For the transistor model depicted in Figure 8.15, plot S_{11} in the frequency range of $0-1$ GHz on a polar coordinate, in $50\,\Omega$ reference impedance system. Put a cross mark on the frequencies 50 MHz, 200 MHz, 500 MHz, 850 MHz, 1000 MHz.

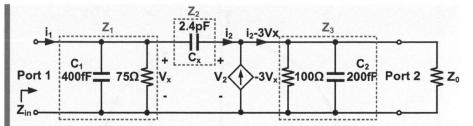

Figure 8.15: Small-signal model of the transistor for S_{11} calculation.

Solution:
To compute S_{11}, we should first calculate the input impedance while the output is loaded by Z_0. It can be found from Figure 8.15 that

$$i_1 = \frac{V_x}{Z_1} + i_2 \tag{8.56}$$

or

$$i_1 = \frac{V_x}{Z_1} + \frac{V_x - V_2}{Z_2} \tag{8.57}$$

Having

$$V_2 = (i_2 - 3V_x)Z_L \tag{8.58}$$

where $Z_L = Z_3 \parallel Z_0$. Substituting the latter equation in the former, we obtain

$$i_1 = V_x \left(\frac{1}{Z_1} + \frac{1}{Z_2} - \frac{\frac{1}{Z_2} - 3}{1 + \frac{Z_2}{Z_L}} \right) \tag{8.59}$$

or

$$Z_{in} = \frac{V_x}{i_1} = \frac{1}{\left(\frac{1}{Z_1} + \frac{1}{Z_2} - \frac{\frac{1}{Z_2} - 3}{1 + \frac{Z_2}{Z_L}} \right)} \tag{8.60}$$

S_{11} can be found as

$$S_{11} = \frac{Z_{in} - Z_0}{Z_{in} + Z_0} = \frac{\dfrac{1}{\frac{1}{Z_1} + \frac{1}{Z_2} - \frac{\frac{1}{Z_2} - 3}{1 + \frac{Z_2}{Z_L}}} - Z_0}{\dfrac{1}{\frac{1}{Z_1} + \frac{1}{Z_2} - \frac{\frac{1}{Z_2} - 3}{1 + \frac{Z_2}{Z_L}}} + Z_0} \tag{8.61}$$

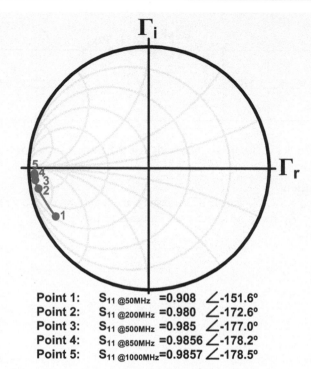

Point 1: $S_{11\ @50MHz} = 0.908\ \angle\text{-}151.6°$
Point 2: $S_{11\ @200MHz} = 0.980\ \angle\text{-}172.6°$
Point 3: $S_{11\ @500MHz} = 0.985\ \angle\text{-}177.0°$
Point 4: $S_{11\ @850MHz} = 0.9856\ \angle\text{-}178.2°$
Point 5: $S_{11\ @1000MHz} = 0.9857\ \angle\text{-}178.5°$

Figure 8.16: The plot of S_{11} of the transistor model from 50 MHz to 1000 MHz.

where

$$Z_1 = \frac{1}{j\omega\,(400\,\text{fF})}\,\|\,75\,\Omega, Z_2 = \frac{1}{j\omega\,(2.4\,\text{pF})}, Z_3 = \frac{1}{j\omega\,(200\,\text{fF})}\,\|\,100\,\Omega \quad (8.62)$$

The S_{11} plot is given as below

■

Example 8.9 For a transistor, the output circuit model is shown in Figure 8.17. Compute S_{22} in the frequency span of $100\,\text{MHz} - 1.5\,\text{GHz}$.

Solution:
S_{22} is given by

$$S_{22} = \frac{\frac{1}{jc_o\omega+\frac{1}{r_o}} - Z_0}{\frac{1}{jc_o\omega+\frac{1}{r_o}} + Z_0} = \frac{1 - Z_0\left(jc_o\omega+\frac{1}{r_o}\right)}{1 + Z_0\left(jc_o\omega+\frac{1}{r_o}\right)} \quad (8.63)$$

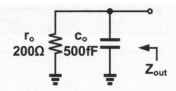

Figure 8.17: Equivalent circuit for S_{22} modeling.

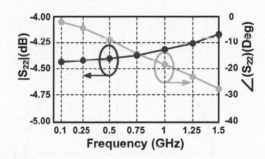

Figure 8.18: The amplitude and phase of S_{22} of the transistor.

S_{22} phase and amplitude behavior is depicted in Figure 8.18 from the above computation

Example 8.10 Two lossless transmission lines have been used to connect two ports of a transistor on a PCB to two SMA connectors for S-parameter measurements (consider the transmission lines as airlines). If the reference plane in test 1 has been calibrated for S_{11} and S_{22}, and assuming $Z_{out} = 25 - j50$ and $Z_{in} = 10$ at 500 MHz, determine the real input and output impedance of the transistor on the Smith chart. (You may use the Smith software to do this.)

Solution:
We first derive the S-parameters considering the reference plane for test 1 as

$$S_{11} = \frac{Z_{in} - Z_0}{Z_{in} + Z_0} = \frac{10 - 50}{10 + 50} = -\frac{2}{3} = 0.67\angle 180° \qquad (8.64)$$

$$S_{22} = \frac{Z_{out} - Z_0}{Z_{out} + Z_0} = \frac{25 - j50 - 50}{25 - j50 + 50} = \frac{1 - j8}{13} = 0.62\angle -82.9° $$

The wavelength can be calculated as $\lambda = c/f = 60$ cm, and hence the length of the line is equal to $\lambda/15$. These two parameters at the test reference plane 2 will become

$$S'_{11} = S_{11}e^{j2\beta l} = 0.67\angle 228° \qquad (8.65)$$

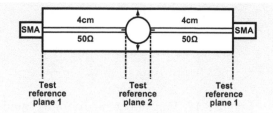

Figure 8.19: The circuit used in Example 8.10.

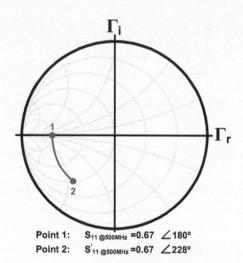

Point 1: $S_{11\ @500MHz} = 0.67\ \angle 180°$

Point 2: $S'_{11\ @500MHz} = 0.67\ \angle 228°$

Figure 8.20: The input reflection coefficient, depicted on the Smith chart [(1): S_{11}, (2): S'_{11}].

and

$$S'_{22} = S_{22}e^{j2\beta l} = 0.62\angle -34.9° \tag{8.66}$$

In another way, these parameters can be calculated as

$$Z_{in} = Z_0 \left(\frac{Z'_{in} + jZ_0 \tan(\beta \ell)}{Z_0 + jZ'_{in} \tan(\beta \ell)} \right) = 50 \left(\frac{Z'_{in} + j50 \tan\left(\frac{2\pi}{15}\right)}{50 + jZ'_{in} \tan\left(\frac{2\pi}{15}\right)} \right) = 10 \tag{8.67}$$

By solving for Z'_{in}, we arrive at $Z'_{in} = 11.89 - j21.19$ and the reflection coefficient becomes

$$S'_{11} = \frac{Z'_{in} - Z_0}{Z'_{in} + Z_0} = 0.67\angle 228° \tag{8.68}$$

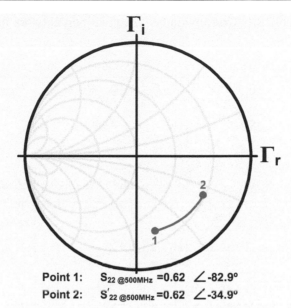

Point 1: $S_{22\,@500\text{MHz}} = 0.62 \angle\text{-}82.9°$

Point 2: $S'_{22\,@500\text{MHz}} = 0.62 \angle\text{-}34.9°$

Figure 8.21: The output reflection coefficient, depicted on the Smith chart [(1): S_{22}, (2): S'_{22}].

The same steps should be taken for the output impedance. Hence,

$$Z_{\text{out}} = Z_0 \left(\frac{Z'_{\text{out}} + jZ_0 \tan(\beta\ell)}{Z_0 + jZ'_{\text{out}} \tan(\beta\ell)} \right) = 50 \left(\frac{Z'_{\text{out}} + j50 \tan\left(\frac{2\pi}{15}\right)}{50 + jZ'_{\text{out}} \tan\left(\frac{2\pi}{15}\right)} \right) = 25 - j50 \quad (8.69)$$

and solving for the output impedance, it yields $Z'_{\text{out}} = 83.77 - j96.6$ and again the output reflection coefficient can be computed as

$$S'_{22} = \frac{Z'_{\text{out}} - Z_0}{Z'_{\text{out}} + Z_0} = 0.62\angle{-34.9°} \quad (8.70)$$

∎

8.3 Conversion of Network Matrices

It is noteworthy that a linear two port can be characterized by either of the impedance matrix, admittance matrix, or scattering parameters. Modern day network analyzers while measuring the scattering parameters can compute online the [Z] and the [Y] matrices as well. In Table 8.1, the conversion of all the three network matrices is represented, so that a user can compute either of the two matrices once he/she has the measurement results of one of the matrices.

Table 8.1: Conversion between two-port network matrices.

	S	Z	Y
S_{11}	S_{11}	$\frac{(Z_{11}-Z_0)(Z_{22}+Z_0)-Z_{12}Z_{21}}{\Delta Z}$	$\frac{(Y_0-Y_{11})(Y_0+Y_{22})+Y_{12}Y_{21}}{\Delta Y}$
S_{12}	S_{12}	$\frac{2Z_{12}Z_0}{\Delta Z}$	$\frac{-2Y_{12}Y_0}{\Delta Y}$
S_{21}	S_{21}	$\frac{2Z_{21}Z_0}{\Delta Z}$	$\frac{-2Y_{21}Y_0}{\Delta Y}$
S_{22}	S_{22}	$\frac{(Z_{11}+Z_0)(Z_{22}-Z_0)-Z_{12}Z_{21}}{\Delta Z}$	$\frac{(Y_0+Y_{11})(Y_0-Y_{22})+Y_{12}Y_{21}}{\Delta Y}$
Z_{11}	$Z_0\frac{(1+S_{11})(1-S_{22})+S_{12}S_{21}}{(1-S_{11})(1-S_{22})-S_{12}S_{21}}$	Z_{11}	$\frac{Y_{22}}{\lvert Y\rvert}$
Z_{12}	$Z_0\frac{2S_{12}}{(1-S_{11})(1-S_{22})-S_{12}S_{21}}$	Z_{12}	$\frac{-Y_{12}}{\lvert Y\rvert}$
Z_{21}	$Z_0\frac{2S_{21}}{(1-S_{11})(1-S_{22})-S_{12}S_{21}}$	Z_{21}	$\frac{-Y_{21}}{\lvert Y\rvert}$
Z_{22}	$Z_0\frac{(1-S_{11})(1+S_{22})+S_{12}S_{21}}{(1-S_{11})(1-S_{22})-S_{12}S_{21}}$	Z_{22}	$\frac{Y_{11}}{\lvert Y\rvert}$
Y_{11}	$Y_0\frac{(1-S_{11})(1+S_{22})+S_{12}S_{21}}{(1+S_{11})(1+S_{22})-S_{12}S_{21}}$	$\frac{Z_{22}}{\lvert Z\rvert}$	Y_{11}
Y_{12}	$Y_0\frac{-2S_{12}}{(1+S_{11})(1+S_{22})-S_{12}S_{21}}$	$\frac{-Z_{12}}{\lvert Z\rvert}$	Y_{12}
Y_{21}	$Y_0\frac{-2S_{21}}{(1+S_{11})(1+S_{22})-S_{12}S_{21}}$	$\frac{-Z_{21}}{\lvert Z\rvert}$	Y_{21}
Y_{22}	$Y_0\frac{(1+S_{11})(1-S_{22})+S_{12}S_{21}}{(1+S_{11})(1+S_{22})-S_{12}S_{21}}$	$\frac{Z_{11}}{\lvert Z\rvert}$	Y_{22}

$\Delta Z = (Z_{11}+Z_0)(Z_{22}+Z_0)-Z_{12}Z_{21}$; $\Delta Y = (Y_{11}+Y_0)(Y_{22}+Y_0)-Y_{12}Y_{21}$; $\lvert Z\rvert = Z_{11}Z_{22}-Z_{12}Z_{21}$; $\lvert Y\rvert = Y_{11}Y_{22}-Y_{12}Y_{21}$; $Y_0 = \frac{1}{Z_0}$

8.4 Conclusion

Scattering parameters are mainly used for high-frequency measurement of active and passive devices. These parameters are measured using the network analyzer. Network analyzers can distinguish between the incident and the reflected waves at the two ports of the Device Under the Test (DUT). The scattering parameters are determined with respect to a reference impedance. This reference impedance in most of the measurement equipment is chosen to be $50\,\Omega$. The input and the output reflection coefficients of a device, i.e., S_{11} and S_{22} can normally be shown on the Smith chart, while the forward gain S_{21} and the reverse gain S_{12} of a two-port are depicted on polar or Cartesian coordinates. It is convenient to transform the scattering parameters into impedance/admittance matrices and vice versa. An RF transistor S-parameters can be measured, using a network analyzer, through a calibration procedure where the effects of the interconnects can be canceled out.

8.5 **References and Further Reading**

1. D. Pozar, *Microwave Engineering*, fourth edition, Hoboken, NJ: J. Wiley & Sons, Inc., 2012.
2. S. Ramo, J.R. Whinnery, T. Van Duzer, *Fields and Waves in Communication Electronics*, New York: J. Wiley & Sons, 1994.
3. G. Gonzalez, *Microwave Transistor Amplifiers: Analysis and Design.*, NJ: Prentice-Hall, 1994.
4. U.L. Rohde, A.M. Pavio, G.D. Vendelin, *Microwave Circuit Design Using Linear and Nonlinear Techniques*, Hoboken, NJ: J. Wiley & Sons, Inc., 2005.

8.6   Problems

Problem 8.1 Determine the S-parameters for the two-port network depicted in Figure 8.22.

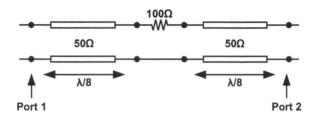

Figure 8.22: The resistive/transmission line circuit for S-parameters determination.

Problem 8.2 The input of a transistor is modeled as in Figure 8.23. Sketch S_{11} in a $50\,\Omega$ system on the Smith chart from 0 to 1 GHz.

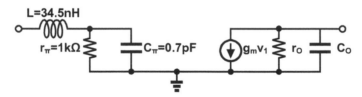

Figure 8.23: The circuit model for a transistor to compute S_{11}.

Problem 8.3 The input of a bipolar transistor at 1 GHz is modeled by an RLC circuit as depicted in Figure 8.24. Here, $S_{11} = 0.63\angle -54°$. A π network is used for impedance matching to $50\,\Omega$, where $L_1 = 7\,\text{nH}$.

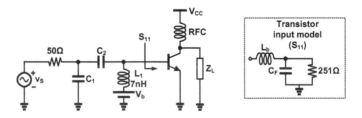

Figure 8.24: The input circuit for a BJT amplifier.

1. Find the proper values of C_F and L_b in the model to give in the stated value of S_{11}.
2. Determine C_1 and C_2 for the required matching.
3. If the input inductor loss is modeled by a series resistance of $5\,\Omega$, and assuming $Q_{C_1} = Q_{C_2} = 100$, estimate the bandwidth of the matching network and calculate the matching circuit loss in this case, in dB.

Problem 8.4 Determine the S-parameters in a $50\,\Omega$ reference system, in the two attenuator networks (a) and (b) shown in Figure 8.25.

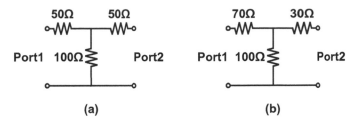

Figure 8.25: Two T-section attenuator circuits.

Problem 8.5 The input of a transistor is modeled as either of the networks depicted in Figure 8.26, plot S_{11} in a $75\,\Omega$ reference system on the Smith chart for each of the three cases shown in the figure for the frequency range of $0-1$ GHz (with steps of 100 MHz). Compare these three cases.

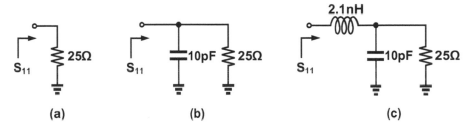

Figure 8.26: The transistors' input equivalent circuits.

Problem 8.6 Considering the bipolar transistor model given in Figure 8.27, determine the corresponding expressions for the S-parameters, in terms of the circuit parameters, with a reference impedance of Z_0. First assume $C_\mu \parallel r_\mu$ is negligible, and then repeat the problem while taking into consideration their effect.

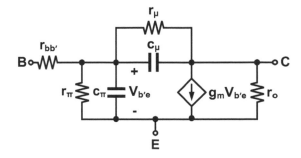

Figure 8.27: The equivalent circuit of the transistor to determine the S-parameters.

Problem 8.7 Compute the S-parameters of the network depicted in Figure 8.28 at 2 GHz.

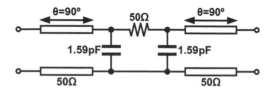

Figure 8.28: The RC section for S-parameters' determination.

Problem 8.8 Find an expression for the two-port S-parameters for a reference impedance of Z_0 as shown in Figure 8.29. For which values of L and C, S_{11} and S_{22} will tend to zero?

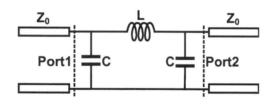

Figure 8.29: The delay cell for S-parameters' determination.

Problem 8.9 A transistor's S-parameters have been measured through 2 mm length line pads at 4 GHz. The substrate effective relative permittivity as in Figure 8.30 has a value of 2. What are the S-parameters of the pure transistor?

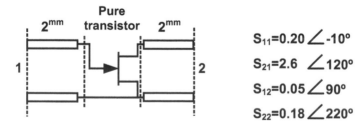

Figure 8.30: The transistor S-parameters for de-embedding.

Problem 8.10 Determine the S-parameter of the cascaded chain of amplifiers as in Figure 8.31.

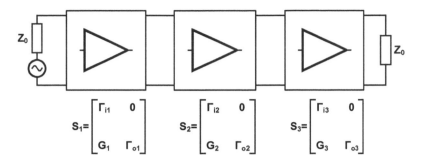

Figure 8.31: The cascaded chain of three unilateral amplifiers.

Problem 8.11 A pair of similar amplifiers are cascaded by two three dB 90° couplers with the specified S-parameters as in Figure 8.32. Determine the overall S-parameters between port 1 and port 2. What is the advantage of this architecture?

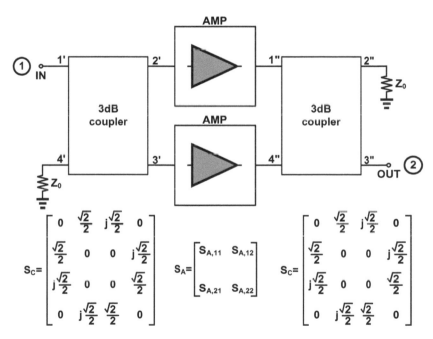

Figure 8.32: A pair of balanced amplifiers cascaded by two 90°, 3 dB couplers.

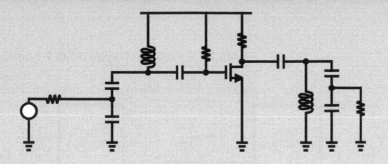

9. Amplifier Design Using S-parameters

9.1 Amplifier Design Using Scattering Parameters

In this chapter, we mainly focus on the analysis and basic design of microwave transistor-based amplifiers using S-parameters usually available from the datasheet. The most important design factors in microwave amplifiers include stability, power gain, bandwidth, noise figure, dc biasing, and power consumption. The first amplifier design step is choosing a proper transistor. Subsequently, using the measured S-parameters, matching, gain, and stability factor can be evaluated accordingly. An unconditionally stable transistor would not oscillate with a passive load and a passive source. However, for potentially unstable transistors, specific measures should be taken into account while terminating them to passive loads/sources using the required stability contours on the Smith chart.

9.2 Specification of Amplifiers

A generic transistor amplifier can be completely characterized using these parameters: (1) stability (absence of unwanted oscillations), (2) maximum power gain, (3) input and output impedances, (4) the transducer gain, (5) optimum load and source impedances, and (6) simultaneous conjugate matching.

It can be easily concluded that all the above parameters are subject to variations with frequency or bias level, complicating the design of wide-band microwave amplifiers. Finding a stable transistor in its proposed quiescent point can be considered as the first step in the design procedure. The desired performance can be estimated using S-parameters. Consider Figure 9.1, where the general case of a microwave amplifier is depicted with its input and output matching circuits.

Contrary to the ordinary analog electronic circuits where the parameters such as the voltage gain and the current gain are mostly evaluated and employed, in RF communication circuits, the power gain is of most importance. The reason is that

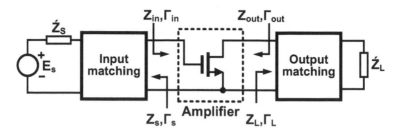

Figure 9.1: Illustration of the general case of a microwave amplifier with input and output matching circuits.

in communication systems, the RF power is the most scarce quantity, and it should compete with noise, so approximately in all RF stages, the power gain is the quantity to be optimized. Let us now introduce the most important property of a microwave amplifier, that is, its power gain. Three separate definitions are considered for the power gain as follows.

(1) Transducer Power Gain is the power delivered to the load, Z_L, divided by the available power of the source, P_{AVS}, i.e.,

$$G_T = \frac{P_L}{P_{AVS}} = \frac{G_L|V_L|^2}{\frac{1}{4G_S}|I_S|^2} = 4G_SG_L\frac{|V_L|^2}{|I_S|^2} \tag{9.1}$$

Here, G_S is the source conductance, G_L is the load conductance, V_L is the load voltage, and I_S is the source's short circuit currents. Here, if the source and the load reflection coefficients are complex conjugate of the reflection coefficients of the input and the output of the transistor, that is,

$$\Gamma_L = \Gamma_{out}{}^* \tag{9.2}$$

then the absorbed power of the load becomes equal to the available power of the output

$$P_L = P_{AVO} \tag{9.3}$$

and if

$$\Gamma_{in} = \Gamma_s{}^* \tag{9.4}$$

then the available power of the source will be absorbed by the input, that is,

$$P_{AVS} = P_{in} \tag{9.5}$$

In this case, the transducer gain will attain its maximum value. This is the simultaneous conjugate match condition.

In terms of device's S-parameters and the source and the load reflection coefficients, the transducer power gain can be expressed as

$$G_T = \frac{|S_{21}|^2 \left(1 - |\Gamma_S|^2\right) \left(1 - |\Gamma_L|^2\right)}{|1 - \Gamma_S \Gamma_{in}|^2 |1 - S_{22} \Gamma_L|^2} \tag{9.6}$$

This is the most realistic definition of the power gain.

(2) Available Power Gain (APG) is the ratio of the output available power to the power available from the source, and can be expressed as

$$G_A = \frac{P_{AVO}}{P_{AVS}} = \frac{\frac{1}{4G_O}|I_O|^2}{\frac{1}{4G_S}|I_S|^2} = \frac{G_S}{G_O}\frac{|I_O|^2}{|I_S|^2} \tag{9.7}$$

Here, G_S is the source conductance, G_O is the output conductance, I_S is the source's short-circuit current, and I_O is the output's short-circuit current. If $\Gamma_{in} = \Gamma_s{}^*$, that is, $P_{AVS} = P_{in}$, then the APG will attain its maximum value.

In terms of device's S-parameters and the source reflection coefficient, the APG can be expressed as

$$G_A = \frac{|S_{21}|^2 \left(1 - |\Gamma_S|^2\right)}{|1 - S_{11}\Gamma_S|^2 \left(1 - |\Gamma_{out}|^2\right)} \tag{9.8}$$

One can generally expect that

$$G_T \le G_A \tag{9.9}$$

(3) Operational Power Gain is the ratio of the power absorbed by the load from the network to the power delivered to the network from the source, which can be written as

$$G_P = \frac{P_L}{P_{in}} = \frac{G_L |V_L|^2}{G_{in} |V_{in}|^2} = \frac{G_L}{G_{in}}|A_V|^2 \tag{9.10}$$

Here, G_{in} is the input conductance, G_L is the load conductance, V_L is the load voltage, V_{in} is the input voltage, and A_V is the voltage gain. If $\Gamma_L = \Gamma_{out}{}^*$, that is, $P_L = P_{AVO}$, then power gain will attain its maximum value. One can generally expect that

$$G_T \le G_P \tag{9.11}$$

In case of simultaneous conjugate match, all these three quantities will converge to the value, that is,

$$G_{P_{max}} = G_{A_{max}} = G_{T_{max}} \tag{9.12}$$

We shortly develop expressions for each one the above triple definitions in terms of S-parameters.

9.3   Performance Parameters of an Amplifier

We have recognized thus far that the transducer power gain is dependent on the input reflection coefficient, output reflection coefficient, and S-parameters, the power gain (operational power gain) is dependent on output reflection coefficient and S-parameters, and finally, the APG is dependent on the input reflection coefficient and S-parameters. Another point of importance is the matching network in amplifiers which match the input and the output impedances to a certain value, say $50\,\Omega$.

9.3.1   Stability

Passive loads, in general, have a reflection coefficient with an absolute value less than unity, whereas for active two-ports (such as biased transistors), S-parameters could be such that the input and/or the output reflection coefficients can have an absolute value larger than unity. In this case, the circuit can become potentially unstable. In other words, this means the input resistance or the output resistance is negative, which can lead to instability. That is $|\Gamma_{in}| > 1$, or $|\Gamma_{out}| > 1$. In case of a unilateral network, for example, $S_{12} = 0$, the unstable case reduces to $|S_{11}| > 1$ and $|S_{22}| > 1$, each revealing the presence of a negative input or output resistances.

Regarding the stability, one could say a two-port is *unconditionally stable* if the input reflection coefficient's absolute value is smaller than unity for all passive loads, and the output reflection coefficient's absolute value is smaller than unity for all passive sources. Or in other words, if the real part of both the input and the output impedances for all passive load impedances and for all passive source impedances are positive. We should bear in mind that all the expressions used here are in the frequency domain, and hence, they are valid only within a certain bandwidth. If a two-port network is not unconditionally stable, there exists a combination of load and source impedances that leads to a negative real part for the input or a negative real part for the output impedances. In terms of the reflection coefficients the stability means, the input reflection coefficient and the output reflection coefficient for passive loads/sources should satisfy the following condition, *i.e.*, for $|\Gamma_s| < 1$ and $|\Gamma_L| < 1$, can be described as

$$|\Gamma_{in}| = \left| S_{11} + \frac{S_{21}\Gamma_L S_{12}}{1 - S_{22}\Gamma_L} \right| \le 1 \quad \text{for all } |\Gamma_L| \le 1 \tag{9.13}$$

and

$$|\Gamma_{out}| = \left| S_{22} + \frac{S_{21}\Gamma_s S_{12}}{1 - S_{11}\Gamma_s} \right| \le 1 \quad \text{for all } |\Gamma_S| \le 1 \tag{9.14}$$

Relations 9.13 and 9.14 suggest that if the load or the source terminations are passive, the input and the output reflection coefficients should remain passive, that is, with an absolute value less than unity. To check the stability condition, relations 9.13 and 9.14 can be solved by putting each of the left-hand sides of the inequalities equal to unity. A two-port network is conditionally stable if there exists passive loads for which the absolute value of the input reflection coefficient is greater than unity ($|\Gamma_{in}| > 1$), and there exists passive sources for which the absolute value of the output reflection coefficient is greater than unity ($|\Gamma_{out}| > 1$). As such the concept of stability using input and output reflection coefficients, the problem of stability can be easily handled, taking into consideration the locus for $|\Gamma_{in}| = 1$, once the load impedance varies, and the locus for $|\Gamma_{out}| = 1$ once the source impedance varies, on the Smith chart. It can be shown that both loci are described by circles given by the following relations (These circles correspond to relations 9.13 and 9.14, respectively, once the equality holds).

The locus in the source plane can be described by

$$|\Gamma_s - C_s|^2 = r_s^2 \qquad (9.15)$$

Equation 9.15 describes a circle on the Smith chart whose center is C_s and whose radius is r_s.
where

$$r_s = \left| \frac{S_{12}S_{21}}{|S_{11}|^2 - |\Delta|^2} \right| \qquad (9.16)$$

$$C_s = \frac{(S_{11} - \Delta S_{22}^*)^*}{|S_{11}|^2 - |\Delta|^2} \qquad (9.17)$$

and the locus in the load plane can be described by

$$|\Gamma_L - C_L|^2 = r_L^2 \qquad (9.18)$$

Equation 9.18 describes a circle on the Smith chart whose center is C_L and whose radius is r_L, where

$$r_L = \left| \frac{S_{12}S_{21}}{|S_{22}|^2 - |\Delta|^2} \right| \qquad (9.19)$$

$$C_L = \frac{(S_{22} - \Delta S_{11}^*)^*}{|S_{22}|^2 - |\Delta|^2} \qquad (9.20)$$

In Equations 9.16 and 9.20, Δ is the S-matrix determinant.

$$\Delta = S_{11}S_{22} - S_{12}S_{21} \qquad (9.21)$$

With S-parameters known at a particular frequency, using Equations 9.16 and 9.17, the circle corresponding to $|\Gamma_{out}| = 1$ can be found on the source impedance plane (Smith chart) and the circle corresponding to $|\Gamma_{in}| = 1$ can be found on the load impedance plane (Smith chart). These two circles are called the source stability circle and the load stability circle, respectively. This means that these circles are the boundary between the stable and the unstable region.

Therefore, either the inside or the outside of the stability circles corresponds to the stable region, in other words the region where $|\Gamma_{out}| < 1$ or $|\Gamma_{in}| < 1$, respectively. If the output is loaded by $\Gamma_L = 0$, then the input reflection coefficient would become $\Gamma_{in} = S_{11}$. The following cases could occur:

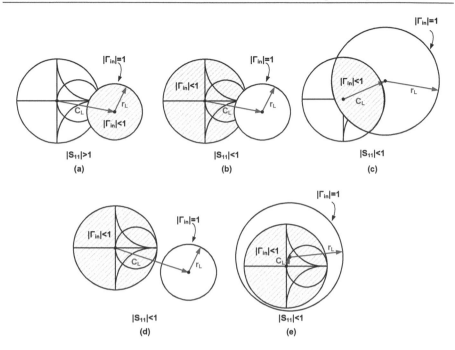

Figure 9.2: Input stability circles on the load plane Smith chart for five cases (the shaded areas correspond to the stable region), (a) for $|S_{11}| > 1$ and conditional stability, (b) for $|S_{11}| < 1$, the stability circle intersecting the chart while not comprising the chart center, and consequently, conditional stability, (c) for $|S_{11}| < 1$, the stability circle intersecting the chart while comprising the chart center, and consequently, conditional stability, (d) for $|S_{11}| < 1$, the stability circle does not intersect the chart, and consequently, unconditional stability, and (e) for $|S_{11}| < 1$, the stability circle comprises the whole chart, and consequently, unconditional stability.

1. If $|S_{11}| > 1$, then obviously there is a load ($\Gamma_L = 0$) for which the two-port becomes unstable, and therefore, the two-port can be conditionally stable (Figure 9.2(a)).
2. If $|S_{11}| < 1$, and if the stability circle intersects the Smith chart, and the center of the chart ($\Gamma_L = 0$) is outside the stability circle, the points inside the chart and outside the stability circle correspond to the stable region. Anyway, the two-port is conditionally stable in this case (Figure 9.2(b)).
3. If $|S_{11}| < 1$, and if the stability circle intersects the Smith chart, and the center of the chart ($\Gamma_L = 0$) is inside the stability circle, the points inside the chart and inside the stability circle correspond to the stable region. Anyway, the two-port is conditionally stable in this case (Figure 9.2(c)).
4. If $|S_{11}| < 1$, and if the stability circle does not intersect the Smith chart, and the center of the chart ($\Gamma_L = 0$) is outside the stability circle, the points inside

the chart and outside the stability circle correspond to the stable region. Consequently, the two-port is unconditionally stable (Figure 9.2(d)).

5. If $|S_{11}| < 1$, and if the stability circle does not intersect the Smith chart but the whole chart is inside the stability circle (evidently the center of the chart is inside the stability circle as well), the points inside the chart and inside the stability circle correspond to the stable region. Consequently, the two-port is unconditionally stable (Figure 9.2(e)).

The same argument can be used for the source plane and S_{22}, as depicted in Figure 9.3. This leads to an important conclusion: if $|S_{11}| < 1$ and $|S_{22}| < 1$, the network is unconditionally stable if and only if the stability circles at the load plane and at the source plane do not intersect the Smith chart. These points are demonstrated in Figures 9.2 and 9.3.

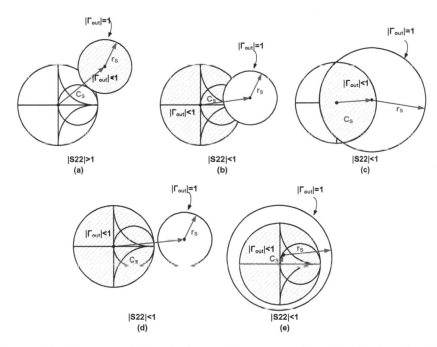

Figure 9.3: Output stability circles on the source plane Smith chart for five cases (the shaded areas correspond to the stable region), (a) for $|S_{22}| > 1$ and conditional stability, (b) for $|S_{22}| < 1$, the stability circle intersecting the chart while not comprising the chart center, and consequently, conditional stability, (c) for $|S_{22}| < 1$, the stability circle intersecting the chart while comprising the chart center, and consequently, conditional stability, (d) for $|S_{22}| < 1$, the stability circle does not intersect the chart and consequently, unconditional stability, and (e) for $|S_{22}| < 1$, the stability circle comprises the whole chart, and consequently, unconditional stability.

As observed in Figure 9.2(b), when $|S_{11}| < 1$, the shaded area includes the origin, while in Figure 9.2(a), when $|S_{11}| > 1$, only a small area within the Smith chart is covered by the stability circle. The same point is illustrated for the output reflection coefficient in Figure 9.3. In Figure 9.3(b), when $|S_{22}| < 1$, the shaded area includes the origin, while in Figure 9.3(a), when $|S_{22}| > 1$, only a small area within the Smith chart is covered by the stability circle.

Now, the necessary and the sufficient conditions for unconditional stability of a two-port are that the stability circles do not intersect the Smith chart at the source and at the load planes. That is

$$||C_L| - r_L| > 1 \quad \text{and} \quad |S_{11}| < 1 \tag{9.22}$$

$$||C_s| - r_s| > 1 \quad \text{and} \quad |S_{22}| < 1 \tag{9.23}$$

Given $|S_{11}| < 1$ and $|S_{22}| < 1$, the above conditions can be translated into the following pair of conditions:

$$k = \frac{1 - |S_{11}|^2 - |S_{22}|^2 + |\Delta|^2}{2|S_{12}S_{21}|} > 1 \tag{9.24}$$

and

$$|\Delta| = |S_{11}S_{22} - S_{12}S_{21}| < 1 \tag{9.25}$$

In Equation 9.24, k is called Rollett's stability factor. Equations 9.24 and 9.25 are the necessary and the sufficient conditions for stability of the two-port, given $|S_{11}| < 1$ and $|S_{22}| < 1$. Or, in other words, the conditions are $k > 1$ and $|\Delta| < 1$.

9.3.2 Maximum APG

It can be shown that in order to obtain the maximum power gain, simultaneous conjugate matching must be realized at the input and the output. This conjugate matching is only possible for the unconditionally stable two-port. Then the maximum available gain can be found as

$$MAG = 10\log \left| \frac{S_{21}}{S_{12}} \right| \times \left| k - \sqrt{k^2 - 1} \right| \tag{9.26}$$

provided $S_{12} \neq 0$ and $k > 1$. Note that for the case where $S_{12} = 0$ one can use the unilateral transducer power gain as in Equation 9.29. Now, consider when we have no constraint on the two-port and the two-port is unconditionally stable, for simultaneous conjugate match, we should have

$$\Gamma_{in} = S_{11} + \frac{S_{12}S_{21}\Gamma_L}{1 - S_{22}\Gamma_L} = \Gamma_s^* \tag{9.27}$$

and

$$\Gamma_{out} = S_{22} + \frac{S_{12}S_{21}\Gamma_s}{1 - S_{11}\Gamma_s} = \Gamma_L{}^* \tag{9.28}$$

The simultaneous resolution of Equations 9.27 and 9.28 which results in maximum APG or the maximum transducer gain is called the simultaneous complex conjugate match (we discuss the solution of these equations in the following). Therefore, for design purposes, we first choose a specific transistor with required power-frequency capability. Then, using the load and the source, proper matching networks can be designed. We should also note that there exists a relationship between the input and the output reflection coefficients through S_{12}, further complicating impedance matching. For the sake of simplicity, we first consider the case where $S_{12} = 0$, which is the property of a unilateral transistor.

Conjugate Matching for a Unilateral Amplifier

As stated earlier, in this case, we have $S_{12} = 0$. As a result, we have $\Gamma_{in} = S_{11}$ and $\Gamma_{out} = S_{22}$. Considering the complex conjugate matching condition, which simply reduces to $\Gamma_s = S_{11}{}^*$ and $\Gamma_L = S_{22}{}^*$, and ultimately, using the transducer power gain expression, we arrive at

$$G_T = \frac{1}{1 - |S_{11}|^2}|S_{21}|^2\frac{1}{1 - |S_{22}|^2} = G_{s,max}|S_{21}|^2 G_{L,max} \tag{9.29}$$

This transducer gain could be translated to three parts. The first part expresses the source matching gain ($G_{s,max}$), the second part, the unmatched transducer gain ($|S_{21}|^2$), and the third part, the load matching gain ($G_{L,max}$).

Conjugate Matching for a Bilateral Amplifier

Here we consider the two-port is no more unilateral, that is, $S_{12} \neq 0$. The complex conjugate matching condition can be deduced from the solution of the following pair of equations for the two unknowns Γ_L and Γ_s:

$$\Gamma_s{}^* = S_{11} + \frac{S_{21}\Gamma_L S_{12}}{1 - S_{22}\Gamma_L} \tag{9.30}$$

and

$$\Gamma_L{}^* = S_{22} + \frac{S_{21}\Gamma_s S_{12}}{1 - S_{11}\Gamma_s} \tag{9.31}$$

If Equations 9.30 and 9.31 are solved for the input and output reflection coefficients simultaneously, two complex second-order equations would result whose solutions are given in Equations 9.33 and 9.38:

$$C_1\Gamma_{s,O}^2 - B_1\Gamma_{s,O} + C_1^* = 0 \tag{9.32}$$

Therefore

$$\Gamma_{s,O} = \frac{B_1 \pm \sqrt{B_1{}^2 - 4|C_1|^2}}{2C_1} \tag{9.33}$$

where we have

$$B_1 = 1 + |S_{11}|^2 - |S_{22}|^2 - |\Delta|^2 \tag{9.34}$$

and

$$C_1 = S_{11} - \Delta S_{22}{}^* \tag{9.35}$$

There exists a solution if and only if $B_1{}^2 - 4|C_1|^2 \geq 0$ which means $k > 1$. In order to determine the sign before the square root term in Equation 9.34, the solution which has an absolute value less than unity would be acceptable. In other words, we look for the sign of B_1, if it is positive, then the square root sign is negative and *vice versa*. The source impedance can be found from

$$Z_s = Z_0 \frac{1 + \Gamma_s}{1 - \Gamma_s} \tag{9.36}$$

Equivalently for the output reflection coefficient, we have

$$C_2 \Gamma_{L,O}^2 - B_2 \Gamma_{L,O} + C_2^* = 0 \tag{9.37}$$

Therefore

$$\Gamma_{L,O} = \frac{B_2 \pm \sqrt{B_2{}^2 - 4|C_2|^2}}{2C_2} \tag{9.38}$$

where

$$B_2 = 1 + |S_{22}|^2 - |S_{11}|^2 - |\Delta|^2 \tag{9.39}$$

and

$$C_2 = S_{22} - \Delta S_{11}{}^* \tag{9.40}$$

There exists a solution if and only if $B_2{}^2 - 4|C_2|^2 \geq 0$ which means $k > 1$. In order to determine the sign before the square root term in Equation 9.38, the solution which has an absolute value less than unity would be acceptable. In other words, we look for the sign of B_2, if it is positive, then the square root sign is negative and *vice versa*. The load impedance can be found from

$$Z_L = Z_0 \frac{1 + \Gamma_L}{1 - \Gamma_L} \tag{9.41}$$

It should be noted that, if one of the source or load reflection coefficients are found, the other one can be found from either through Equations 9.31 or 9.30.

Example 9.1 The S-parameters at the given bias are measured for a bipolar transistor at 200 MHz and at $V_{CE} = 10V, I_C = 10\,mA$ as $S_{11} = 0.4\angle 162°, S_{12} = 0.04\angle 60°, S_{21} = 5.2\angle 63°, S_{22} = 0.35\angle -39°$. Assume that the amplifier is terminated to a $50\,\Omega$ impedance at both input and the output. Perform complex conjugate matching such that the maximum transducer power gain occurs.

Solution:
First, we examine the stability conditions using the provided S-parameters from Equations 9.24 and 9.25 as

$$\Delta = S_{11}S_{22} - S_{12}S_{21} = 0.4\angle 162° (0.35\angle -39°) - 0.04\angle 60° (5.2\angle 63°) \tag{9.42}$$

$$= 0.068\angle -57°$$

Now, we compute k as

$$k = \frac{1 - |S_{11}|^2 - |S_{22}|^2 + |\Delta|^2}{2|S_{12}S_{21}|} - \frac{1 - 0.4^2 - 0.35^2 + 0.068^2}{2(0.04)(5.2)} - 1.74 > 1 \tag{9.43}$$

As we have $k > 1$, $|\Delta| < 1$, $|S_{11}| < 1$, and $|S_{22}| < 1$, the two-port network is unconditionally stable. The MAG can be calculated from section 9.3.2 as

$$B_1 = 1 + |S_{11}|^2 - |S_{22}|^2 - |\Delta|^2 = 1 + 0.4^2 - 0.35^2 - 0.068^2 = 1.03 > 0 \tag{9.44}$$

It follows that

$$MAG = 10\log\left|\frac{S_{21}}{S_{12}}\right| + 10\log\left|k - \sqrt{k^2 - 1}\right| \tag{9.45}$$

$$= 10\log\frac{5.2}{0.04} + 10\log\left|1.74 - \sqrt{1.74^2 - 1}\right| = 16.1\,dB$$

For example, if the desired maximum gain was larger than 16.1 dB, this transistor could not be used. We can find the load reflection coefficient assuming the complex conjugate; we have

$$C_2 = S_{22} - \Delta S_{11}^* = (0.35\angle -39°) - (0.068\angle -57°)(0.4\angle -162°) = 0.377\angle -39° \tag{9.46}$$

and

$$B_2 = 1 + |S_{22}|^2 - |S_{11}|^2 - |\Delta|^2 = 1 + 0.35^2 - 0.4^2 - 0.068^2 = 0.958 \tag{9.47}$$

which results in

$$\Gamma_{L,O} = \frac{B_2 \pm \sqrt{B_2^2 - 4\,|C_2|^2}}{2C_2} = \frac{0.958 - \sqrt{0.958^2 - 4(0.377)^2}}{2\,(0.377\angle{-39°})} = 0.487\angle{+39°}$$

(9.48)

Therefore, we have $\Gamma_{L,O} = 0.487\angle{39°}$, which is equivalent to $Z_L = 79.5 + j64$. The source reflection coefficient can be found to be

$$\begin{aligned}
\Gamma_{s,O} &= \left(S_{11} + \frac{S_{21}\Gamma_L S_{12}}{1 - S_{22}\Gamma_L} \right)^* \\
&= \left(0.4\angle{162°} + \frac{5.2\angle{63°}\,(0.487\angle{39°})\,0.04\angle{60°}}{1 - 0.35\angle{-39°}\,(0.487\angle{39°})} \right)^* \\
&= 0.522\angle{-162°}
\end{aligned}$$

(9.49)

This reflection coefficient is equivalent to $Z_s = 16 - j7$.

Now, we show the reflection coefficients on the Smith chart.

We can now match the $50\,\Omega$ impedance to the desired $\Gamma_{s,O}$ and $\Gamma_{L,O}$, using the circuit topology shown in Figure 9.7. We begin at the origin and add a parallel capacitor to move clockwise along the constant conductance contour. Then, a series inductor is added so that we move clockwise, on the constant resistance contour, to reach the desired source reflection coefficient, $\Gamma_{s,O}$. These steps are demonstrated in Figure 9.5.

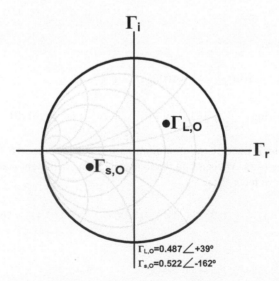

Figure 9.4: Input and output reflection coefficients on the Smith chart.

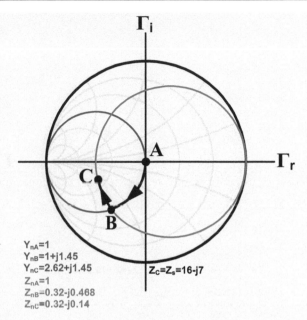

Figure 9.5: Steps for input matching on the Smith chart.

The values of the capacitor and the inductor can be found to be

$$C = \frac{|\text{Im}\{Y_B\} - \text{Im}\{Y_A\}|}{\omega} = \frac{1.45 \times 20 \times 10^{-3}}{2\pi \left(200 \times 10^6\right)} = 23pF \tag{9.50}$$

and

$$L = \frac{|\text{Im}\{Z_C\} - \text{Im}\{Z_B\}|}{\omega} = \frac{0.328 \times 50}{2\pi \left(200 \times 10^6\right)} = 13nH \tag{9.51}$$

respectively.

The same procedure can be carried out in order to find the output matching network. We first choose a series capacitor and move counterclockwise, on the constant resistance contour, from the origin to point B in Figure 9.6, then by choosing a parallel inductor, we move counterclockwise, on the constant conductance contour, to reach $\Gamma_{L,O}$ point. These steps are depicted in Figure 9.6.

The inductor and capacitor values can be similarly found as

$$L = \frac{1}{\omega |\text{Im}\{Y_C\} - \text{Im}\{Y_B\}|} = \frac{1}{2\pi \left(200 \times 10^6\right) 0.79 \times 20 \times 10^{-3}} = 50.3nH \tag{9.52}$$

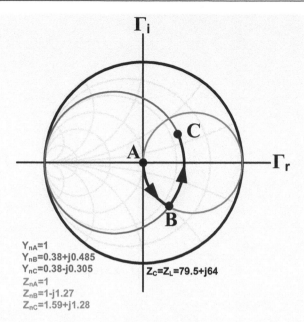

Figure 9.6: Steps for output matching on the Smith chart.

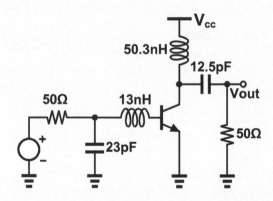

Figure 9.7: The overall matching network.

and

$$C = \frac{1}{\omega \left| \text{Im}\{Z_B\} - \text{Im}\{Z_A\} \right|} = \frac{1}{2\pi \left(200 \times 10^6 \right) 1.23 \times 50} = 12.5 pF \quad (9.53)$$

The overall matching networks in the transistor circuit are depicted in Figure 9.7. ∎

We now proceed to study the design of two-port amplifiers for a specific gain.

9.4 Power Gain Contours

When the value of $|S_{12}|$ is not negligible (the device is not unilateral), chances are that the device becomes potentially unstable and the solution for simultaneous conjugate matching does not exist. A routine and conventional method is to design the circuit for a specific operational power gain (OPG), in this case, one normally uses the constant power gain contours. Remind that the OPG is independent of the source impedance. Therefore, constant OPG contours can be drawn for both unconditional and conditional stability on the load plane.

9.4.1 OPG Contours for Bilateral Unconditionally Stable Amplifiers

Our purpose in this section is to design a two-port amplifier with a specific gain. Now, we follow the design procedure by just considering a single-stage amplifier. The straightforward way to specify the amplifier power gain is to use constant gain contours on the Smith chart. We observe that there exists certain loci of constant gain loads on the load plane. It can be shown that

$$G_p = \frac{|S_{21}|^2 \left(1 - |\Gamma_L|^2\right)}{\left(1 - \left|\frac{S_{11} - \Delta\Gamma_L}{1 - S_{22}\Gamma_L}\right|^2\right)|1 - S_{22}\Gamma_L|^2} = |S_{21}|^2 g_p \qquad (9.54)$$

where

$$g_p = \frac{G_p}{|S_{21}|^2} = \frac{1 - |\Gamma_L|^2}{1 - |S_{11}|^2 + |\Gamma_L|^2 \left(|S_{22}|^2 - \Delta^2\right) - 2\operatorname{Re}(\Gamma_L C_2)} \qquad (9.55)$$

where C_2 is given by Equation 9.40. In the above equations, G_p and g_p are a function of S-parameters and the load reflection coefficient. It can be shown that those values of load reflection coefficient which yield a constant g_p lie on a circle, which we refer to as constant OPG contour from now on. The equation of this contour on the load reflection coefficient plane is given by (9.56):

$$|\Gamma_L - C_p| = r_p \qquad (9.56)$$

where C_p is the center and r_p is the radius of the circle which are given by

$$C_p = \frac{g_p C_2{}^*}{1 + g_p \left(|S_{22}|^2 - |\Delta|^2\right)} \qquad (9.57)$$

and

$$r_p = \frac{\sqrt{1 - 2k|S_{12}S_{21}|g_p + |S_{12}S_{21}|^2 g_p{}^2}}{\left|1 + g_p \left(|S_{22}|^2 - |\Delta|^2\right)\right|} \qquad (9.58)$$

Equation 9.57 suggests that the distance from the origin to the center of the circle is equal to $|C_p|$ and its angle can be found by the angle of $C_2{}^*$. The maximum OPG occurs where r_p is equal to zero. Now, if we equate r_p to zero, it follows from Equation 9.58 that

$$g^2_{p,max}|S_{12}S_{21}|^2 - 2k|S_{12}S_{21}|g_{p,max} + 1 = 0 \qquad (9.59)$$

where $g_{p,max}$ is the maximum value of g_p. Solving for g_p for the unconditionally stable case, we obtain

$$g_{p,max} = \frac{1}{|S_{12}S_{21}|}\left(k - \sqrt{k^2 - 1}\right) \qquad (9.60)$$

This is indeed the case where the conjugate matched condition occurs at the output. Now, substituting Equation 9.60 into Equation 9.55, we have

$$G_{p,max} = \frac{|S_{21}|}{|S_{12}|}\left(k - \sqrt{k^2 - 1}\right) \qquad (9.61)$$

Provided $|S_{12} \neq 0|$ and $k > 1$. The minimum value for g_p is equal to zero, which means $G_p = 0$ as well. We can observe from Equation 9.55 that $G_p = 0$, if only if the magnitude of the load reflection coefficient is equal to unity ($\Gamma_L = 1$). In other words, OPG is equal to zero if all the power is reflected by the load. For a given power gain, the load reflection coefficient can be deduced from constant OPG contours. The maximum gain occurs when the load reflection coefficient lies at the distance $|\Gamma_{L,O}|$ and where $g_{p,max} = G_{p,max}/|S_{21}|^2$. The maximum output power occurs when we have complex conjugate matching at the input, i.e., $\Gamma_s = \Gamma_{in}^*$. We can say equivalently that if $\Gamma_s = \Gamma_{in}^*$, the input power is equal to the maximum available input power. Therefore, in this case, the maximum transducer gain is equal to the maximum OPG. The constant power contour can be drawn as follows: (1) For a given power gain, we draw the desired circle from Equation 9.56; (2) We choose a desired load reflection coefficient on the contour; (3) For the given load reflection coefficient, the maximum output power is obtained when we have complex conjugate matching at the input, i.e., $\Gamma_s = \Gamma_{in}^*$. The resulting source reflection coefficient gives an OPG equal to the transducer power gain.

Example 9.2 The S-parameters at the given bias are measured for a bipolar transistor at 250 MHz and at $V_{CE} = 5V, I_C = 5$ mA as $S_{11} = 0.277\angle - 59°$, $S_{12} = 0.078\angle 93°$, $S_{21} = 1.92\angle 64°$, $S_{22} = 0.848\angle - 31°$. Assume that the input and the output terminations are $Z_s = 35 - j60$ and $Z_L = 50 - j50$, respectively. Design a two-port amplifier such that the gain of 9 dB is achieved at 250 MHz.

Solution: The stability factor is:

$$k = \frac{1 - |S_{11}|^2 - |S_{22}|^2 + |\Delta|^2}{2|S_{12}S_{21}|} = 1.033 \qquad (9.62)$$

where

$$\Delta = S_{11}S_{22} - S_{12}S_{21} = (0.277\angle - 59°)(0.848\angle - 31°)$$
$$- (0.078\angle 93°)(1.92\angle 64°) = 0.324\angle - 64.8° \qquad (9.63)$$

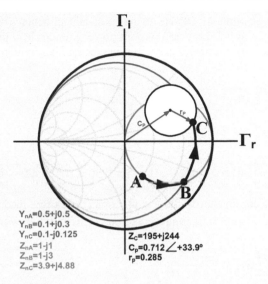

$Y_{nA}=0.5+j0.5$
$Y_{nB}=0.1+j0.3$
$Y_{nC}=0.1-j0.125$
$Z_{nA}=1-j1$
$Z_{nB}=1-j3$
$Z_{nC}=3.9+j4.88$

$Z_{c}=195+j244$
$C_{p}=0.712\angle+33.9°$
$r_{p}=0.285$

Figure 9.8: The load matching using a series capacitor and a parallel inductor to achieve 9 dB gain.

The amplifier is unconditionally stable with $k = 1.033$ and $\Delta < 1$. It can be shown that the maximum gain is equal to 12.9 dB. We have

$$C_2 = S_{22} - \Delta S_{11}{}^* = (0.848\angle - 31°) - (0.324\angle - 64.8°)(0.277\angle 59°)$$
$$= 0.768\angle - 33.9° \tag{9.64}$$

and

$$g_p = \frac{G_p}{|S_{21}|^2} = \frac{10^{0.9}}{(1.92)^2} = \frac{7.94}{(1.92)^2} = 2.15 \tag{9.65}$$

Now, we can write

$$C_p = \frac{g_p C_2{}^*}{1 + g_p\left(|S_{22}|^2 - |\Delta|^2\right)} = \frac{2.15(0.768\angle 33.9°)}{1 + 2.15\left(0.848^2 - 0.324^2\right)} = 0.712\angle 33.9° \tag{9.66}$$

and

$$r_p = \frac{\sqrt{1 - 2k|S_{12}S_{21}|g_p + |S_{12}S_{21}|^2 g_p{}^2}}{\left|1 + g_p\left(|S_{22}|^2 - |\Delta|^2\right)\right|} = 0.285 \tag{9.67}$$

The constant gain contour is shown in Figure 9.8. In order to arrive at a point on the contour from the starting point, A ($Z_{nA} = 1 - j1$), we move counterclockwise on a constant resistance circle ($R = 50\,\Omega$) with a series capacitor to point B and then we move counterclockwise on the constant conductance circle, using a parallel inductor, so as to intersect the constant power gain circle at point C.

The values of the series capacitor and the parallel inductor can be found as

$$C = \frac{1}{\omega \left| \text{Im}\{Z_B\} - \text{Im}\{Z_A\} \right|} = \frac{1}{2\pi \left(250 \times 10^6\right) 2 \times 50} = 6.4 pF \qquad (9.68)$$

and

$$L = \frac{1}{\omega \left| \text{Im}\{Y_C\} - \text{Im}\{Y_B\} \right|} = \frac{1}{2\pi \left(250 \times 10^6\right) 0.425 \times 0.02} = 75 nH \quad (9.69)$$

For complex conjugate matching at the input, the chosen load reflection coefficient is $\Gamma_L' = 0.82\angle 14.2°$. The input reflection coefficient then would be

$$\Gamma_s' = \left(S_{11} + \frac{S_{21}\Gamma_L'S_{12}}{1 - S_{22}\Gamma_L'} \right)^*$$

$$= \left(0.277\angle -59° + \frac{(1.92\angle 64°)(0.82\angle 14.2°)(0.078\angle 93°)}{1 - (0.848\angle -31°)(0.82\angle 14.2°)} \right)^* \qquad (9.70)$$

$$= 0.105\angle 160°$$

The input matching is shown in Figure 9.9.

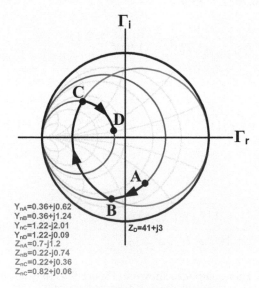

$Y_{nA}=0.36+j0.62$
$Y_{nB}=0.36+j1.24$
$Y_{nc}=1.22-j2.01$
$Y_{nD}=1.22-j0.09$
$Z_{nA}=0.7-j1.2$
$Z_{nB}=0.22-j0.74$
$Z_{nC}=0.22+j0.36$
$Z_{nC}=0.82+j0.06$

$Z_D=41+j3$

Figure 9.9: Source matching.

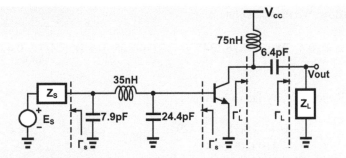

Figure 9.10: The input and output matching networks for complex source and load impedances.

We turn clockwise from point A, the normalized input impedance ($Z_{nA} = 0.7 - j1.2$), with a parallel capacitor on the constant conductance circle to point B. Then, using a series inductor, we turn clockwise on the constant resistance circle to arrive at point C, and finally by adding a parallel capacitor, we turn clockwise on the constant conductance circle to arrive at the desired source reflection coefficient. The values of the parallel capacitors and the series inductor are given by

$$C = \frac{|\text{Im}\{Y_B\} - \text{Im}\{Y_A\}|}{\omega} = \frac{0.62 \times 0.02}{2\pi \left(250 \times 10^6\right)} = 7.9 pF \tag{9.71}$$

and

$$L = \frac{|\text{Im}\{Z_C\} - \text{Im}\{Z_B\}|}{\omega} = \frac{1.1 \times 50}{2\pi \left(250 \times 10^6\right)} = 35 nH \tag{9.72}$$

and

$$C = \frac{|\text{Im}\{Y_D\} - \text{Im}\{Y_C\}|}{\omega} = \frac{1.92 \times 0.02}{2\pi \left(250 \times 10^6\right)} = 24.4 pF \tag{9.73}$$

The overall matching network is depicted in Figure 9.10.   ∎

While in Example 9.2 the load and the source impedances were not equal to $50\,\Omega$, normally this is not the case and every stage is normally matched to the reference impedance, $e.g.$, $50\,\Omega$. In the next example, we take into consideration both the stability and the power gain contours.

Example 9.3 S-parameters at 200 MHz and bias information for transistor 2N5179 are given as $V_{CE} = 6V$, $I_C = 5\,\text{mA}$, $S_{11} = 0.4\angle 280°$, $S_{12} = 0.048\angle 65°$, $S_{21} = 5.4\angle 103°$, $S_{22} = 0.78\angle 345°$. Also assume $50\,\Omega$ terminations. Design the two-port amplifier such that 12 dB power gain is achieved at 200 MHz.

Solution: We first calculate Δ as

$$\Delta = (0.4\angle 280°)\,(0.78\angle 345°) - (0.048\angle 65°)\,(5.4\angle 103°) = 0.429\angle - 58.18°$$
(9.74)

The Rollet stability factor becomes

$$k = \frac{1 - 0.4^2 - 0.78^2 + 0.429^2}{2\,(0.048)\,(5.4)} = 0.802$$
(9.75)

Therefore, the two-port is conditionally stable ($k < 1$). Recalling Equation 9.35, we now derive the input and the output stability circles as

$$C_1 = S_{11} - \Delta S_{22}{}^* = 0.4\angle 280° - (0.429\angle - 58.18°)\,(0.78\angle - 345°)$$
$$= 0.241\angle - 136.6°$$
(9.76)

and from Equation 9.40

$$C_2 = S_{22} - \Delta S_{11}{}^* = 0.78\angle 345° - (0.429\angle - 58.18°)\,(0.4\angle - 280°) = 0.65\angle - 24°$$
(9.77)

The center and the radius of the locus of the unity output reflection coefficient (input stability circle) can be found from Equation 9.17:

$$C_s = \frac{C_1{}^*}{|S_{11}|^2 - |\Delta|^2} = \frac{0.241\angle 136.6°}{0.4^2 - 0.429^2} = 10\angle -43.4°$$
(9.78)

and recalling Equation 9.16

$$r_s = \left| \frac{S_{12}S_{21}}{|S_{11}|^2 - |\Delta|^2} \right| = \left| \frac{(0.048\angle 65°)\,(5.4\angle 103°)}{0.4^2 - 0.429^2} \right| = 10.78$$
(9.79)

The center and the radius of the locus of the unity input reflection coefficient (output stability circle) can be found from Equation 9.20:

$$C_L = \frac{C_2{}^*}{|S_{22}|^2 - |\Delta|^2} = \frac{0.65\angle 24°}{0.78^2 - 0.429^2} = 1.53\angle 24°$$
(9.80)

and recalling Equation 9.19

$$r_L = \left| \frac{S_{12}S_{21}}{|S_{11}|^2 - |\Delta|^2} \right| = \left| \frac{(0.048\angle 65°)\,(5.4\angle 103°)}{0.78^2 - 0.429^2} \right| = 0.610$$
(9.81)

We then proceed to find the center and the radius corresponding to the desired gain in the example. It follows from Equation 9.55:

$$g_p = \frac{G_p}{|S_{21}|^2} = \frac{10^{1.2}}{5.4^2} = \frac{15.85}{5.4^2} = 0.543 \qquad (9.82)$$

and then, recalling Equation 9.57, we have

$$C_p = \frac{g_p C_2^*}{1 + g_p \left(|S_{22}|^2 - |\Delta|^2\right)} = \frac{0.543\,(0.65\angle 24°)}{1 + 0.543\,(0.78^2 - 0.429^2)} = 0.287\angle 24° \quad (9.83)$$

and from Equation 9.58

$$r_p = \frac{\sqrt{1 - 2k|S_{12}S_{21}|\,g_p + |S_{12}S_{21}|^2 g_p^2}}{\left|1 + g_p \left(|S_{22}|^2 - |\Delta|^2\right)\right|} = 0.724 \qquad (9.84)$$

The areas of interest are shown in the Smith chart in Figure 9.11. We should choose a point on the constant gain contour such that it lies outside the output stability circle. Let's choose $\Gamma_L = (0.724 - 0.287)\angle(24 + 180)° = 0.437\angle 204°$ to be sure of the output stability. Now we choose a proper value for Γ_s such that the transducer gain becomes equal to the operating power gain (the input is matched to the source), so it can be calculated as follows

$$\Gamma_s = \left(S_{11} + \frac{S_{12}S_{21}\Gamma_L}{1 - S_{22}\Gamma_L}\right)^* = 0.409\angle 68° \qquad (9.85)$$

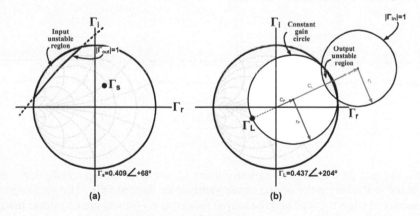

Figure 9.11: (a) The input stability circle and the corresponding source reflection coefficient. (b) The output stability circle and the chosen load reflection coefficient.

As it can be seen in Figure 9.11, in the computed Γ_s is outside the unstable source region. So now a standard technique can be used to realize the specified values of Γ_s and Γ_L out of a 50 Ω source and a 50 Ω load using an L-section matching network. So an amplifier with a 12 dB gain is designed with a conditionally stable device. ∎

9.4.2 APG Contours for Bilateral Conditionally Stable Amplifiers

In a similar way, the constant APG contours can be obtained on the source plane:

$$G_A = \frac{|S_{21}|^2 \left(1 - |\Gamma_s|^2\right)}{\left(1 - \left|\frac{S_{22} - \Delta\Gamma_s}{1 - S_{11}\Gamma_s}\right|\right)|1 - S_{11}\Gamma_s|^2} = |S_{21}|^2 g_a \tag{9.86}$$

where the normalized APG is

$$g_a = \frac{G_A}{|S_{21}|^2} = \frac{1 - |\Gamma_s|^2}{1 - |S_{22}|^2 + |\Gamma_s|^2 \left(|S_{11}|^2 - \Delta^2\right) - 2real\{\Gamma_s C_1\}} \tag{9.87}$$

and

$$C_1 = S_{11} - \Delta S_{22}^* \tag{9.88}$$

In a similar way to those expressions already derived earlier for OPG, the center and the radius of constant APG circles which are drawn on the input reflection coefficient plane can be, respectively, found as

$$C_a = \frac{g_a C_1^*}{1 + g_a \left(|S_{11}|^2 - |\Delta|^2\right)} \tag{9.89}$$

and

$$r_p = \frac{\sqrt{1 - 2k|S_{12}S_{21}|g_a + |S_{12}S_{21}|^2 g_a^2}}{\left|1 + g_a \left(|S_{11}|^2 - |\Delta|^2\right)\right|} \tag{9.90}$$

At the end, the circles can be plotted using C_a and r_p on the input reflection coefficient plane with every point on the circle yielding the desired gain. The maximum gain is achieved when the load and the output reflection coefficient are conjugate matched, in which case the transducer power gain (TPG) and APG become equal. Interestingly, as constant noise contours are also drawn on the input reflection coefficient plane, a compromise can be made between the available gain and the minimum noise figure on this plane. As a result for the amplifiers where the operating power gain is critical, we use the procedure described in Example 9.3, and for the amplifiers where the APG and the noise figure are of more importance, we use the procedure described in the next section.

9.5 Noise Behavior of a Two-Port Network

The noise figure (NF) for a two-port network can be defined by the quantity through which it decreases the output SNR with respect to the input SNR. In many receiver applications, minimum noise is the most important design parameter. As minimum noise and maximum power gain do not coincide, a compromise is made by plotting both the constant noise and the constant APG contours, and choosing the proper source reflection coefficient. This trade-off can be well extended to noise, stability, matching, and power gain. If the objective is to design a wideband amplifier, then the most important condition to satisfy would be achieving a flat power gain response with minimal distortion, all of which is achieved by using a compensated matching network, may be using a negative feedback and a balanced architecture for the amplifier. It is important to note that using small-signal S-parameters is valid as long as the transistor operates in the linear regime, its output power remaining below saturation in linear region. It should be noted that the more we move into the nonlinear regime, the more the input/output impedances or the reflection coefficients become dependent on the operating power level.

9.5.1 Noise In a Two-Port

In RF amplifiers, even in the absence of the signal and the interference at the input, a small voltage can be recorded at the output. This negligibly small value is referred to as the amplifier output noise voltage. We recognize that the total output noise power is the combination of the intrinsic noise of the amplifier, resulting from the transistor itself or other noisy components, and the input noise amplified by the amplifier. Depicted in Figure 9.12 is the noise model of an RF or a microwave two-port amplifier.

The input noise can be modeled by an equivalent noise voltage source (V_n) related to the source resistance, a series noise voltage source (e_n), and a parallel noise current source (i_n), the two latter pertaining to the two-port referred to its input. The source resistance generates a thermal or Johnson noise. In fact, the noise is generated by the random movements of electrons due to thermal excitations. This noise can be found for a specific bandwidth as

$$\overline{V^2}_{n,rms} = 4kTBR_S \tag{9.91}$$

where k is the Boltzmann constant of a value of $1.374 \times 10^{-23} \frac{J}{\circ k}$, T is the absolute temperature of the resistor in Kelvins, and B is the operating bandwidth. As suggested by Equation 9.91, the noise power is a direct function of its bandwidth and the bandwidth is normally limited value in an RF system. Thermal noise is widely known

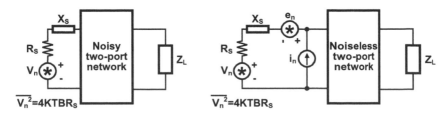

Figure 9.12: The noise model of a two-port amplifier.

as a *white noise*, as it contains all frequency components and has a flat spectral behavior over the frequency range. The available noise power of a resistor can be derived as

$$P_N = \frac{\overline{V^2}_{n,rms}}{4R_S} = KTB \tag{9.92}$$

Note that noise voltage has a random value, its instantaneous value is not known, but its rms value can be estimated. The wider the bandwidth, the narrower and larger the instantaneous voltage spikes.

> **Example 9.4** Find the available noise power which a resistor generates at the standard temperature, *i.e.*, $T_0 = 290°$k in a bandwidth of 1 Hz. Then, calculate the noise voltage and power for a $2\,M\Omega$ resistor in a bandwidth of 5 kHz and at the standard temperature.
>
> **Solution:** We have from Equation 9.92,
>
> $$P_N = KTB = \left(1.374 \times 10^{-23}\right)(290)(1) = 3.985 \times 10^{-21} W \tag{9.93}$$
>
> which if written in dBm, we have
>
> $$P_N\,(dBm) = 10\log\frac{P_N}{10^{-3}} = 10\log 3.985 \times 10^{-18} = -174\,dBm \tag{9.94}$$
>
> Now, in order to obtain the noise voltage for a 5 kHz bandwidth, we write
>
> $$v_{n,rms} = \sqrt{4\left(1.374 \times 10^{-23}\right)(290)(5000)\left(2 \times 10^6\right)} = 12.6\,\mu V \tag{9.95}$$
>
> and finally, the noise power can be calculated from Equation 9.93 as
>
> $$P_N\,(dBm) = 10\log\frac{P_N}{10^{-3}} = 10\log\frac{\left(12.6 \times 10^{-6}\right)^2}{4\left(2 \times 10^6\right)} \times 10^3$$
>
> $$= 10\log\left(19.9 \times 10^{-15}\right) = -137\,dBm \tag{9.96}$$

Noise figure is a quantitative measure of noise behavior in an RF or a microwave amplifier. The noise figure is defined as the ratio of total available noise power at the output of the amplifier to the available output noise power resulting from the same noiseless two-port connected to a resistive source termination, say R, at standard temperature, at the input. This definition can be written as

$$F = \frac{P_{N_O}}{P_{N_i}G_A} \tag{9.97}$$

In Equation 9.97, P_{N_O} is the total noise power at the output of the amplifier, $P_{N_i} = KT_0B$ is the available noise power resulting from the termination resistor, R, at standard temperature, $T_0 = 290°k$, B is the bandwidth, and G_A is the APG, which is given by

$$G_A = \frac{P_{S_O}}{P_{S_i}} \tag{9.98}$$

In Equation 9.98, P_{S_O} is the output signal power and P_{S_i} is the input signal power. Therefore, Equation 9.97, can be rewritten as

$$F = \frac{P_{S_i}/P_{N_i}}{P_{S_O}/P_{N_O}} = \frac{SNR_i}{SNR_O} \tag{9.99}$$

In other words, the noise figure can be defined as the ratio of the SNR at the input to the SNR at the output. Given the fact that the noise figure is always greater than unity, the output SNR is always smaller than the input SNR, so it is evident that in an amplifier one cannot improve the SNR, as such an RF engineer would rather strive not to degrade it. The curious student might ask why do we ever amplify the signal in a receiver where at every stage the SNR would be degraded within the amplifier chain? The reason is that to be able to further process the signal, it is necessary that the signal should attain a certain required level (a few dBm's, for example) for detection. In an amplifier, in order to find the minimum noise figure, the choice of a proper source reflection coefficient is important. In Figure 9.13, a model is provided for calculation of noise figure in cascaded amplifiers. As depicted in Figure 9.13, P_{N_i} is the input noise power, G_{A1} and G_{A2} are the APGs of the first and the second stage, and P_{N1} and P_{N2} represent the available noise powers at the output of each amplifier which result from its own intrinsic noise. Hence, the total noise at the output can be written as

$$P_{N_O} = G_{A2}\left(G_{A1}P_{N_i} + P_{N1}\right) + P_{N2} \tag{9.100}$$

Using the definition provided earlier, noise figure can be calculated as

$$F = \frac{P_{N_O}}{P_{N_i}G_{A1}G_{A2}} = 1 + \frac{P_{N1}}{P_{N_i}G_{A1}} + \frac{P_{N2}}{P_{N_i}G_{A1}G_{A2}} \tag{9.101}$$

Equation 9.101 can also be rewritten as

$$F = F_1 + \frac{F_2 - 1}{G_{A1}} \tag{9.102}$$

Since we have

$$F_1 = 1 + \frac{P_{N1}}{P_{N_i}G_{A1}} \tag{9.103}$$

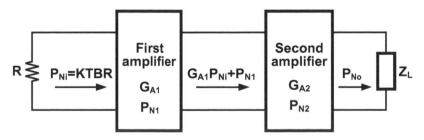

Figure 9.13: Model for noise figure calculation in a cascade of two stages.

and

$$F_2 = 1 + \frac{P_{N2}}{P_{N_i} G_{A2}} \tag{9.104}$$

F_1 and F_2 are the noise figures corresponding to each one of the stages. Equation 9.102 suggests that the noise of the second stage is diminished by a factor of G_{A1}. We then realize that the noise of the second stage cannot affect the overall noise figure significantly. This leads to an important observation from the design point of view: the noise of the first stage is the most important part of the overall noise figure, given the fact that the power gain of the preceding amplifiers masks the noise of the following stages, and therefore it is appropriate to diminish the noise figure of the first stage at the price of losing a little bit of the power gain. In fact, we can target for the minimum noise figure for a lesser power gain at the first stage. A trade-off always exists between NF and the APG in any design. The design can be made such that the minimum NF is obtained. Consider two amplifiers with NF's and gains of F_1, F_2, G_{A1}, and G_{A2}, respectively. If the first amplifier precedes the second amplifier, the overall NF denoted by F_{12} can be written as

$$F_{12} = F_1 + \frac{F_2 - 1}{G_{A1}} \tag{9.105}$$

and if the second amplifier comes first, F_{21} is given by

$$F_{21} = F_2 + \frac{F_1 - 1}{G_{A2}} \tag{9.106}$$

To have a lower NF in the first case, i.e., $F_{12} < F_{21}$, we should have

$$F_1 + \frac{F_2 - 1}{G_{A1}} < F_2 + \frac{F_1 - 1}{G_{A2}} \tag{9.107}$$

which can be rewritten as

$$\frac{F_1 - 1}{1 - \frac{1}{G_{A1}}} < \frac{F_2 - 1}{1 - \frac{1}{G_{A2}}} \tag{9.108}$$

Equation 9.108 can be simplified as

$$M_1 < M_2 \tag{9.109}$$

where the quantity M is defined as the noise measure of the amplifier:

$$M = \frac{F - 1}{1 - \frac{1}{G_A}} \tag{9.110}$$

M is an alternative quantity to represent the noise performance of an amplifier, including noise figure and the power gain. Equation 9.109 suggests that, if the noise measure of the first stage, M_1, is smaller than that of the second stage, M_2, the overall noise figure in this structure will be smaller. Hence, we predict that, in order to have a minimal NF in a cascade of stages, the amplifier with the lower noise measure, M, should precede the one with larger noise measure. It can be shown that NF for a cascade of many stages is given by

$$F = F_1 + \frac{F_2 - 1}{G_{A1}} + \frac{F_3 - 1}{G_{A1} G_{A2}} + \frac{F_4 - 1}{G_{A1} G_{A2} G_{A3}} + \cdots \quad (9.111)$$

which is called the *"Friis' relation"* for the noise figure. This equation for the special case of $F_1 = F_2 = \cdots F_n$ and $G_{A1} = G_{A2} = \cdots = G_{An}$ and, assuming an infinite chain of identical amplifiers, reduces to

$$F = 1 + \frac{F_1 - 1}{1 - \frac{1}{G_{A1}}} = 1 + M_1 \quad (9.112)$$

This means the noise figure of an infinite chain of identical amplifiers tends to a limited value (the noise measure plus unity).

9.6 Constant Noise Figure Contours

It can be shown that the NF of a two-port amplifier can be expressed by

$$F = F_{\min} + \frac{R_n}{G_s} |Y_s - Y_{opt}|^2 \quad (9.113)$$

where R_n is the equivalent noise resistance of the two-port, or in a normalized notation $r_n = R_n / Z_0$, $Y_S = G_s + jB_s$ is the source admittance of the two-port, and $Y_{opt} = G_{opt} + jB_{opt}$ denotes the optimum source admittance which results in the minimum noise figure, $F_{\min}$. The source admittance and the optimum noise admittance can be readily expressed in terms of their corresponding reflection coefficient as

$$Y_S = \frac{1 - \Gamma_s}{1 + \Gamma_s} Y_0 \quad (9.114)$$

and

$$Y_{opt} = \frac{1 - \Gamma_{opt}}{1 + \Gamma_{opt}} Y_0 \quad (9.115)$$

Substituting Equations 9.114 and 9.115 into Equation 9.113, we obtain

$$F = F_{\min} + \frac{4 r_n |\Gamma_s - \Gamma_{opt}|^2}{\left(1 - |\Gamma_s|^2\right) |1 + \Gamma_{opt}|^2} \quad (9.116)$$

This relation is a function of $F_{\min}$, r_n, and Γ_{opt}. These three parameters, two real and one complex (or equivalently four real parameters), are widely known as the noise parameters and are provided either in the datasheet by the vendor or can be found by measurement. The input reflection coefficient can be altered and the noise figure can be measured accordingly. Multiple measurements with different source reflection coefficients (measured using a network analyzer), and the noise figure measured by a noise figure meter will allow the engineer to extract the four required noise parameters. Here F_{min} is a function of the bias current or the bias voltage of the device and the operating frequency. Therefore, only a single Γ_{opt} corresponds to every value of F_{min}. Equation 9.116 can be rewritten such that for a specific input reflection coefficient, Γ_s, the resulting NF is, say $F = F_i$. Hence

$$\frac{\left|\Gamma_s - \Gamma_{opt}\right|^2}{1 - \left|\Gamma_s\right|^2} = \frac{F_i - F_{min}}{4r_n}\left|1 + \Gamma_{opt}\right|^2 \tag{9.117}$$

As suggested by Equation 9.117, for a given noise figure, F_i, the right-hand side of the equation is a constant. Therefore, if we define a factor, N_i as

$$N_i = \frac{F_i - F_{min}}{4r_n}\left|1 + \Gamma_{opt}\right|^2 \tag{9.118}$$

and, it follows that

$$\frac{\left|\Gamma_s - \Gamma_{opt}\right|^2}{1 - \left|\Gamma_s\right|^2} = N_i \tag{9.119}$$

Equation 9.119 can be rewritten as

$$\left|\Gamma_s\right|^2 - \frac{2}{1+N_i}Re\left[\Gamma_s\Gamma_{opt}{}^*\right] + \frac{\Gamma_{opt}{}^2}{1+N_i} = \frac{N_i}{1+N_i} \tag{9.120}$$

This represents a contour on the input reflection coefficient plane. Equation 9.120 can also be rewritten alternatively as

$$\left|\Gamma_s - \frac{\Gamma_{opt}}{1+N_i}\right|^2 = \frac{N_i{}^2 + N_i\left(1 - \left|\Gamma_{opt}\right|^2\right)}{(1+N_i)^2} \tag{9.121}$$

This describes a circle on the input reflection coefficient plane with the center, $C_{F,i}$, and the radius, $r_{F,i}$, as follows

$$C_{F_i} = \frac{\Gamma_{opt}}{1+N_i} \tag{9.122}$$

and

$$r_{F_i} = \frac{1}{1+N_i}\sqrt{N_i{}^2 + N_i\left(1 - \left|\Gamma_{opt}\right|^2\right)} \tag{9.123}$$

Using Equation 9.118, N_i can be obtained for a given value of F_i. Ultimately, having both the centers and the radii of constant noise figure contours, they can all be plotted

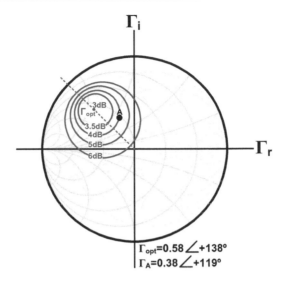

$\Gamma_{opt}=0.58 \angle +138°$
$\Gamma_A=0.38 \angle +119°$

Figure 9.14: Constant NF contours on the plane of input reflection coefficient.

on the input reflection coefficient plane. Equations 9.118 through 9.123 suggest that, when $F_i = F_{min}$, N_i is equal to zero, and the center corresponds to Γ_{opt} and the radius would be zero (F_{min} contour lies at the center point, Γ_{opt}, with a radius of zero). We realize from Equation 9.120 that the centers of all other constant noise figure contours lie on a line connecting the center of the chart to the Γ_{opt} point; in other words, they lie on a line passing through the center with $\angle\Gamma_{opt}$. A typical group of constant noise figure contours are depicted in Figure 9.14. As it is obvious from Figure 9.14, the minimum noise figure, F_{min}, corresponds to $\Gamma_s = \Gamma_{opt} = 0.58\angle138°$ and it is equal to 3 dB. Any other neighboring point will have a higher noise figure. At point A, for instance, with $\Gamma_s = 0.38\angle119°$, we have $NF = 4$ dB.

In practical designs, there is always an unwanted discrepancy between the targeted NF value and the NF derived from the measurement which stems from the matching network imperfection and also inaccuracies in the transistors noise parameters' measurement. Typically speaking, this may amount to a few 0.1 dB's to 1 dB change in NF.

In a practical low-noise amplifier design, there exists a trade-off between the APG, the noise figure, and VSWR, or equivalently, matching. The trade-off between the noise and the power gain is illustrated in Figure 9.15, which the transistor of choice is unilateral and a group of constant NF and constant gain contours are plotted. As it is obvious from Figure 9.15, the maximum power gain and the minimum NF points do not coincide in general. The normalized power gain in Figure 9.15 is $G_s = 3$ dB, which occurs with $\Gamma_s = 0.7\angle110°$ and results in $F_i = 4$ dB. The minimum NF, $F_{min} = 0.8$ dB is achieved with $\Gamma_s = 0.6\angle40°$, and corresponds to normalized $G_s = -1$ dB.

Using the above constant gain and the constant noise figure contours, one can easily make a trade-off between the gain and the noise figure. As to say for a given noise figure, the point at which a constant gain circle (with maximum possible gain) is

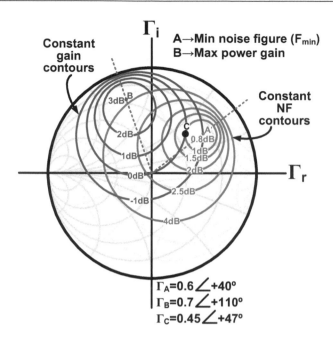

Figure 9.15: Constant NF and the normalized available power gain contours on the input reflection coefficient plane at 6 GHz.

tangent to it gives the best compromise, otherwise for a given power gain the point at which a constant noise figure circle (with minimum possible noise figure) is tangent to it gives the best compromise. For example, if we choose to have a NF of 1 dB, we would then choose point C where the 0 dB power gain circle is tangent to it and the corresponding Γ_S is $0.45\angle47°$.

9.7 Design of a Single-Stage Low-Noise Amplifier

If our objective is to design a minimum NF stage, the source impedance and the bias points must be chosen such that the minimum NF is achieved for the device. This can be done either by performing a set of measurements on the device or using the noise datasheet provided by the device vendor. The input matching network can then be easily designed. For design considerations, the condition $k > 1$ should be satisfied, so that the transistor is stable. Once the optimum input reflection coefficient is realized for the minimum noise, the load reflection coefficient should be chosen such that the load is matched to output. That is

$$\Gamma_L = \left(S_{22} + \frac{S_{21}S_{12}\Gamma_s}{1 - S_{11}\Gamma_s} \right)^* \qquad (9.124)$$

Example 9.5 clarifies the design procedure.

Example 9.5 The optimum noise bias point for a given bipolar transistor is $V_{CE} = 10\,V$ and $I_C = 5\,mA$. The optimum noise input reflection coefficient at 200 MHz is $\Gamma_s = 0.7\angle 140$. S-parameters at 200 MHz (in a 50 Ω measurement system) are given as $S_{11} = 0.4\angle 168°$, $S_{12} = 0.04\angle 60°$, $S_{21} = 5.2\angle 63°$, $S_{22} = 0.35\angle - 39°$. Design an LNA at 200 MHz with the source and load impedances of 75 Ω and 100 Ω. Then, determine what the transducer power gain is expected of this design.

Solution:
We first evaluate the Rollet stability factor as

$$k = \frac{1 - |S_{11}|^2 - |S_{22}|^2 + |\Delta|^2}{2|S_{12}S_{21}|} = \frac{1 - 0.4^2 - 0.35^2 + 0.068^2}{2(0.04)(5.2)} = 1.74 \quad (9.125)$$

Considering $k > 1$ and $|\Delta| < 1$ the transistor is unconditionally stable. As such, the amplifier is stable. We then design the input matching network for a 75 Ω source impedance. We depict the optimum noise source impedance on the Smith chart at point C, $\Gamma_s = 0.7\angle 140$ corresponding to $Z_{nC} = 0.2 + j0.35$. Using Figure 9.16, starting from point A ($Z_{nA} = 1.5 + j0$), we turn clockwise on the constant conductance contour ($G = 0.667$) to intersect the constant resistance contour ($R - 0.2$) at point B. So for the parallel capacitor, we would have

$$C - \frac{|\text{Im}\{Y_B\} - \text{Im}\{Y_A\}|}{\omega} = \frac{1.7 \times 0.02}{2\pi (200 \times 10^6)} = 27pF \quad (9.126)$$

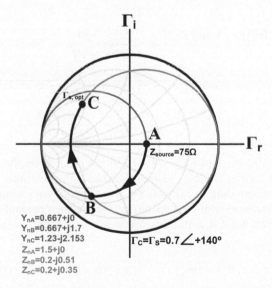

Figure 9.16: Input matching on the Smith chart.

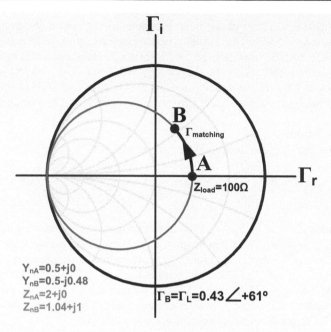

Figure 9.17: Output matching on the Smith chart.

Starting from point B ($Z_{nB} = 0.2 - j0.51$), we turn clockwise on the constant resistance contour ($R = 0.2$) to arrive at point C. So for the series inductor, we would have

$$L = \frac{|\text{Im}\{Z_C\} - \text{Im}\{Z_B\}|}{\omega} = \frac{50 \times 0.86}{2\pi \left(200 \times 10^6\right)} = 34nH \qquad (9.127)$$

For the output matching network, we have from Equation 9.124

$$\Gamma_L = \left(0.35\angle - 39° + \frac{(5.2\angle 63°)(0.7\angle 140°)(0.04\angle 60°)}{1 - (0.4\angle 168°)(0.7\angle 140°)}\right)^* = 0.43\angle 61° \quad (9.128)$$

In Figure 9.17, we depict $\Gamma_L = 0.43\angle 61°$ at point B on the Smith chart which corresponds to $Y_{nB} = 0.5 - j0.48$. We depict the required load impedance at point A corresponding to $Y_{nA} = 0.5 + j0$. As such, with adding a single parallel inductance, we can turn counter clockwise, on the constant conductance circle ($G = 0.5$) from point A to point B. We would have then

$$L = \frac{1}{\omega |\text{Im}\{Y_C\} - \text{Im}\{Y_B\}|} = \frac{1}{2\pi \left(200 \times 10^6\right) 0.48 \times 0.02} = 83\,\text{nH} \quad (9.129)$$

The 330 pF capacitors at the input and the output are for DC decoupling purpose and their AC impedances are negligible.

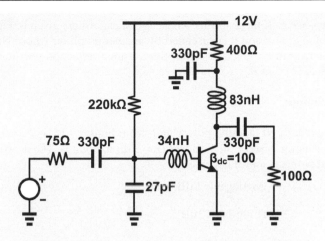

Figure 9.18: The overall matching network.

The overall matching network is depicted in Figure 9.18.

For the transducer gain (once we put the expression for Γ_{in} from Equation 8.30 in Equation 9.6) we have

$$G_T = \frac{|S_{21}|^2 \left(1 - |\Gamma_L|^2\right)\left(1 - |\Gamma_s|^2\right)}{|(1 - S_{11}\Gamma_s)(1 - S_{22}\Gamma_L) - S_{21}\Gamma_L S_{12}\Gamma_s|^2} = 23 \tag{9.130}$$

Or

$$10\log 23 = 13.6\,\text{dB} \tag{9.131}$$

∎

Now that we have learned how to design a single-stage amplifier, we move on to design a two-stage low noise amplifier and understand the design procedure.

9.8 Design of Two-Stage Amplifiers

Consider the NF of a cascade of multiple stages of amplifiers, which is referred to as the Friis' NF equation, as

$$F = F_1 + \frac{F_2 - 1}{G_{A1}} + \frac{F_3 - 1}{G_{A1}G_{A2}} + \frac{F_4 - 1}{G_{A1}G_{A2}G_{A3}} + \cdots \tag{9.132}$$

In order to minimize the overall NF, the NF of the first stage must be minimized and its gain maximized. The input reflection coefficient must be chosen such that it results in the minimum overall NF. As we discussed earlier, the maximum gain point and the minimum noise figure point are normally distinct on the source plane Smith chart. It is possible to draw a line connecting the two point on the chart, and then choose the reflection coefficient which results in the lowest noise along this line.

Example 9.6 If the amplifier whose specifications are given in Figure 9.15 has a value of $|S_{21}| = 8$ and it is cascaded by another amplifier whose NF is $F_2 = 5\,dB$, find the overall NF once the amplifier's source reflection coefficient is either at point A or at point B.

Solution:

We record NF and the power gain from Figure 9.15 as $F = 0.8\,dB$ and $G = -1\,dB$ at point A and $F = 4\,dB$ and $G = 3\,dB$ at point B. The denormalized power gains at points A and B become

$$G_A = -1 + 10\log 8 = 8\,dB \tag{9.133a}$$

$$G_B = 3 + 10\log 8 = 12\,dB \tag{9.133b}$$

We have from Equation 9.132

$$A : F_T = F_1 + \frac{F_2 - 1}{G_1} = 10^{0.08} + \frac{10^{0.5} - 1}{10^{0.8}} = 1.545 \quad \text{or} \quad 10\log 1.545 = 1.89\,dB \tag{9.134}$$

and

$$B : F_T = F_1 + \frac{F_2 - 1}{G_1} = 10^{0.4} + \frac{10^{0.5} - 1}{10^{1.2}} = 2.65 \quad \text{or} \quad 10\log 2.65 = 4.23\,dB \tag{9.135}$$

■

This design procedure for a specific power gain and the minimum possible NF consists of three steps: (1) The contours corresponding to the desired APG should be drawn; (2) The NF contour which is tangent to the required power gain contour should be drawn next; and (3) The optimum reflection coefficient lies on the specified loci where the two circles are tangent.

Example 9.7 The optimum bias points along with S-parameters and noise parameters for a transistor are provided at $4\,GHz$: $V_{CE} = 10V$, $I_C = 4\,mA$, $S_{11} = 0.552\angle 169°$, $S_{12} = 0.049\angle 23°$, $S_{21} = 1.681\angle 26°$, $S_{22} = 0.839\angle -67°$, $F_{min} = 2.5\,dB$, $\Gamma_{opt} = 0.475\angle 166°$, $r_n = 3.5\,\Omega$. Design the amplifier such that the overall noise figure of this stage followed by another stage with $NF = 7\,dB$ is minimized.

Solution:
We first evaluate the stability of the transistor.

$$|\Delta| = S_{11}S_{22} - S_{12}S_{21} = 0.419 < 1 \tag{9.136}$$

$$k = \frac{1 - |S_{11}|^2 - |S_{22}|^2 + |\Delta|^2}{2|S_{12}S_{21}|} = 1.012 > 1 \tag{9.137}$$

As a result, the amplifier is unconditionally stable. The constant APG and the constant NF contours are plotted in Figure 9.19 in the plane of the source reflection coefficient. Now, we choose a set of points with the specified noise figure and the maximum possible APG as follows, and we compute the overall noise figure consequently (from Equation 9.132).

$$A_1 : F_1 = 2.5\,\text{dB}, G_1 = 11dB \Rightarrow F_T = 2.095(3.21\,\text{dB}) \tag{9.138}$$

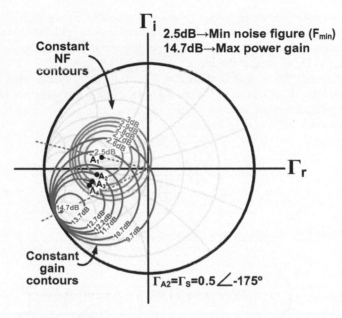

Figure 9.19: Constant power gain and constant NF contours for the given transistor on the Smith chart.

$$A_2 : F_1 = 2.6\,\text{dB}, G_1 = 12.2\,\text{dB} \Rightarrow F_T = 2.06(3.14\,\text{dB}) \; ; \; \Gamma_S = 0.524\angle +186° \tag{9.139}$$

$$A_3 : F_1 = 2.7\,\text{dB}, G_1 = 12.7\,\text{dB} \Rightarrow F_T = 2.077(3.17\,\text{dB}) \tag{9.140}$$

$$A_4 : F_1 = 2.8\,\text{dB}, G_1 = 13\,\text{dB} \Rightarrow F_T = 2.16(3.35\,\text{dB}) \tag{9.141}$$

As such, the input reflection coefficient (at point A_2) is chosen as

$$\Gamma_s = 0.524\angle +186° \tag{9.142}$$

For the output matching, the load reflection coefficient can be calculated as

$$\Gamma_L = \left(S_{22} + \frac{S_{21}S_{12}\Gamma_s}{1 - S_{11}\Gamma_s}\right)^* = 0.871\angle +70° \tag{9.143}$$

It is noteworthy that all the procedure which has been described in terms of S-parameters formulation can be repeated in terms of other circuit parameters such as Y-parameters. The following example illustrates this approach.

Example 9.8 Admittance parameters for a field effect transistor at 1 GHz are given as $Y_{11} = 8.79\,\text{mj}\mho, Y_{12} = -2.5\,\text{mj}\mho, Y_{21} = 1\,\text{m}\mho - 2.5\,\text{mj}\mho, Y_{22} = 0.33\,\text{m}\mho + 3.77\,\text{mj}\mho$.
(a) Determine the circuit model values as depicted in Figure 9.20.
(b) Does this transistor lead to a stable amplifier design? Determine the value of the required parallel resistor at the output to provide unconditionally stability if a parallel $1\,\text{k}\Omega$ is added at the input.
(c) Assuming the output is short circuited, for matching the input to $50\,\Omega$, a π section is used, as shown in Figure 9.21. Determine the values of the capacitors and the inductor with an input quality factor of 10.
(d) Calculate the matching bandwidth.

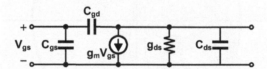

Figure 9.20: Equivalent circuit model of the transistor in Example 9.8.

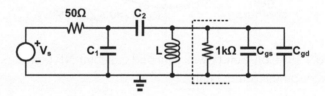

Figure 9.21: The π matching network employed in Example 9.8.

Solution:
(a) The equivalent model of the transistor can be shown as in Figure 9.20. The values can be found using the provided Y-parameters as

$$Y_{12} = -j\omega C_{gd} \Rightarrow C_{gd} = 398\,\text{fF} \tag{9.144}$$

$$Y_{11} = j\omega\left(C_{gs} + C_{gd}\right) \Rightarrow C_{gs} = 1\,\text{pF} \tag{9.145}$$

$$Y_{21} = g_m - j\omega C_{gd} \Rightarrow g_m = 1\,\text{m}\mho \tag{9.146}$$

$$Y_{22} = g_{ds} + j\omega\left(C_{gd} + C_{ds}\right) \Rightarrow g_{ds} = 330\,\mu\mho, C_{ds} = 202\,\text{fF} \tag{9.147}$$

(b) The Rollet's stability factor in terms of Y-parameters can be expressed as [5]

$$k = \frac{2g_{11}g_{22}}{|Y_{12}Y_{21}| + Re\{Y_{12}Y_{21}\}} \tag{9.148}$$

Given the fact that $g_{11} = 0$ then $k = 0$ which suggests instability. Adding a 1 kΩ resistor at the input, the Rollet's stability factor can be rewritten as [5]

$$k = \frac{2(g_{11} + G_S)g_{22}}{|Y_{12}Y_{21}| + Re\{Y_{12}Y_{21}\}} = \frac{0.66}{6.73 - 6.25} = 1.375 \tag{9.149}$$

This indicates that the circuit is unconditionally stable in this case, and no parallel resistor is needed at the output.

(c) The value of the components can be found as

$$Q = \frac{n^2}{G_S}C_{eq}\omega = 10 \tag{9.150}$$

Let, $\frac{n^2}{G_S} = 1k$, then $C_{eq} = 1.59\,\text{pF}$, where

$$n = \frac{C_1 + C_2}{C_2} = \sqrt{\frac{1000}{50}} = 4.47 \tag{9.151}$$

and

$$C_{eq} = \frac{C_1 C_2}{C_1 + C_2} \tag{9.152}$$

Therefore, from the above

$$C_1 = 7.08\,\text{pF}, \quad C_2 = 2.05\,\text{pF} \tag{9.153}$$

For matching, we should have

$$\frac{1}{L\omega} = C_{eq}\omega + b_{11} = 10\,\text{m} + 8.79\,\text{m} = 18.79\,\text{m}\mho \tag{9.154}$$

$$L = 8.47\,\text{nH} \tag{9.155}$$

(d) For estimating the matching network bandwidth, we calculate the total equivalent resistance in parallel with the inductor

$$R_{total} = 1\text{k} \parallel 1\text{k} = 500\,\Omega \tag{9.156}$$

$$Q_{total} = \frac{R_{total}}{L\omega} = 9.4 \tag{9.157}$$

$$BW = \frac{f_0}{Q} = 106\,\text{MHz} \tag{9.158}$$

Example 9.9 Assume that the bandwidth of a receiver is equal to 30 kHz, and the maximum allowable noise figure at the input is 7 dB. If the required $SNR_{min} = 6$ dB, compute the sensitivity of the receiver.

Solution:
We have

$$P_{sen} = (KT_0BF)(SNR)_{min} \qquad (9.159)$$

Note $10\log KT_0 = -174$ dBm/Hz.
The sensitivity in decibles becomes:

$$P_{sen} = \left(-174\,\text{dBm} + 10\log\left(30 \times 10^3\right) + 7\,\text{dB}\right) + 6\,\text{dB} = -116.2\,\text{dBm} \quad (9.160)$$

■

9.9 Conclusion

In this chapter, we discussed the design of RF/microwave amplifiers. We investigated the stability condition at the input and the output in terms of S-parameters. We learned about the radii and the centers of the stability circles at the source and at the load reflection coefficient planes. After resolving the stability problem, we presented three distinct definitions for the power gain of a two-port amplifier which were the operating power gain, the APG, and the transducer power gain. Then, we introduced the constant operating power gain contours on the load plane and the constant APG contours on the source plane over the Smith chart. Subsequently, we provided the definition of the noise figure and introduced constant NF contours on the source plane. Finally, a 3-step design procedure was provided by means of which the trade-off between noise figure and the power gain can be achieved, leading to the low-noise design with proper power gain.

9.10 References and Further Reading

1. D. Pozar, *Microwave Engineering*, fourth edition, Hoboken, NJ: J. Wiley & Sons, Inc., 2012.
2. G. Gonzalez, *Microwave Transistor Amplifiers: Analysis and Design.*, NJ: Prentice-Hall, 1994.
3. U.L. Rohde, A.M. Pavio, G.D. Vendelin, *Microwave Circuit Design Using Linear and Nonlinear Techniques*, Hoboken, NJ: J. Wiley & Sons, Inc., 2005.
4. R. Chi-Hsi Li, *RF Circuit Design*, Hoboken, NJ: J. Wiley & Sons, Inc., 2009.
5. R.S. Carson, *High Frequency Amplifiers*, New York, NY: J. Wiley & Sons, Inc., 1982.
6. J. Everard, *Fundamentals of RF Circuit Design with Low Noise Oscillators*, United Kingdom: J. Wiley & Sons, Inc., 2000.
7. P.R. Gray, P.J. Hurst, S.H. Lewis, R.G. Meyer, *Analysis and Design of Analog Integrated Circuits*, fifth edition, Hoboken, NJ: J. Wiley & Sons, Inc., 2009.

8. T.H. Lee, *The Design of CMOS Radio-Frequency Integrated Circuits*, second edition, Cambridge: Cambridge University Press, 2003.

9. M.C. Albuquerque, F. Farzaneh, J. Obregon, "Définition théorique des paramètres en fort niveau d'un multipôle actif non linéaire," *Annales des Télécommunications,* Vol. 40, Issue 3, pp. 106–110, March 1985.

10. J. Obregon, F. Farzaneh, "Definition of Nonlinear Reflection Coefficient of a Microwave Device Using Describing Function Formalism," *IEEE Transactions on Microwave Theory and Techniques,* Vol. 32, Issue 4, pp. 452–455, May 1984.

9.11 Problems

Problem 9.1 Prove that in the case where the source and the load impedances are equal to the characteristic impedance, the transducer power gain (G_T) can be written as $G_T = |S_{21}|^2$. Then calculate the operating power gain (G_P) and the APG (G_A) in terms of the S-parameters of the transistor.

Problem 9.2 Consider the circuit shown in Figure 9.22. Compute G_T, G_A, and G_P, with the following parameters:
$\Gamma_s = 0.49\angle - 150°, \Gamma_L = 0.56\angle 90°$
$S_{11} = 0.54\angle 165°, S_{12} = 0.09\angle 20°, S_{21} = 2\angle 30°, S_{22} = 0.5\angle - 80°$

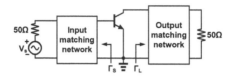

Figure 9.22: The amplifier circuit for determining various power gains.

Problem 9.3 The S-parameters for three transistors are given. Comment on their stability by drawing the stability circles at the source and the load planes.

$$
\begin{array}{lll}
S_{11} = 0.674\angle - 152° & S'_{11} = 0.385\angle - 55° & S''_{11} = 0.7\angle - 50° \\
S_{12} = 0.075\angle 6.2° & S'_{12} = 0.045\angle 90° & S''_{12} = 0.27\angle 75° \\
S_{21} = 1.74\angle 36.4° & S'_{21} = 2.7\angle 78° & S''_{21} = 5\angle 120° \\
S_{22} = 0.6\angle - 92.6° & S'_{22} = 0.89\angle - 26.5° & S''_{22} = 0.6\angle 80°
\end{array}
$$

Problem 9.4 Show that as S_{12} approaches zero, the centers and the radii of the input and output stability circles can be estimated as $C_S \approx \frac{1}{S_{11}}, r_S \approx 0, C_L \approx \frac{1}{S_{22}}, r_L \approx 0$. Given $|S_{11}| < 1$ and $|S_{22}| < 1$, what would you deduce from this?

Problem 9.5 Two different amplifiers with the specified S-parameters are cascaded as shown in Figure 9.23. Compute the overall S-parameters of these cascaded amplifiers in terms of their corresponding S-parameters (Hint: use the concept of loaded two-port S-parameters).

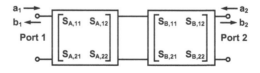

Figure 9.23: Cascaded amplifiers to determine the overall S-parameters.

Problem 9.6 Consider the circuits illustrated in Figure 9.24. Determine how resistive loading affects the overall S-parameters and the stability of the two-port network. The transistor's S-parameters are:
$S_{11} = 0.69\angle - 78°, S_{12} = 0.033\angle 41.4°, S_{21} = 5.67\angle 123°, S_{22} = 0.84\angle - 25°$

Hint: (1) Use the results of problem 9.5, (2) For the series resistance two-port S-parameters, one can easily show that $S_{21} = 1 - S_{11}$, and for the parallel resistance two-port $S_{21} = 1 + S_{11}$.

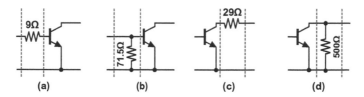

(a)   (b)   (c)   (d)

Figure 9.24: A transistor cascaded by either of series or parallel resistances.

Problem 9.7 Prove that maximum transducer power gain, G_T, occurs in a unilateral amplifier once we have $\Gamma_s = S_{11}{}^*, \Gamma_L = S_{22}{}^*$.

Problem 9.8 S-parameters for a transistor in a $50\,\Omega$-system are given as:
$S_{11} = 2.3\angle -135°, S_{12} = 0, S_{21} = 4\angle 60°, S_{22} = 0.8\angle -60°$
Comment on the stability of this transistor. Draw the circle corresponding to $G_A = 4\,\mathrm{dB}$. Then, design the matching network such that with $G_A = 4\,\mathrm{dB}$, $|\Gamma_{out}|$ is minimum.

Problem 9.9 Design a transistor amplifier in a $50\,\Omega$-system with maximum G_T. The S-parameters are given as:
$S_{11} = 0.277\angle -59°, S_{12} = 0.078\angle 93°, S_{21} = 1.92\angle 64°, S_{22} = 0.848\angle -31°$.

Problem 9.10 Considering the following S-parameters, first plot the stability circles. Secondly, determine G_P where $\Gamma_s = 0.2\angle 145°, \Gamma_L = 0$. Finally, determine the maximum value of G_P.
$S_{11} = 0.5\angle 45°, S_{12} = 0.4\angle 145°, S_{21} = 4\angle 120°, S_{22} = 0.4\angle -40°$.

Problem 9.11 Consider an amplifier with a silicon transistor as depicted in Figure 9.25. First determine the required resistances R_C and R_B for the quiescent point of $V_{CE} = 10V, I_C = 5\,\mathrm{mA}$. Secondly, compute the input and output impedances and consequently the source and the load reflection coefficients. The operating frequency is 300 MHz.

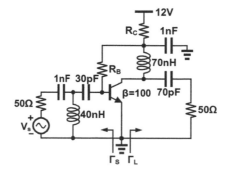

Figure 9.25: A transistor amplifier with corresponding load and source impedances and the bias circuitry.

Problem 9.12 The S-parameters and the noise parameters of a transistor are given at 1 GHz as:

$S_{11} = 0.6\angle 170°, S_{12} = 0.05\angle 16°, S_{21} = 2\angle 30°, S_{22} = 0.5\angle -95°$
$F_{min} = 2.5\,dB, \Gamma_{opt} = 0.5\angle 145°, R_n = 5\,\Omega$

Verify the stability and determine the maximum G_A. Then, plot the constant power gain contour having a gain 3 dB lower than $G_{A,max}$. Furthermore, plot the constant NF contours for $NF = 3\,dB$ and $NF = 4\,dB$. Finally, derive the NF of the amplifier at the maximum power gain point (in the source impedance plane).

Problem 9.13 Consider Figure 9.26. We wish to design a 2 GHz amplifier having $NF = 2\,dB$ with maximum possible G_T. First determine the required Γ_S and corresponding G_A. Then compute the required Γ_{out}. What would be the value of G_T then? The S-parameters are given at 2 GHz as:

$S_{11} = 0.646\angle 172°, S_{12} = 0.051\angle 13.5°, S_{21} = 3.042\angle 47.9°, S_{22} = 0.642\angle -64°$

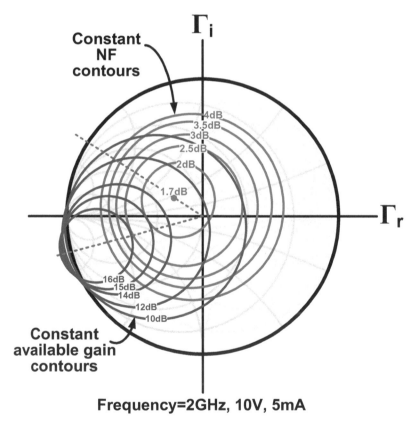

Figure 9.26: The Smith chart to design an amplifier with specific NF and G_A.

Problem 9.14 Consider the cascade of amplifiers/mixer depicted in Figure 9.27. Calculate the overall NF and the APG, G_A, of the chain.

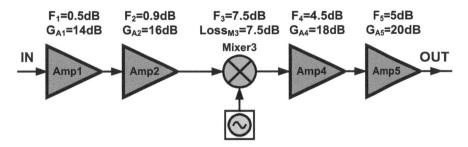

Figure 9.27: The cascade amplifiers/mixer for determination of the overall noise figure.

Problem 9.15 For the circuit depicted in Figure 9.28, calculate the input and the output power in dBm as well as the output voltage in dBVolts.

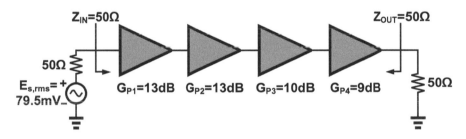

Figure 9.28: A cascade of amplifiers for the output power calculations.

Problem 9.16 For the MOS amplifier operating at $2\,\text{GHz}$ whose S-parameters are $S_{11} = 0.3\angle 160°, S_{12} = 0.03\angle 62°, S_{21} = 6.1\angle 65°, S_{22} = 0.4\angle -38°$, and depicted in Figure 9.29

1. Neglecting S_{12} determine the input and the output matching loads and calculate G_p. Assuming ideal inductors and capacitors, determine their required values.
2. If L_3 is replaced by a short-circuited $50\,\Omega$ stub, calculate the electrical length of the line as a fraction of the wavelength.

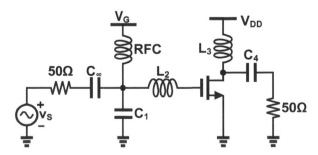

Figure 9.29: The transistor MOS amplifier for input and output matching.

Problem 9.17 Consider a transistor with the following parameters at 2 GHz:
$S_{11} = 0.55\angle 170°, S_{12} = 0.01\angle 23°, S_{21} = 1.68\angle 26°, S_{22} = 0.84\angle - 67°$
$F_{min} = 2\,dB, \Gamma_{opt} = 0.48\angle 165°, R_n = 6.25\,\Omega$

1. Comment on the stability and determine the maximum G_T.
2. Design the amplifier for the minimum NF and the maximum possible power gain. Find the proper values of Γ_S and Γ_L, determine the maximum possible power gain. Design the input and the output matching networks using a pair of lumped elements (capacitors and inductors).
3. In a second attempt design the amplifier for the maximum G_T, determine the required Γ_S and Γ_L in this case. Determine the corresponding NF in this case.
4. If the amplifier is followed by a mixer with $NF = 7\,dB$, calculate the overall NF in either of the cases in parts 2 and 3. In which case the receiver would be more sensitive?

Problem 9.18 Consider Figure 9.30. Our objective is to design a two-stage amplifier. The second stage being an amplifier with $G_2 = 14\,dB$ and $NF = 5\,dB$ at 1.6 GHz, preceded by a SiGe internally matched LNA, *BGU*7007 manufactured by NXP (see the datasheet at *www.nxp.com*). Determine the required bias voltage and the bias current of the LNA for an overall gain $G_T \geq 30\,dB$ and the total noise figure $NF_T \leq 1.2\,dB$.

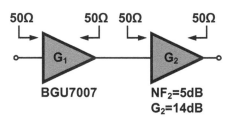

Figure 9.30: The two-stage amplifier to be designed using a *BGU*7007 internally matched LNA.

Problem 9.19 Assume a transistor with the following S-parameters at 1 GHz:
$S_{11} = 0.6\angle - 180°, S_{12} = 0.01\angle - 80°, S_{21} = 2.5\angle 30°, S_{22} = 0.6\angle - 83°$
Design an amplifier using this transistor with the power gain, G_P of 9.5 dB (plot the constant power gain contour in the load plane and choose the minimum $|\Gamma_L|$ point). Now determine the appropriate Γ_S such that the input is matched to it. Design the matching network using T-lines (on a substrate of $\varepsilon_{eff} = 3.3$) and open/short stubs.

Problem 9.20 The S-parameters and the noise parameters of a transistor operating at 4 GHz with $V_{CE} = 10V, I_C = 4mA$, are given as:
$S_{11} = 0.552\angle 169°, S_{12} = 0.049\angle 23°, S_{21} = 1.681\angle 26°, S_{22} = 0.839\angle - 67°$
$F_{min} = 2.5\,dB, \Gamma_{opt} = 0.475\angle 166°, R_n = 3.5\,\Omega$

1. Using the constant noise and the constant gain contours given in Figure 9.31, provided that the following stage has a NF of 7 dB. Choose the source reflection coefficient (impedance) such that the overall NF becomes at most 3.2 dB.
2. If the minimum required gain is equal to 10.7 dB, regardless of the second stage,

choose a point on the contours depicted in Figure 9.31 to have the minimum noise figure in this case. Determine the required load reflection coefficient to match the output, and then design the matching network in a 50Ω system at both the input and the output using air-filled T-lines ($\varepsilon_r = 1$) and open/short stub.

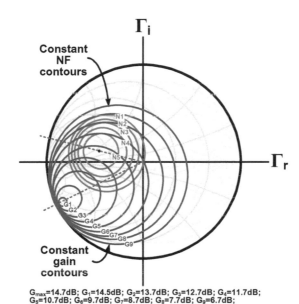

$$G_{max}=14.7dB; G_1=14.5dB; G_2=13.7dB; G_3=12.7dB; G_4=11.7dB;$$
$$G_5=10.7dB; G_6=9.7dB; G_7=8.7dB; G_8=7.7dB; G_9=6.7dB;$$

$$N_1=3dB; N_2=2.9dB; N_3=2.8dB; N_4=2.7dB; N_5=2.6dB;$$

Figure 9.31: The constant noise and the constant available gain circles at the source plane of the transistor at 4 GHz.

Problem 9.21 Calculate the maximum G_T, if stable, for the Motorola silicon bipolar transistor with part number $MRF962$ at 700 MHz at different biases of $V_{CE} = 5V, I_C = 10mA, 25mA, 50mA$ where the S-parameters data are as follows:

Table 9.1: Part of $MRF962$ datasheet.

V_{CE}	I_C	f	S_{11}		S_{21}		S_{12}		S_{22}	
(Volts)	(mA)	(MHz)	$|S_{11}|$	$\angle\phi$	$|S_{21}|$	$\angle\phi$	$|S_{12}|$	$\angle\phi$	$|S_{22}|$	$\angle\phi$
5	10	700	0.78	−176	3.16	77	0.071	26	0.23	−117
5	25	700	0.80	178	3.82	78	0.055	40	0.31	−158
5	50	700	0.81	176	4.09	78	0.048	50	0.38	−169
5	25	1500	0.81	164	1.82	59	0.086	42	0.34	−167

Problem 9.22 Evaluate stability at $1.5GHz$ and for $V_{CE} = 5V, I_C = 25mA$, and design the matching network for maximum transducer gain at this frequency. Use Table. 9.1.

Problem 9.23 In the given amplifier depicted in Figure 9.32, match the input and the output at the frequency of 318.3 MHz. First determine the input and the output admittances of the transistor at the operating frequency, and then find the required reactive components to match the input to 50 Ω and the output to 20 Ω.

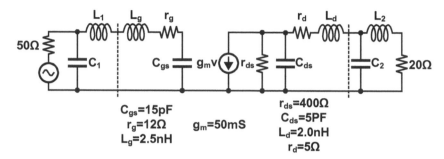

$C_{gs}=15pF$
$r_g=12Ω$ $g_m=50mS$
$L_g=2.5nH$

$r_{ds}=400Ω$
$C_{ds}=5PF$
$L_d=2.0nH$
$r_d=5Ω$

Figure 9.32: The equivalent circuit of a FET transistor and the associated matching circuits.

Problem 9.24 The Y-parameters of the cascode MOSFET stage at 159 MHz are depicted in Figure 9.33. Determine the reactive matching components for the maximum transducer power gain and compute the mentioned power gain.

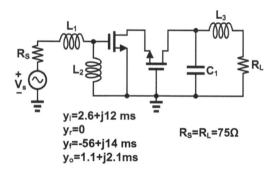

$y_i=2.6+j12$ ms
$y_r=0$
$y_f=-56+j14$ ms
$y_o=1.1+j2.1ms$

$R_S=R_L=75Ω$

Figure 9.33: A cascode MOS stage amplifier.

Problem 9.25 The Y-parameters of a *RF* transistor are as depicted in Figure 9.34. Moreover, one of the constant noise circles is given on the source admittance plane. First determine the optimum noise admittance (Hint: transform the relation 9.161 into a constant noise figure circle equation and from there determine G_o and B_o). Then, determine the required output admittance for the conjugate match condition. Finally, design the required LC circuits to match the input and the output to the 50 Ω reference impedance.

$$F = F_m + \frac{R_n}{G_S} \left[(B_s - B_o)^2 + (G_S - G_o)^2 \right] \tag{9.161}$$

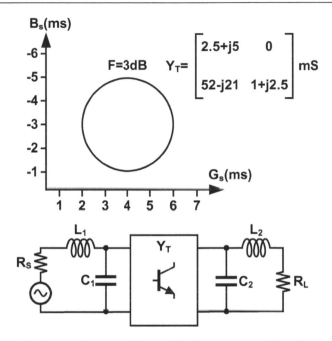

Figure 9.34: The low-noise amplifier with the corresponding input and output matching circuits.

Problem 9.26 In a low-noise amplifier, a noise figure of $F = 3$ (or 5 dB) has been measured for three consecutive source admittances, Y_{S1}, Y_{S2}, and Y_{S3} as depicted in Figure 9.35. Draw the corresponding constant noise figure circle and from there determine the corresponding optimum noise admittance. Use Equation 9.161.

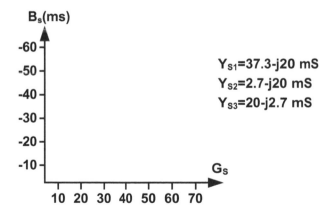

Figure 9.35: The measured source admittances for a noise figure $F = 3$ (source admittance plane).

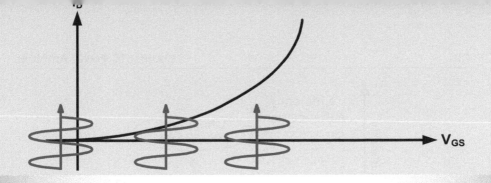

10. Power Amplifier

RF power amplifiers (PAs) consume the highest amount of power among all the transmitter blocks. While advancements in technology have resulted in aggregating all transmitter blocks into one single integrated circuit, the PA block is still integrated separately in many applications. In a transmitter chain, the data signal modulates properly the carrier signal and is then upconverted from the IF to the RF frequency. Afterward, a PA provides the necessary RF power level to transmit the signal according to the standard of interest, and the signal is radiated into the air by the antenna. By virtue of the power amplification, the PA mostly operates in a nonlinear or large-signal regime which mandates careful considerations regarding both its design and its simulation. One of the most important trade-offs in a PA is that of efficiency and linearity, as shown in Figure 10.1. In customary PAs, more linearity is expected to result in less efficiency and vice versa, while better linearity and good efficiency are both required to obtain a high data rate and low power consumption, respectively.

PA nonlinearity stems from the large-signal behavior of the active devices. An important issue that must be taken into consideration is the presence of the signal harmonics as well as IM products which have a potentially adverse effect on the adjacent channels. For this reason, the standards stipulate a measure called adjacent channel power ratio (ACPR) to regulate this issue. Furthermore, efficiency considerations impose lower limits on the supply voltage and active device architecture. In other words, the voltage source must be capable of providing sufficiently high currents and the active device should have a high voltage swing without entering the breakdown region. For driving purpose, a predriver stage is often used before the PA to provide the required signal level at the PA input. Noting the fact that the antenna impedance is in the order of $50\,\Omega$ and the breakdown voltage limit in active devices, an impedance matching circuit is utilized to match the required load impedance to the antenna impedance level. Considering the limited supply voltage for higher powers, we would need higher current swings which means lower impedance levels at the PA output. Impedance matching circuits generally introduce a finite loss due to use of low-Q inductors or capacitors, and consequently causing a poorer efficiency.

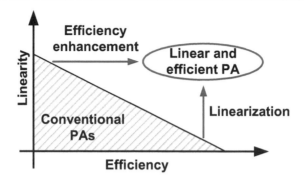

Figure 10.1: A graphical representation of the compromise between the linearity and the efficiency in a typical power amplifier.

In this chapter, PA design specifications are initially discussed. Then, different PA classes are introduced, followed by assessing a number of linearization techniques.

10.1 PA Specification

In this section, the most vital design specifications of PAs are introduced and studied. These specifications consist of the efficiency, the output power, the in-band noise, the gain, the linearity (AM to AM and AM to PM distortion, ACPR, and error vector magnitude (EVM)), and the stability. In general, power amplification can be discussed in two categories dependent on the linearity of the operating region. Assume the PA as a block with a single-tone input as

$$x(t) = A\cos(\omega_c t + \theta) \tag{10.1}$$

Then, the output signal is given by Equation 10.2

$$y(t) = |G|A\cos(\omega_c t + \theta + \angle G) \tag{10.2}$$

where G is the PA gain. Unlike the linear amplification, the output signal in a nonlinear amplifier will take the form of

$$y(t) = M(A(t))\cos(\omega_c t + \theta + \angle A(t)) \tag{10.3}$$

$M(A(t))$ and $\angle A(t)$ show AM to AM and AM to PM conversion, respectively. If the input contains multiple tones, linear amplification is preferred so that no IM products are generated, although nonlinear amplification yields a higher efficiency in certain cases.

10.1.1 PA Efficiency

Efficiency is the most critical parameter in PA design. A PA with 50% efficiency delivering $1W$ power (or $30\,dBm$) to a $50\,\Omega$ load, for example, also dissipates $1\,W$

power in the circuit, and thus having a total power consumption of 2 Watts. The dissipated power generates heat and necessitates specific measures in the circuit implementation, also shortening the battery life. In the context of PA design, power-added-efficiency (PAE) and efficiency are defined as

$$PAE = \frac{P_{out,RF} - P_{in,RF}}{P_{total,DC}} \qquad (10.4)$$

$$\eta = \frac{P_{out,RF}}{P_{total,DC} + P_{in,RF}} \cong \frac{P_{out,RF}}{P_{total,DC}} \qquad (10.5)$$

In Equation 10.4, the numerator accounts for the difference between high-frequency output power and the input power, and the denominator represents the total DC power dissipation. Equation 10.4 gives the power transferred to the load minus the input power divided by the DC power, and is called PAE for that matter. As studied in previous chapters, the output power exhibits a compressive behavior as a function of the input power as shown in Figure 10.2. In other words, owing to nonlinear amplification, the output signal no longer increases in proportion to the input signal when the input amplitude exceeds a certain value. The rate of change of the numerator in Equation 10.4 becomes smaller, leading to a compression in PAE as shown in Figure 10.3.

As an example for the importance of PAE in cell phones, the higher the PAE of a cell phone's PA, the longer would be the battery lifetime and the call durations.

10.1.2 PA Output Power

Consider Figure 10.4 where a PA is shown followed by a band-pass filter (or a selective impedance matching network).

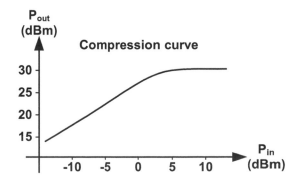

Figure 10.2: Typical compression curve of a power amplifier.

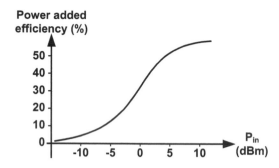

Figure 10.3: Compression curve of the PAE of an amplifier.

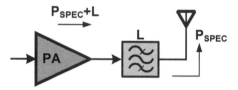

Figure 10.4: Power amplifier followed by a matching network.

The band-pass filter (BPF) is used to suppress the spurs and the undesired harmonics generated by the PA, also providing a proper load impedance for the amplifier. Nevertheless, the ultimate efficiency is degraded due to the insertion loss of the filter. To alleviate this issue, the PA must be designed for higher output power and better efficiency. For Figure 10.4, PAE can be written as

$$PAE = \frac{P_{\text{out}} * L - P_{\text{in}}}{P_{\text{total,DC}}} \tag{10.6}$$

Here, L is the insertion loss of the output filter. Note that the numerical value of L should be used in the above equation instead of its dB value. Each communication standard allows a specified amount of power to be transmitted. The effective radiated power (ERP) can be defined as the product of the power supplied to the antenna by the antenna gain relative to a half-wave dipole in a given direction. Another definition also exists as equivalent isotropically radiated power (EIRP) which is the product of the power supplied to the antenna by the antenna gain in a specific direction relative to an isotropic antenna. In general

$$EIRP \approx ERP + 2.2\,\text{dB} \tag{10.7}$$

where the 2.2 dB term is the gain of half-wave dipole antenna with respect to an isotropic antenna.

Table 10.1 shows various specifications of output power for a number of communication standards.

Another parameter of the PA is characterized with its probability density function (PDF). To be adapted to the output power requirements (e.g., in a CDMA system), the

Table 10.1: Output power specifications for a few communication systems in their different classes.

Wireless Standard		Maximum mobile station output power		
CDMA (IS-95)	PCS band (ERIP)	Class I	Class II	Class III
		28–33 dBm	22–30 dBm	18–27 dBm
	Cellular band (ERP)	Class I	Class II	Class III
		31–38 dBm	27–34 dBm	23–30 dBm
WCDMA		27 dBm		
AMPS (ERP)		Class I	Class II	Class III
		36 dBm	32 dBm	28 dBm
GSM		Class 2	Class 3	Class 4
		39 dBm	37 dBm	33 dBm
DCS-1800		Class 1	Class 2	Class 3
		30 dBm	24 dBm	36 dBm
4G LTE		23 dBm (maximum possible output power)		
IEEE802.11b		20 dBm		
Bluetooth		Class 1	Class 2	Class 3
		20 dBm	4 dBm	0 dBm

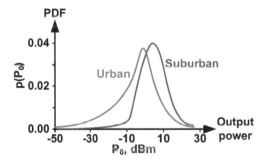

Figure 10.5: PDF of the transmitted power of the PA for a mobile set in an urban or suburban area.

mobile unit mostly transmits a power inferior to the maximum value. For this reason, in urban areas with more base stations, the chances of successful transmissions are higher even at low power levels, while power should be necessarily high in suburban areas. The PDF depicts the probability distribution of the power transmitted from a CDMA transmitter unit, a sample of which is shown in Figure 10.5 for an amplifier in an urban area or in a suburban area.

10.1.3 Receive-band Noise

Consider the full-duplex system of Figure 10.6. In full-duplex systems such as CDMA, the PA output at the transmitter degrades the sensitivity of the receiver system. The

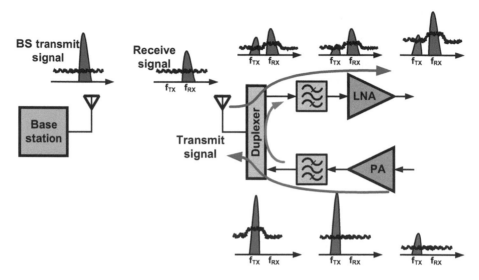

Figure 10.6: The effect of the TX-band noise at the receiver in a full-duplex system.

duplexer provides about 30 to 40 dB isolation between the receive and the transmit paths the reason for which the transmitted signal could degrade the receiver sensitivity by leaking to it. In fact, the PA amplifies the noise in the receiver band as well as the desired signal in the transmitter band. The PA noise itself will be added to the receiver, as such degrading its performance. Figure 10.6 demonstrates this process. In Figure 10.6, the base station transmits the signal with a large output level and the attenuated version of the original signal is received at the receiving antenna. The PA amplifies the signal in the transmitter band, but it leaks to the receive path through the duplexer (due to finite isolation between the two paths). In addition, the PA noise in the receiver band is only slightly attenuated by the duplexer and is added to the receive path. Consequently, detecting the received signal will be tougher, i.e, the receiver sensitivity will be degraded.

10.1.4 PA Gain

In practice, the minimum required PA gain is affected by four parameters:
- Maximum output power
- Loss after the PA
- Maximum driver output power
- Loss after the driver

Now, consider the system in Figure 10.7 in which a driver precedes a PA and there are interstage matching networks with a certain loss.

The overall minimum gain of the power amplifier in Figure 10.7 then can be obtained as

$$Gain_{min}(dB) = P_{SPEC,out}(dBm) + L(dB) + Ld(dB) - P_{max\,Driver}(dBm) \quad (10.8)$$

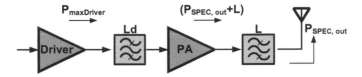

Figure 10.7: A PA with the PA driver and the interstage matching networks.

The maximum permissible gain of the PA is determined by parameters such as the driver noise in the receive band and the input–output isolation which prevents probable oscillation in the amplifier chain. The practical PA gain is typically within the range of 10 to 30 dB. Figures 10.8 and 10.9 show the compression effect in a PA.

10.1.5 Linearity Considerations in PA

Another important parameter to assess a PA's performance is its linearity. Generally, there are two types of nonlinearities to be considered in PAs:

- Amplitude nonlinearity (AM to AM distortion)

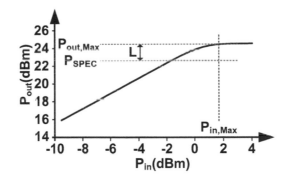

Figure 10.8: Typical compression of the output power versus the input power of a PA.

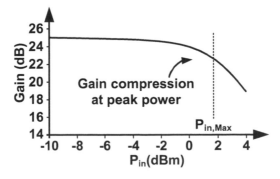

Figure 10.9: Typical gain compression in a PA.

- Phase nonlinearity (AM to PM distortion)

These nonlinearities have adverse effects on system performance which are stated below:

- Production of undesired harmonic components
- Spectral regrowth (degradation in ACPR)
- Degradation in noise power ratio (or reduction in SNR)
- Increase in error vector magnitude

Amplitude nonlinearity (AM to AM distortion): The degree of amplitude nonlinearity can be defined by parameters such as 1 dB compression point (P_{1dB}), third-order intercept point (IP_3), and carrier to intermodulation ratio (C/I). It is instructive to restate the important parameters regarding the nonlinearity:

P_{1dB} is the point where the nonlinear output gain of the system drops by 1 dB below the linear output gain. This point is shown on the curve of the output power in terms of the input power in Figure 10.10.

IP_3 is the point where the linear output power curve of the system intersects the third-order IM curve (in a two-tone excitation), as shown in Figure 10.10.

C/I is defined as the ratio of the amplitude of the desired signal at the system output to the maximum intermodulation distortion term (IMD) (typically, $C/I > 30dB$).

Phase nonlinearity (AM to PM distortion): This type of nonlinearity usually occurs once there is a nonlinear reactance like a voltage-dependent capacitor in the system (e.g, junction capacitors). Consider the circuit in Figure 10.11 where a resistor and a voltage-dependent capacitor are part of the circuit with a sinusoidal excitation current. Imagine a nonlinear capacitance defined by the following dynamic equation

$$q = C_1 v + C_2 v^2 + C_3 v^3 \tag{10.9}$$

Now, consider this nonlinear capacitance is driven by a sinusoidal voltage

$$v(t) = V_1 cos(\omega t) \tag{10.10}$$

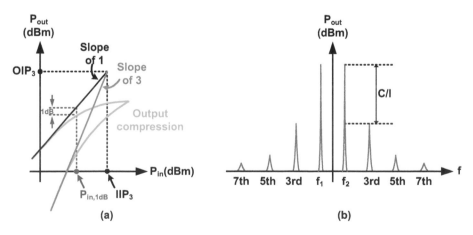

Figure 10.10: (a) P_{1dB} and IP_3 points in a nonlinear system, (b) Output spectrum in a nonlinear system with a two-tone input.

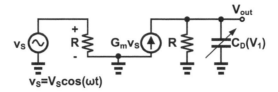

Figure 10.11: An amplifier with a nonlinear capacitance at its load.

Then, the capacitance's charge would have the following form

$$q = \left(C_1 + \frac{3}{4}C_3 V_1^2\right) V_1 \cos(\omega t) + \frac{C_2 V_1^2}{2}(1 + \cos(2\omega t)) + \frac{C_3 V_1^3}{4}\cos(3\omega t) \quad (10.11)$$

Now, retaining the fundamental term and ignoring the DC and higher harmonic terms, we can define a large-signal dynamic capacitance as

$$C_D(V_1) = C_1 + \frac{3}{4}C_3 V_1^2 \quad (10.12)$$

Considering this large-signal dynamic capacitance, one can obtain the output as

$$v_{out}(t) = \frac{G_m V_s R}{\sqrt{1 + (R\omega C_D(V_1))^2}}\cos(\omega t + \phi(t)) \quad (10.13)$$

where

$$V_1 = \frac{G_m V_s R}{\sqrt{1 + (R\omega C_D(V_1))^2}} \quad (10.14)$$

By substituting the value of $C_D(V_1)$ from Equation 10.12, one can rewrite Equation 10.14 in the following form

$$\left(\frac{3}{4}R\omega C_3\right)^2 V_1^6 + \frac{3}{2}(R\omega)^2 C_3 C_1 V_1^4 + \left(1 + (R\omega C_1)^2\right)V_1^2 - (G_m V_s R)^2 = 0 \quad (10.15)$$

By resolving the above equation, one can obtain the value of V_1^2, and therefore obtain the phase value as

$$\phi(t) = -\tan^{-1}(R\omega C_D(V_1)) \quad (10.16)$$

According to Equation 10.16, the output phase is a function of the input voltage amplitude. Figure 10.12 shows a sample of phase shift in the output of a system as the input signal amplitude increases. As the input signal power is raised, the output phase begins to change significantly after passing a threshold power.

It should be noted that phase modulation results in spectral regrowth. To gain a better understanding of spectral regrowth, consider the output signal of a phase-modulated single-tone sinusoid as

$$\begin{aligned} Vout(t) &= A\cos(\omega_c t + k\sin(\omega_b t)) \\ &= A\cos(\omega_c t)\cos(k\sin(\omega_b t)) - A\sin(\omega_c t)\sin(k\sin(\omega_b t)) \quad (10.17) \end{aligned}$$

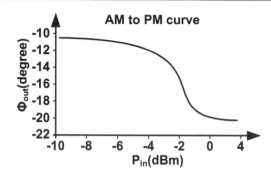

Figure 10.12: A typical AM to PM characteristics for a nonlinear amplifier.

Assuming $k \ll 1$, Equation 10.17 can be rewritten as below

$$Vout(t) \approx A\cos(\omega_c t) - Ak\sin(\omega_c t)\sin(\omega_b t)$$
$$= A\cos(\omega_c t) + \tfrac{Ak}{2}\cos((\omega_c + \omega_b)t) - \tfrac{Ak}{2}\cos((\omega_c - \omega_b)t) \qquad (10.18)$$

Figure 10.13(a) shows the one-sided frequency spectrum of Equation 10.18 for $k \ll 1$, only the first harmonic of the baseband will appear about the carrier. In the case where $k \approx 1$ or $k > 1$ (Figure 10.13(b)), the harmonics of ω_b will also appear in both sidebands and we can observe a certain spectral regrowth where in Equation 10.17, the carrier multipliers $\cos(k\sin(\omega_b t))$ and $\sin(k\sin(\omega_b t))$ can be expressed in terms of Bessel functions of even and odd order, respectively

$$\cos(k\sin(\omega_b t)) = J_0(k) + 2J_2(k)\cos(2\cos(\omega_b t)) + \ldots \qquad (10.19)$$
$$\sin(k\sin(\omega_b t)) = 2J_1(k)(\sin(\omega_b t)) + 2J_3(k)(\sin(3\omega_b t)) + \ldots \qquad (10.20)$$

As such, the even and the odd harmonics of the baseband signal will also modulate the in-phase and the quadrature components of the carrier and the signal spectrum will grow. In the following sections, some effects of nonlinearity will be briefly discussed:

Generation of undesired harmonic components: High gain and high output voltage swing limitations lead the PA to perform in a large-signal regime. This results

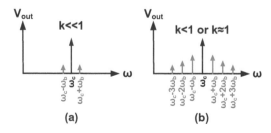

Figure 10.13: Output spectrum of Figure 10.11 due to PM conversion, (a) $k \ll 1$, and (b) $k < 1$ or $k \approx 1$.

in distortions in the output signal and production of undesired harmonic components as depicted in Figure 10.14(a). These components will grow as the input signal power is raised. To determine the influence of the undesired components, a parameter called the total harmonic distortion (THD) is defined as the square root of ratio of the summation of all undesired components to the main harmonic in percentage:

$$\text{THD} = 100 \times \sqrt{\frac{\sum\limits_{n=2}^{\infty} I_{d_n}^2}{I_{d_1}^2}} \tag{10.21}$$

where I_{d_n} is the current component of the nth harmonic of the main frequency. Furthermore, if two or more blockers are present at the input of the nonlinear system, the IM causes more spurious components to be generated (Figure 10.14(b)). This is why using filters in the output of PAs is common to attenuate the undesired harmonic components.

Spectral regrowth: The nonlinear performance of the system results in the generation of undesired components in the spectrum, having adverse effects on the adjacent channel. Both AM to AM and AM to PM conversions cause the spectrum to regrow at the output of the amplifier. Adjacent channel power ratio (ACPR) represents the amount of spectral regrowth in a PA, thus being a criterion of its linearity. Imagine a nonlinear tuned amplifier as depicted in Figure 10.15 where its nonlinear dynamic transconductance is described as follows

$$i = \alpha v + \beta v^3 + \gamma v^5 \tag{10.22}$$

Here

$$v = V_{S1} \cos(\omega_1 t) + V_{S2} \cos(\omega_2 t) \tag{10.23}$$

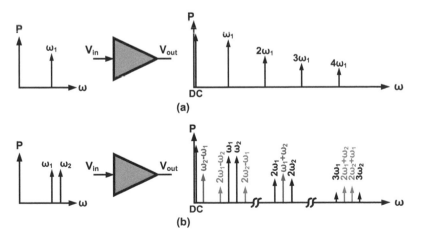

Figure 10.14: Generation of undesired harmonic and intermodulation components at the output of a nonlinear PA.

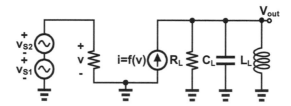

Figure 10.15: A nonlinear tuned amplifier corresponding to its nonlinear transconductance.

Then

$$i = \alpha \left(v_{S1} + v_{S2} \right)$$
$$+ \beta \left(v_{S1}^3 + 3v_{S1}^2 v_{S2} + 3v_{S1} v_{S2}^2 + v_{S2}^3 \right)$$
$$+ \gamma \left(v_{S1}^5 + 5v_{S1}^4 v_{S2} + 10v_{S1}^3 v_{S2}^2 + 10v_{S1}^2 v_{S2}^3 + 5v_{S1} v_{S2}^4 + v_{S1}^5 \right) \qquad (10.24)$$

Putting the sinusoidal voltage terms in Equation 10.24, we can compute the output fundamental voltages as well as the third and the fifth IM terms which happen to be within the output bandwidth. The third and fifth IM currents will have the following forms

$$I_{2\omega_1 - \omega_2} = \frac{3\beta V_{S1}^2 V_{S2}}{4} \cos\left((2\omega_1 - \omega_2) t \right) \qquad (10.25a)$$

$$I_{2\omega_2 - \omega_1} = \frac{3\beta V_{S1} V_{S2}^2}{4} \cos\left((2\omega_2 - \omega_1) t \right) \qquad (10.25b)$$

$$I_{3\omega_1 - 2\omega_2} = \frac{5\gamma V_{S1}^3 V_{S2}^2}{8} \cos\left((3\omega_1 - 2\omega_2) t \right) \qquad (10.26a)$$

$$I_{3\omega_2 - 2\omega_1} = \frac{5\gamma V_{S1}^2 V_{S2}^3}{8} \cos\left((3\omega_2 - 2\omega_1) t \right) \qquad (10.26b)$$

The output voltage including the IM terms (considering flat impedance, R_L, for all of the terms) can be described by the following expression.

$$v_{out} \simeq \alpha R_L \left(V_{S1} \cos(\omega_1 t) + V_{S1} \cos(\omega_2 t) \right)$$
$$+ \frac{3\beta R_L}{4} \left\{ V_{S1}^2 V_{S2} \cos\left((2\omega_1 - \omega_2) t \right) + V_{S1} V_{S2}^2 \cos\left((2\omega_2 - \omega_1) t \right) \right\}$$
$$+ \frac{5\gamma R_L}{8} \left\{ V_{S1}^3 V_{S2}^2 \cos\left((3\omega_1 - 2\omega_2) t \right) + V_{S1}^2 V_{S2}^3 \cos\left((3\omega_2 - 2\omega_1) t \right) \right\}$$
$$(10.27)$$

As it is seen here when there is at least two carrier frequency components at the input, the output spectrum will be widened by a factor of three due to the third-order IM and it will be widened by a factor of five due to the fifth-order IM. This results in a spectral regrowth at the sidebands of the carriers as seen in Figure 10.16. Also, Figure 10.17 depicts spectral regrowth and the resulting distortion in the time domain.

Note that while the fundamental terms grow with a slope of 10 dB/decade, the

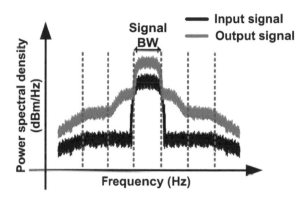

Figure 10.16: Spectral regrowth in a nonlinear PA as seen in the frequency domain.

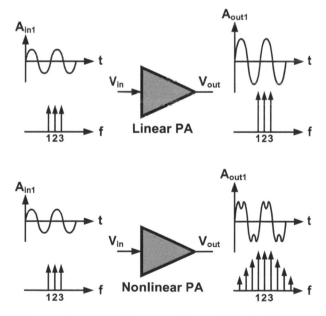

Figure 10.17: Spectral regrowth in a nonlinear PA, single tone shown in the time domain.

third-order IM terms grow with a slope of 30 dB/decade, and the fifth-order IM terms grow with a slope of 50 dB/decade. This point indicates that the input of the power amplifiers should not be more than a specified level in order to maintain the required signal to IM ratio or the required ACPR.

For instance, in CDMA (IS-95), ACPR is defined as the ratio of the available power at 1250 kHz offset frequency (with respect to the carrier) in a 30 kHz bandwidth to the available power in the main channel (at the center frequency) within the same bandwidth, as shown in Figure 10.18. Therefore, $ACPR_{Lo}$ and $ACPR_{Hi}$ can be defined as

$$ACPR_{Lo} = \frac{P_{Lo}}{P_o} \tag{10.28a}$$

$$ACPR_{Hi} = \frac{P_{Hi}}{P_o} \tag{10.28b}$$

The simplest method to measure ACPR in simulations is to excite the PA with a modulated input. If the fast Fourier transform (FFT) of the output is calculated, ACPR will be obtained. Simulation time is expected to be very long since the envelope frequency is mainly much lower than the carrier frequency. Therefore, another method is used to calculate ACPR which is believed to be more efficient. With a high difference between the modulation frequency and the carrier frequency, this method can help to estimate the ACPR through AM to AM and AM to PM characteristics. It can be performed in the following steps:

1. Measuring or calculating the nonlinear PA gain function (G) to form the AM to AM curve (AM to AM $\Rightarrow G(A)$).
2. Measuring or calculating the nonlinear PA phase function (P) to form the AM to PM curve (AM to PM $\Rightarrow P(A)$).
3. Generating a baseband modulated input signal as $S_i(t) = A(t)\angle\phi(t)$.
4. Estimating the baseband modulated output signal as $S_o(t) = G(A(t))\angle[\phi(t) + P(A(t))]$.
5. Computing the FFT of the signal resulted in the previous step.

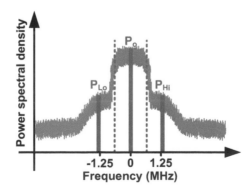

Figure 10.18: ACPR definition in CDMA (IS-95) standard.

Care must be taken with regard to the above method as it is completely an approximative one. So its result might underestimate the spurious generated in the adjacent channels (sidebands), because these unwanted components are mainly generated due to a dynamic IM process.

Decreasing the noise power ratio: The nonlinear performance of PAs increases the in-band noise power especially at the carrier frequency as well as generating undesired components at the adjacent channels. This causes a lower SNR and thus lower sensitivity in the receiver. In other words, the transmitted data will be noisy and hard to be detected by the target receiver.

One of the ways to determine in-band distortion is measuring the noise power ratio (NPR). For this purpose, the input signal is initially passed through a very sharp notch filter to ensure that its in-band spectrum has no components at the carrier frequency. Then, the resulted signal spectrum is applied to the nonlinear system as an input. The nonlinear performance of the system will lead to spectral regrowth and increasing the intermodulation noise floor level at the carrier frequency. Ultimately, the ratio of the obtained IM noise power at the carrier frequency to the output signal power in the band is calculated as NPR, as shown in Figure 10.19.

Increase in Error Vector Magnitude: Error vector magnitude (EVM) is defined as the normalized distance between the desired and actual signal vectors as depicted in Figure 10.20. Due to the nonlinear performance of the system, the yielded vector is not expected to coincide with the desired signal vector in practice.

EVM is reported in both rms and peak values. Typical values range from 7% to 12% in the former and from 22% to 33% in the latter.

Increases in EVM can be interpreted as distortion in the data constellation at the output signal. Undesired changes in phase and amplitude bring about errors in detecting the modulated data, causing problems in the data transmission process. This issue becomes more significant when the modulation level is higher, such as in QAM. Figure 10.21 depicts a sample of data constellation distortion at the output for 16-QAM modulation.

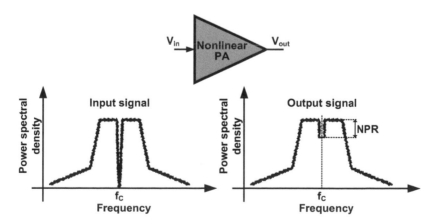

Figure 10.19: Degradation of in-band *SNR* due to nonlinearity.

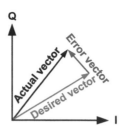

Figure 10.20: Representation of error vector magnitude (EVM) in a sampling $I - Q$ constellation.

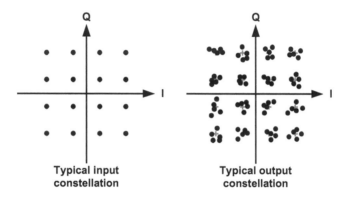

Figure 10.21: A comparison of the input and the output constellation for a 16-QAM modulation for a nonlinear power amplifier in between.

10.1.6 PA Stability Considerations

Ideally, the PA is supposed to be strictly stable, i.e, it should maintain its stability for any passive load or source impedances. In practice, PAs are stable for the VSWR≤ 6 (in all phases). Stability must be guaranteed even when the preceding and the following stages are not perfectly matched. Some of the factors causing instability in PAs include:
- Existing of feedback between various amplifier stages
- Coupling between device terminals or wire-bonds
- Feedback due to device parasitics (e.g., C_μ or C_{gd})
- Feedback paths through supply voltage and ground lines

Another factor is the gain peaking at subharmonics (e.g., at half the fundamental frequency or one-third of the fundamental frequency) which could result in subharmonic spurs and finally bias circuit oscillations.

10.2 PA Topologies

The PA topologies have evolved through the recent decades. Differences in linearity and efficiency are the main reasons that impose various topologies. Another important parameter that makes a significant difference in PA performance is the conduction angle which varies depending on the operation point of the active device. In addition, as the efficiency is of great importance in the design of a PA, a method must be sought to enable minimum power dissipation by the active devices. Harmonic termination can be helpful in some PA classes. In the next sections, the details of the operation classes of PA will be studied as well as the relations regarding the efficiency of each class.

10.2.1 Class A Power Amplifier

In class A amplifiers, the quiescent point of the active device is chosen in such a way that the amplifier is always on, regardless of the input signal; hence, the conduction angle is 360° and a good linearity is expected. In this configuration, the maximum efficiency at the output is 50% and the power capability is $0.125V_{d_{max}}.I_{d_{max}}$. It must be noted that BPFs with high Q are often used in the output to suppress all high-order harmonics of the output voltage (and current). Figure 10.22 shows a sample of the class A amplifier. Figure 10.23(a) shows the quiescent point determination curve for the amplifier of Figure 10.22. As it is clear in Figure 10.23(a), the quiescent point is set at the middle of the load line to guarantee that the device remains on in all circumstances, resulting in a 360° conduction angle. Figure 10.23(b) shows the drain voltage and the drain current of the device and its power loss as a function of time. Considering Figure 10.22, the following equations can be written for the output current and voltage.

$$i_D(t) = I_{DC} - I_{ac}\sin(\omega t) \tag{10.29}$$
$$v_D(t) = V_{DC} + V_{ac}\sin(\omega t) \tag{10.30}$$

Given the fact that in class A operation, we should have

$$I_{ac} \leq I_{DC} \tag{10.31}$$
$$V_{ac} \leq V_{DC} \tag{10.32}$$

The maximum output power would be

$$P_{ac,max} = \frac{1}{2}V_{DC}I_{DC} \tag{10.33}$$

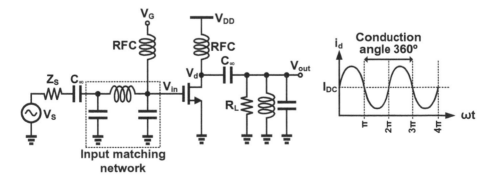

Figure 10.22: Class A power amplifier.

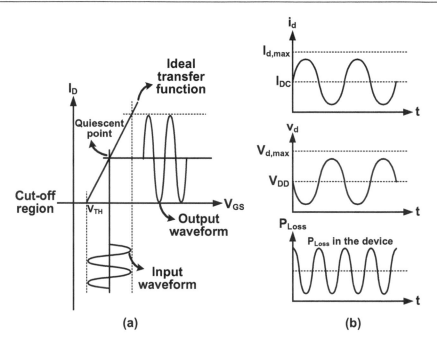

(a)   (b)

Figure 10.23: (a) Quiescent point determination in class A power amplifiers and (b) Drain voltage, drain current, and device power loss of the circuit depicted in Figure 10.22.

The definition of efficiency in PA yields,

$$\eta = \frac{P_{ac}}{P_{DC}} = \frac{\frac{1}{2}V_{ac}I_{ac}}{V_{DD}I_{DC}} \tag{10.34}$$

Therefore, for the maximum efficiency, we will have

$$\eta_{max} = \frac{\frac{1}{2}V_{DC}I_{DC}}{V_{DC}I_{DC}} = 50\% \tag{10.35}$$

Note that, for the maximum efficiency, one should choose the load resistance as

$$R_L = \frac{V_{DC}}{I_{DC}} \tag{10.36}$$

Therefore, reminding that the drain voltage in Figure 10.22 can maximally swing from zero to twice the voltage source ($2V_{DD}$), and the current can swing from zero to twice the DC operating current, the power capability can be defined as

$$\text{Power capability} = \frac{P_{ac,max}}{V_{d,max} \times i_{d,max}} = \frac{\frac{1}{2}V_{DD} \times I_{DC}}{2V_{DD} \times 2I_{DC}} = 0.125 \tag{10.37}$$

10.2.2 Class B Power Amplifier

In class B amplifiers, the quiescent point of the active device is set at the threshold of the device (or approximately zero bias) turning on so that it provides the output current only in half the signal cycle and remains off in the other half-cycle. The conduction angle is consequently 180°, and the maximum efficiency at the output would be $\pi/4 \approx 78.5\%$. This class has the privilege of significant improvement in efficiency besides a fairly appropriate linearity. The power capability is the same as class A if one device is used, but with a push–pull configuration at the output stage, this parameter would be increased to $0.25V_{d_{max}}.I_{d_{max}}$. Furthermore, class-B PAs necessitate the usage of BPFs with high Q at the output to suppress all high-order harmonics of the output voltage (especially the second harmonic voltage). Figure 10.24 depicts a sample class B power amplifier with zero input bias. Figure 10.25(a) demonstrates the quiescent point determination curve for the amplifier of Figure 10.24. Figure 10.25(b) shows the voltage and drain current of the device and its power loss as a function of time. To derive the efficiency relations for the class B amplifiers, assume that the DC current flowing from the voltage source is a half-cycle sinusoidal signal. The average current can be hence obtained by integrating over one period. Consider the instantaneous drain current as

$$i_d(t) = \begin{cases} A\sin(\omega t) & 0 \le \omega t \le \pi \\ 0 & \pi \le \omega t \le 2\pi \end{cases} \tag{10.38}$$

Then, by taking the time average, we obtain

$$I_0 = I_{av} = \frac{1}{T}\int_0^T i_d(t)\,dt = \frac{1}{T}\int_0^{T/2} A\sin(\omega t)\,dt = \frac{A}{\pi} \tag{10.39}$$

$$\tag{10.40}$$

The fundamental component of the output current can be obtained as

$$I_1 = \frac{2}{T}\int_0^T i_d(t)\sin(\omega t)\,dt = \frac{A}{2} \tag{10.41}$$

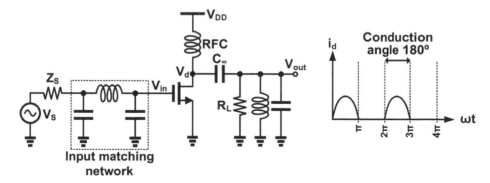

Figure 10.24: A class B power amplifier with the corresponding output current waveform.

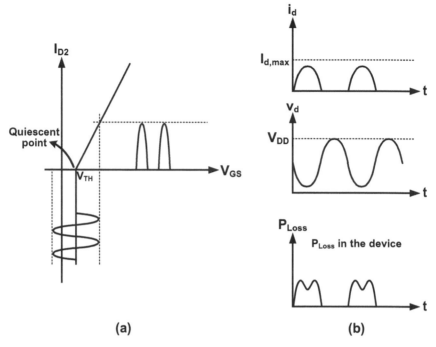

(a)   **(b)**

Figure 10.25: (a) The quiescent operating point and the corresponding waveforms for the class B PA and (b) The output current, the output voltage, and the
power loss waveforms for a single-device class B power amplifier.

Finally, noting that the drain voltage in Figure 10.24 can maximally swing from zero to
the supply voltage, the maximum efficiency is obtained by the ratio of the fundamental
(main harmonic) power to the consumed DC power by the source:

$$\eta_{max} = \frac{P_{ac,max}}{P_{DC}} = \frac{\frac{1}{2}\frac{V_{DD}}{2}\frac{I_{max}}{2}}{\frac{V_{DD}}{2}\frac{I_{max}}{\pi}} = \frac{\pi}{4} \approx 78.5\% \tag{10.42}$$

Note that in this case, for maximum efficiency, the load resistance should be chosen as

$$R_L = \frac{\frac{V_{DD}}{2}}{\frac{I_{max}}{2}} = \frac{V_{DD}}{I_{max}} \tag{10.43}$$

In this case, power capability can be calculated as

$$\text{Power capability} = \frac{P_{ac,max}}{V_{d,max} \times i_{d,max}} = \frac{\frac{1}{2}\frac{V_{DD}}{2} \times \frac{I_{max}}{2}}{V_{DD} \times I_{max}} = 0.125 \tag{10.44}$$

Class B push–pull case

The above derivation was performed for a single-transistor class B amplifier. For the push–pull case, the same procedure can be used with the NMOS conducting in the positive half-cycle and the PMOS conducting in the negative half-cycle. Consequently, the fundamental current will be doubled and the output power will be doubled. Furthermore, the DC power consumption will be doubled and as such the efficiency would remain the same as single-transistor class B power amplifier. However, the harmonic distortion will be reduced and the linearity will be improved. Furthermore, the power capability will be doubled to $0.25V_{d_{max}}.I_{d_{max}}$. Figure 10.26 depicts a sample of the class B PA with push–pull output stage. Note that in this case, for maximum efficiency, the load resistance should be chosen as

$$R_L = \frac{V_{DD}}{2I_{max}} \tag{10.45}$$

Therefore, the load impedance in the class B push–pull case would be half the load impedance in a single-ended class B power amplifier (with the same device and the same bias).

10.2.3 Class AB Power Amplifier

In class AB amplifiers, the quiescent point of the active device is set higher than the threshold of the device's turning on voltage and lower than the middle point of the load line. So, the device provides the output current in more than half a cycle, and less than a full cycle. The conduction angle is hence between 180° and 360°, denoting that the efficiency of this class exceeds that of class A but cannot reach that of class B,

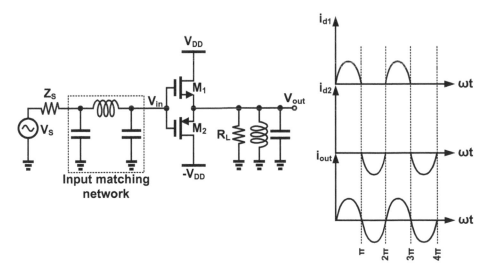

Figure 10.26: A class B push–pull power amplifier with the output current shown as a summation of the positive and the negative half-cycle currents.

while its linearity is better than class B and worse than class A. Figure 10.27 shows the quiescent point determination curve for class AB amplifiers in general.

10.2.4 Class C Power Amplifier

In class C amplifiers, the quiescent point of the active device is set below the threshold of turning on. Therefore, the device provides the output current in less than half a cycle, leading to a conduction angle of less than 180°. In case the conduction angle tends to zero, the efficiency tends to 100%, though the output power will also tend to zero. Indeed, the lower the conduction angle, the more the efficiency and the less the linearity will be. In general, class C efficiency exceeds that of class A and class B power amplifiers while its linearity is worse than the other two. Figure 10.28 shows a sample of the class C amplifier, and Figure 10.29 depicts the quiescent point determination curve for the amplifier of Figure 10.28.

To calculate the power efficiency of the amplifier in Figure 10.28, consider the instantaneous current in the active device as

$$i_d(t) = \begin{cases} I_p \sin(\omega t) - I_D & \theta_1 \leq \omega t \leq \theta_2 \\ 0 & \text{otherwise} \end{cases} \tag{10.46}$$

Thus, the average current (DC current) will be obtained by integrating the current waveform over one period:

$$I_0 = I_{av} = \frac{1}{T} \int_0^T i_d(t) \, dt = \frac{1}{2\pi} \left(\int_{\theta_1}^{\theta_2} (I_p \sin(\omega t) - I_D) \, d(\omega t) \right) \tag{10.47}$$

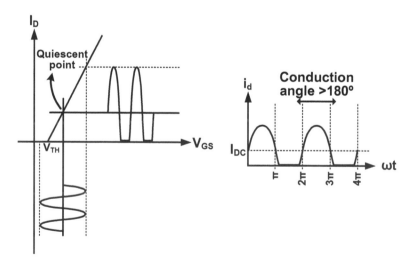

Figure 10.27: Class AB PA's quiescent point and the corresponding current waveform.

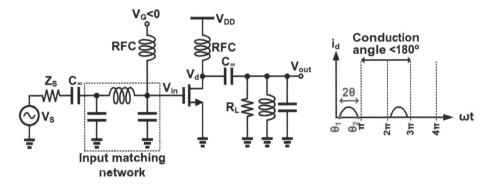

Figure 10.28: A class C power amplifier and its corresponding output current waveform.

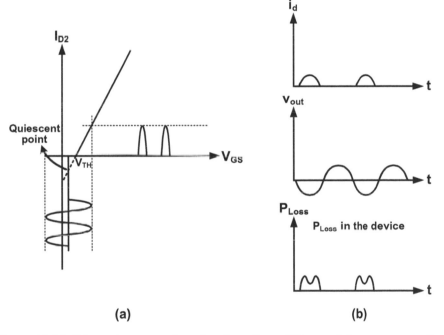

| (a) | (b) |

Figure 10.29: The quiescent point of a class C amplifier and the corresponding output current, output voltage, and the power loss waveforms.

Now, take the following notations into consideration

$$I_p \sin \theta_1 = I_D \tag{10.48a}$$

$$\theta_2 - \theta_1 = 2\theta \tag{10.48b}$$

$$\theta_1 = \frac{\pi}{2} - \theta \tag{10.48c}$$

$$\theta_2 = \frac{\pi}{2} + \theta \tag{10.48d}$$

where 2θ equals the conduction angle (α) of the PA. The total DC power consumption will be given by

$$P_{DC} = V_{DD}I_0 = V_{DD}\frac{I_P}{\pi}\left(\sin(\theta) - \theta\cos(\theta)\right) \tag{10.49}$$

To obtain the AC power delivered to the load, the fundamental current is calculated as

$$I_1 = \frac{1}{\pi}\left(\int_{\frac{\pi}{2}-\theta}^{\frac{\pi}{2}+\theta}\left(I_p\sin(\omega t) - I_D\right)\sin(\omega t)\,d(\omega t)\right) = \frac{I_p}{2\pi}\left(2\theta - \sin(2\theta)\right) \tag{10.50}$$

Finally, assuming the peak output AC voltage as V_{DD}, the efficiency of the class C amplifier is yielded as

$$\eta_{max} = \frac{P_O}{P_{DC}} = \frac{\frac{1}{2}V_{DD}\frac{I_p}{2\pi}\left(2\theta - \sin(2\theta)\right)}{V_{DD}\frac{I_p}{\pi}\left(\sin(\theta) - \theta\cos(\theta)\right)} = \frac{2\theta - \sin(2\theta)}{4\left(\sin(\theta) - \theta\cos(\theta)\right)} \tag{10.51}$$

where $0 < \theta < \pi$. It should be noted that for maximum AC power and efficiency, we should choose the following value for the load impedance.

$$R_L = \frac{V_{DD}}{I_1} = \frac{V_{DD}}{\frac{I_p}{2\pi}\left(2\theta - \sin(2\theta)\right)} = \frac{V_{DD}}{I_p}\frac{2\pi}{\left(2\theta - \sin(2\theta)\right)} \tag{10.52}$$

It is helpful to note that Equation 10.51 applies for the efficiency of A, B, and AB classes as well with the corresponding conduction angles.

10.2.5 Comparison Between Class A, Class B, Class AB, and Class C Amplifiers

One of the most important reasons for PA's classification is the difference in the efficiency and power capability of the classes. Figure 10.30 presents a comparison between A, B, C, and AB classes based on their conduction angles. According to Equation 10.51, the efficiency approaches 100% as θ (half the conduction angle, α) tends to zero, although no power is transferred to the output, for zero conduction angle. In the class A amplifier, θ will reach its maximum value (180°), leading to an efficiency of 50%. To modify and control the conduction angle, the current amplitude can be kept constant while the bias current or the quiescent point is changed. A similar method is performed by acting the other way around. In both cases, the conduction angle can be altered as desired.

10.2.6 Class D Power Amplifier

Other PA topologies such as classes D, E, F, S, etc., have been presented and discussed in various references, *e.g.,* [2] and [3], which can theoretically provide 100% efficiency. In these classes, the active device dissipates theoretically zero power. Transistors act as switches at RF rate in class D amplifiers. Consider Figure 10.31 where the input differential signal has been applied to two NMOS transistors. This architecture actually consists of two class B power amplifiers which have been put in a push–pull configuration. During the positive half-cycle, the upper transistor is turned on,

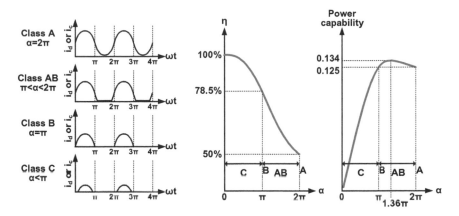

Figure 10.30: The output current waveforms alongside the efficiencies and the power capabilities corresponding to different classes of power amplifiers as a function of the conduction angle, α.

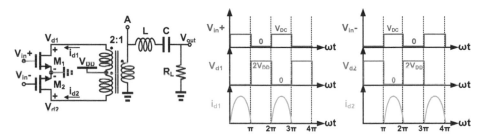

Figure 10.31: A typical class D power amplifier topology and the corresponding voltage and current waveforms.

providing the positive half-cycle transformer current. In the negative half-cycle, the lower transistor is turned on and the negative half-cycle current flows through the transformer. Finally, the currents are superimposed to make up the appropriate full-cycle output signal. A band-pass RLC circuit with high Q suppresses all the higher-order voltage harmonics at the output. The maximum possible efficiency in class D amplifiers is 100% and the maximum power capability equals $(1/\pi)V_{d_{max}} \cdot I_{d_{max}} \approx 0.32V_{d_{max}} \cdot I_{d_{max}}$. This configuration has a better performance at frequencies far lower than the unity gain frequency of the transistors, as the transistors are better switches in this range. Figure 10.31 depicts the corresponding waveforms of class D amplifier. One of the drawbacks of class D PA's is that it is not applicable for linear modulations such as AM, SSB, or DSB in normal circumstances.

As Figure 10.31 shows, the drain voltage of the transistors can maximally swing up to twice the voltage source value. Assuming that the transistors act as ideal switches, the voltage at node A can be considered as a rectangular wave swinging symmetrically between $\pm V_{DD}$. Writing the Fourier series expansion of the voltage at node A, we have

$$v_A(t) = \frac{4}{\pi} \sum_{n=0}^{n=+\infty} \frac{\sin((2n+1)\,\omega t)}{2n+1} \tag{10.53}$$

The band-pass circuit at the output selects only the main harmonic of v_A. Consequently, assuming that the amplitude of the first harmonic is $4V_{DD}/\pi$ at node A, the output power will be

$$P_O = \frac{V_O^2}{2R_L} = \frac{\left(\frac{4V_{DD}}{\pi}\right)^2}{2R_L} = \frac{8V_{DD}^2}{\pi^2 R_L} \tag{10.54}$$

In addition, in order to obtain the average current of each transistor in a period, reminding that the output current is the superposition of the currents of each transistor at consecutive half-cycles, one can write

$$I_{ave} = \frac{1}{T} \int_0^{\frac{T}{2}} \frac{4V_{DD}}{R_L \pi} \sin(\omega t)\,dt = \frac{4V_{DD}}{\pi^2 R_L} \tag{10.55}$$

Ultimately, due to the presence of two active devices, the total power consumption will be calculated

$$P_{DC} = 2 \times V_{DD} \frac{4V_{DD}}{\pi^2 R_L} = \frac{8V_{DD}^2}{\pi^2 R_L} \tag{10.56}$$

which has exactly the same expression as Equation 10.54. Hence, in the class D amplifier, the efficiency is ideally 100%. The above relations are based on the assumption that the switches are ideal while they might have a limited on-resistance which decreases the efficiency of the PA. If this resistance is not ignored, the output power of Equation 10.54 will change as follows

$$P_O = \frac{8V_{DD}^2}{\pi^2 R_L} \left(\frac{R_L}{r_{on} + R_L}\right)^2 \tag{10.57}$$

The total power consumption in this case will be calculated as

$$P_{DC} = \frac{8V_{DD}^2}{\pi^2 R_L} \times \frac{R_L}{r_{on} + R_L} \tag{10.58}$$

Finally, the maximum efficiency of the PA considering the on-resistance will be

$$\eta_{max} = \frac{R_L}{r_{on} + R_L} \tag{10.59}$$

The on-resistance might cause the device to have a saturation voltage when turned on, noted by V_{sat}. The drain voltage of neither of the devices will thus reach to the ground level, as they enter the saturation region. Therefore, V_{DD} should be substituted by $V_{DD} - V_{sat}$, and the output power equation will change as

$$P_O = \frac{8(V_{DD} - V_{sat})^2}{\pi^2 R_L} \tag{10.60}$$

For the total power consumption, we have

$$P_{DC} = 2 \times V_{DD} \frac{4\left(V_{DD} - V_{sat}\right)}{\pi^2 R_L} = \frac{8\left(V_{DD} - V_{sat}\right) V_{DD}}{\pi^2 R_L} \tag{10.61}$$

Finally, the maximum efficiency of the PA considering the transistor saturation voltage will be

$$\eta_{max} = \frac{V_{DD} - V_{sat}}{V_{DD}} \tag{10.62}$$

In case where the output switches have both a turn-on resistance and a saturation voltage, the maximum efficiency of a class D amplifier can be expressed as

$$\eta_{max} = \frac{V_{DD} - V_{sat}}{V_{DD}} \times \frac{R_L}{r_{on} + R_L} \tag{10.63}$$

Class 1/D power amplifier
The class 1/D amplifier can be considered in a dual mode of what has been described earlier. In a sense that the current is switched through a parallel RLC circuit. Consequently, the current waveform will be rectangular and the voltage waveform will be sinusoidal. Therefore, this mode of operation is somehow the dual of the class D mode described earlier. For this reason, this mode is categorized as class 1/D, or current mode. Notice the circuit topology and the corresponding waveforms in Figure 10.32.

10.2.7 Class E Power Amplifier
In this PA class, the active device acts as a switch again. The design goal is to minimize the device on-resistance, thus there will be no dissipations in the active device. The device performance in class E power amplifiers is in such a way that no current flows through the device when a finite voltage is developed across its output. Furthermore, the voltage across the device is quite small when current flows from it

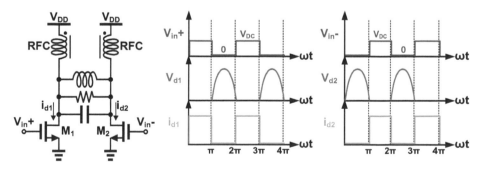

Figure 10.32: A typical circuit topology and current waveforms of a class 1/D power amplifier.

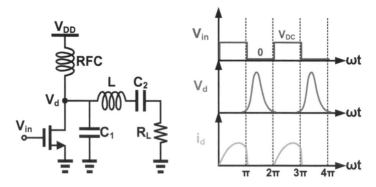

Figure 10.33: A typical class E power amplifier topology and the corresponding voltage and current waveforms.

to the load. Figure 10.33 shows a class E power amplifier. The device voltage and current waveforms are also demonstrated. As shown in Figure 10.33, the drain voltage and its first derivative will be zero when the device is turned on. This configuration has the ability to provide 100% efficiency while its power capability is approximately $0.098 V_{d_{max}} \cdot I_{d_{max}}$.

In the design of class E power amplifiers, the transistor parasitic capacitances are crucial due to the frequency limit they impose on the circuit. In the design of the matching network, the following relations can be used [9]

$$L = \frac{QR_L}{\omega} \tag{10.64a}$$

$$C_1 = \frac{1}{\omega R_L((\pi^2/4)+1)(\pi/2)} \approx \frac{1}{5.447\omega R_L} \tag{10.64b}$$

$$C_2 \approx C_1 \left(\frac{5.447}{Q}\right)\left(1+\frac{1.42}{Q-2.08}\right) \tag{10.64c}$$

In addition, the maximum output power will be calculated by

$$P_{o,max} \approx 0.577 \frac{V_{DD}^2}{R_L} \tag{10.65}$$

10.2.8 Class F Power Amplifier

In class F amplifiers, the harmonic behavior of the output is utilized in its best way by using multiresonance circuits so that the current waveform of the transistor drain gets close to an ideal rectangular pulse. This can also minimize the drain-source voltage and the drain current overlap through time, leading to better efficiency. Figure 10.34 shows a class F power amplifier. The device voltage and current waveforms are also demonstrated.

In Figure 10.34, the L_1C_1 and L_3C_3 resonant circuits are adjusted at f_0 (the desired input frequency) and $3f_0$, respectively. Consequently, the drain voltage will contain a third harmonic component as well as the first.

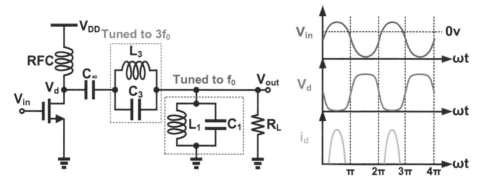

Figure 10.34: A typical class F power amplifier topology and the corresponding voltage and current waveforms.

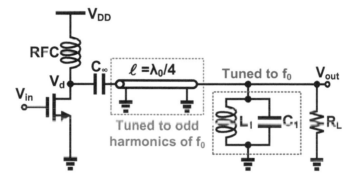

Figure 10.35: A typical class F power amplifier topology using a quarter-wavelength transmission line.

Figure 10.35 is another representation of the class F amplifier in which a transmission line is used instead of the first resonant circuit at the third harmonic. The transmission line has the capability to present the inverse impedance (turn a low impedance into a high impedance) when it is a quarter wavelength long ($\ell = \lambda/4$), hence having a proper input impedance to pass the desired current. It acts equivalently to LC resonant circuits at odd harmonics of the fundamental frequency (f_0), and therefore helps the transistor drain voltage to approach a rectangular wave (for even harmonics the transmission line acts as a half-wave length one, and therefore, no impedance inversion occurs, and the even harmonics of the drain voltages are suppressed). Ultimately, the output band-pass circuit (the resonant circuit at the fundamental harmonic, f_0) suppresses the higher-order harmonics at the output. In fact, class F amplifiers can be treated as a type of class D amplifier in single-ended mode. In this class, the maximum efficiency is 100% and power capability equals to $(1/2\pi) V_{d_{max}} . I_{d_{max}} \approx 0.16 V_{d_{max}} . I_{d_{max}}$ [12].

10.2.9 Class S Power Amplifier

As observed in section 10.2.6, the active devices act as a switch in class D amplifiers. The closer the device performance to an ideal switch, the better the resulting efficiency. One of the methods to ameliorate the switching performance of the device is to increase the applied voltage to the gate (V_{GS}). It helps reduce the on-resistance of the device, leading to increased efficiency. Figure 10.36 shows a general schematic of a class S power amplifier. A signal converter turns the sinusoidal input signal into a pulse width modulation (PWM) signal. The PWM signal is applied to the active device in a class D amplifier, improving its switching performance. Finally, the output signal passes through a high Q band-pass filter which converts the PWM signal into a sinusoid (higher-order harmonics are suppressed).

Figure 10.37 shows a typical signal converter. A comparator compares the input signal to a triangular wave signal which renders a PWM output.

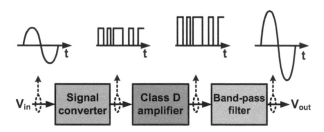

Figure 10.36: Block diagram of a class S power amplifier and its corresponding waveforms.

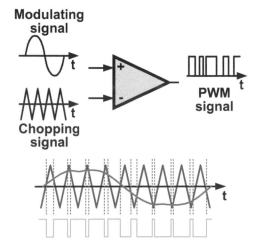

Figure 10.37: A typical class sinusoidal to PWM signal converter.

10.2.10 PA's Performance Comparison

Considering what was mentioned in previous sections regarding power amplifiers' classification, Table 10.2 provides a comparison between the performances of different classes of PAs.

10.3 Linearization Techniques in Power Amplifiers

Since linearity is of great importance in PA design, several techniques to mitigate nonlinear effects are introduced and studied in this chapter. These techniques can be classified into four groups as follows

- Nonlinearity correction at the input by means of back-off, predistortion, Cartesian feedback, and polar feedback.
- Nonlinearity correction at the output by means of feedforward and linear amplification with nonlinear components (LINC).
- Nonlinearity correction at the supply by means of envelope elimination and restoration, pulsewidth modulation (PW and AM), and pulse deletion modulation (PDM).
- Nonlinearity correction at the load by means of switchable amplifier chain.

These methods have been shown schematically in Figure 10.38.

As mentioned before, AM to AM conversion can pose problems to data modulation using signal amplitude; however, AM to PM conversion plays the basic role in modulations based on frequency or phase. In general, comparing different linearization techniques, it can be said that feedback methods impose instability threats or bandwidth problems although their closed-loop nature results in robust linearity. Conversely, feedforward techniques have better stability and a more acceptable bandwidth. In the next subsections, the course of improving stability using the mentioned techniques are briefly explained.

10.3.1 Back-Off

Nonlinearity clearly reduces the gain and increases the IM at the maximum output power. So, the simplest way to linearize the PA operation is to force a back-off from its maximum power. The required back-off depends on the distortion caused by AM to AM and AM to PM conversions. This method benefits from low cost and no extra complexity but requires a device with higher power rating (a bigger device). Back-off

Table 10.2: Performance comparison among different classes of power amplifiers.

Class	Gain	Linearity	Output Power	α	$\eta_{typ}\%$	$\eta_{max}\%$	$V_{d,max}$ (Normalized)	$i_{d,max}$ (Normalized)	Power Capability
A	Large	Best	Moderate	360°	35	50	2	2	0.125
B	Moderate	Good	Moderate	180°	60	78	2	3.14	0.125
C	Small	Bad	Small	<180°	70	78–100	2		<0.125
AB	Moderate	Good	Moderate	>180°	35–60	50–78	2		>0.125
D	Small	Bad	Large	180°	75	100	2	1.57	0.318
E	Small	Bad	Large	180°	80	100	3.6	2.86	0.098
F	Small	Bad	Large	180°	75	100	2	3.14	0.159

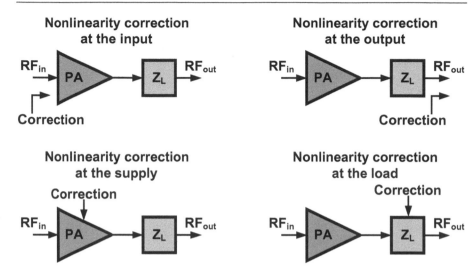

Figure 10.38: Different linearization methods in power amplifiers.

phenomenon is shown in Figure 10.39. The drawback of back-off is that the power
efficiency drops significantly as the back-off is increased.

10.3.2 Predistortion

We can linearize a nonlinear system, knowing its nonlinear transfer function, by
applying the input signal to a system, with an amplitude proportional to the inverse
characteristics of the nonlinear system amplitude response. Ideally, the gain is expected
to remain constant and the phase would vary linearly with frequency (the group
delay would remain constant), and the total system is supposed to remain linear.

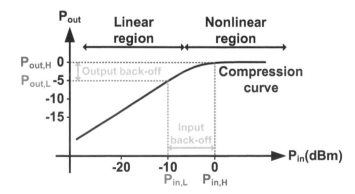

Figure 10.39: A typical output power versus input power curve in decibels depicting the input back-off and the output back-off with respect to the saturation
point.

Predistortion procedure is shown in Figure 10.40. Let $F(u) = G^{-1}(ku)$, therefore $y = G(G^{-1}(ku)) = ku$.

A predistorter conveys the input analog values to the corresponding predistorted values. A look-up table is perfectly suited for open-loop predistortion. It must be noted that correcting the distortion caused by the fabrication process, temperature variations, and aging is difficult. Therefore, it is constructive to upgrade this method to adaptive predistortion (Figure 10.41). In this method, a demodulator reads the output signal, followed by an A/D which digitizes it. In the next step, the adaptor block updates the look-up table data by comparing the input and output signals which have been read.

Finally, a digital predistorter applies the required predistortion to the signal using the look-up table data. The adaptor loop keeps working until the output signal is fully corrected. It can be stated that the distortion owing to AM to AM or AM to PM conversions can be compensated via applying an adaptive predistorter. The drawbacks of this method include the adaptor loop linearity and delay flatness, and the problems the look-up table preparation and updating may impose.

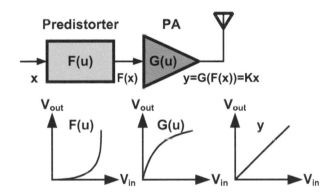

Figure 10.40: A schematic view of predistortion technique .

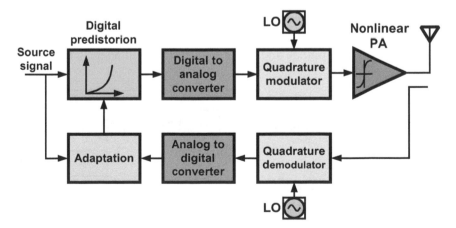

Figure 10.41: The block diagram of an adaptive predistorter technique.

10.3.3 Polar Modulation Feedback

The modulated signal can be shown as a complex number using both polar representation $(V = re^{j\theta})$ and Cartesian representation $(V = V_I + jV_Q)$. Figure 10.42 depicts the polar representation of a complex number.

The polar feedback method utilizes a negative feedback to correct the system nonlinearity. In this method, the first feedback loop determines the phase (θ) by means of a PLL, whereas the second loop senses the amplitude using an envelope detector, and the correction signal is fed to a variable-gain amplifier (VGA). In brief, the feedbacks are split between the phase and the amplitude. The overall structure is shown in Figure 10.43.

This architecture can improve the efficiency at higher output power, but it has not proven very useful when power levels are low. It is necessary to note that the presence of two negative feedback loops causes instability problems, imposing particular problems regarding the gain, the bandwidth, and the phase error. As a rule of thumb, it can be stated that the loop envelope detector bandwidth should be at least 10 times higher than the envelope's maximum frequency. Generally speaking, this linearization method increases the cost and the complexity of the system. Of course, in case where the AM to PM conversion effects are neglected, the phase feedback loop can be omitted resulting in reduced complexity.

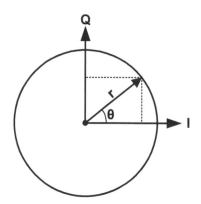

Figure 10.42: Amplitude and phase representation of a signal in the I–Q plane.

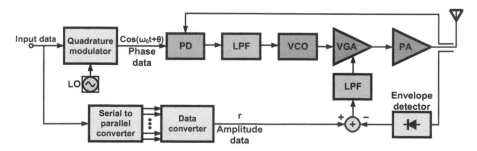

Figure 10.43: Block diagram of linearization with polar feedback.

10.3.4 Cartesian Modulation Feedback

A similar linearization method is Cartesian modulation feedback which is demonstrated in Figure 10.44. The PA output drives a quadrature demodulator to sense the degree of nonlinearity in the baseband data. Then, the restored signal is compared to the initial baseband information and passed through a low-pass filter in order to drive a quadrature modulator. Eventually, the signal is transmitted by the antenna after being amplified. The local oscillator is followed by a phase adjustment circuit so as to deliver the corrected signal for the quadrature modulator.

In a same way as the previous method, the Cartesian feedback increases the cost and the complexity of the circuit, and the stability is still a concern. The architecture is also sensitive to the phase of the local oscillator. Consequently, automatic adjustment is required to compensate for the process and temperature variations. Efficiency is boosted at high output powers only. The loop bandwidth of this structure is higher than that of the polar feedback, making it more effective for higher bandwidth applications.

10.3.5 Feedforward Method

To avoid the instability problem involved in the aforementioned linearization techniques, the feedforward method has been presented. Consider the configuration shown in Figure 10.45. The PA output is coupled to an attenuator. The coupler and the attenuator, in sum, introduce a loss equal to the amplifier gain. A delay line in the feedforward path (delay1) applies a delay equivalent to the sum of the PA delay and the attenuator delay to the input signal. Subtracting the two output signals of the attenuator and the delay line gives an error signal generated due to nonlinearity (mainly the IM products). This error signal is amplified by a supplementary linear amplifier with equal gain to the main amplifier. A second delay line (delay2) is implemented at the output of the power amplifier to compensate the delay of the linear amplifier at the feedforward path. At last, the signals of both paths are subtracted. Ideally, the main path signal distortion must be eliminated by the error signal amplified in the feedforward path.

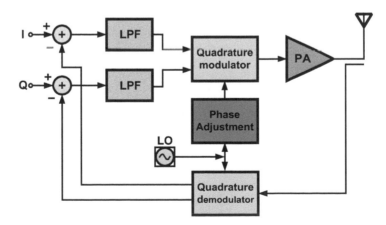

Figure 10.44: Block diagram of linearization with Cartesian feedback.

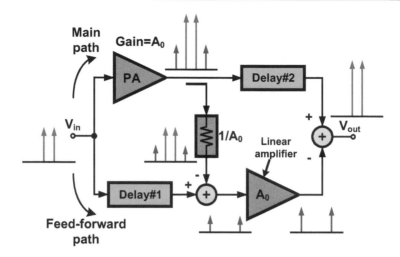

Figure 10.45: Block diagram of linearization with feedforward.

The main advantage of this method is its superior stability compared to the previous ones. However, it requires a supplementary linear amplifier, adding to design complexity even though it is dealing with a small-signal amplitude. Furthermore, delay lines should both provide matching and low loss. Notice that the power amplifier's gain and the coupler plus attenuator's path loss are necessitated to be equal with high precision. High sensitivity to parameters such as fabrication process, temperature, and aging can be mentioned as this system's drawback.

10.3.6 Linear amplification with nonlinear components

Linear amplification with nonlinear components is based on converting the amplitude information into phase and amplifying constant-envelope signals. This method provides capability of linear amplification at high output powers and it is shown in Figure 10.46.

Since the efficiency is proportional to the average output power, this configuration is not efficient for modulations with large peak to average ratio (PAR). To improve linearity, envelope feedback can be used. In the topology depicted in Figure 10.46,

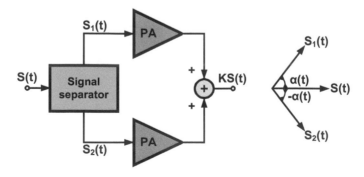

Figure 10.46: Block diagram of the LINC linearization method.

consider the input signal as

$$s(t) = b(t)\cos(\omega t + \phi(t)) \qquad (10.66)$$

Consequently, the separator must generate the following signals

$$s_1(t) = A\cos(\omega t + \phi(t) + \alpha(t)) \qquad (10.67a)$$
$$s_2(t) = A\cos(\omega t + \phi(t) - \alpha(t)) \qquad (10.67b)$$

The parameter added to phase in the output can be defined as

$$\alpha(t) = \cos^{-1}(b(t)/2A) \qquad (10.68)$$

The amplitude information is hence placed within the signal phase at the output signal of the separator and the signal envelope will take a constant value, leading nonlinearity to have a less adverse impact on the data. The total output signal would have the following form

$$s_{\text{out}}(t) = 2AK\cos(\alpha(t))\cos(\omega t + \phi(t)) - Kb(t)\cos(\omega t + \phi(t)) \qquad (10.69)$$

One of the drawbacks of this method is the analog implementation of the input signal separator which could be demanding.

10.3.7 Envelope Elimination and Restoration

As mentioned in Chapter 5, a limiter circuit eliminates the amplitude variations but keeps the changes lying in the phase. Consider the architecture shown in Figure 10.47 where two distinct paths to the output are separated by a separator. The first path utilizes a limiter circuit to extract the phase-modulated information solely. The other path contains an envelope detector to obtain the amplitude modulated data and transfer it to the voltage source (power supply) modulation circuit. In other words, the PA has a modulating voltage source if this technique is used. Altogether, the outputs of these two paths reconstruct the amplified input signal.

This configuration necessitates an efficient source modulator with low loss for which switched capacitor DC/DC converters are an appropriate choice. The method has the capability of providing 100% efficiency at both low and high output powers, and hence it is proper for power amplifiers whose output power is saturated (classes C, D, E, and F). Envelope feedback can also help improve linearity in this technique. The disadvantages of envelope elimination and restoration include AM to PM distortion due to source variations as well as unequal delay between the two paths that can cause distortion.

10.3.8 Pulse Amplitude and Width Modulation

Amplitude modulation can be performed by controlling the width of the pulse applied to the device or the conduction angle. Indeed, the conduction angle determines the length of the time when the amplifier is on, and it flows current to the load. Consider Figure 10.48.

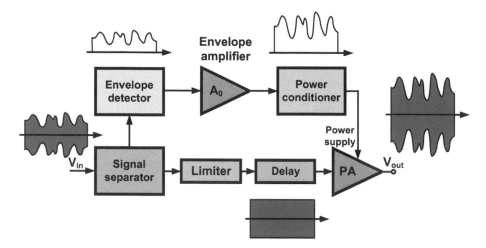

Figure 10.47: Block diagram of envelope elimination and restoration linearization method.

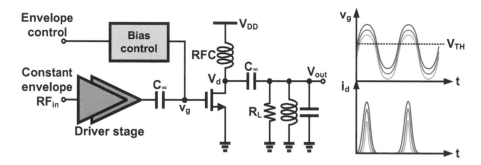

Figure 10.48: Pulse amplitude and width modulation circuit.

As shown in Figure 10.48, the signal with constant envelope and the bias voltage
have been both applied to the transistor gate. The voltage bias value is calculated
through an envelope control circuit. By applying this input signal, square sine-wave-
tips currents are generated in the drain that can reduce the device loss and thus increase
its efficiency. The impedance matching network is placed at the output. In the design of
the bias circuit, it must be noted that the bandwidth must be higher than the envelope's
maximum frequency. The efficiency of this architecture is higher than class A and
class AB, and lower than class C. Applying the envelope feedback can result in further
improvements in linearity.

10.3.9 Switching Parallel Amplifiers

Consider Figure 10.49 where three power amplifiers are preceding a switch array.

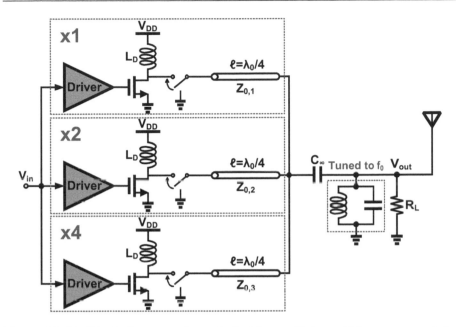

Figure 10.49: Block diagram of a switchable parallel amplifier array used for power control.

Each PA has been designed for a certain value of output power. The output of these PAs is connected to a transmission line with characteristic impedance, $Z_{0,i}$ with $\lambda/4$ length which isolates either of them once its switch is short-circuited. The amplifier array is connected to the load by means of a 3-bit switch control. The weighted combination of the amplifiers' powers appears at the load depending on the switch states [11]. As such, there would be seven ($2^3 - 1$) levels of controllable power which can be interpreted as a linearity improvement as a function of the input. Furthermore, the efficiency would be ameliorated as well. Switches must have low loss to guarantee desirable performance. This architecture has a moderate design complexity.

10.4 Conclusion

In this chapter, we discussed the general requirements and characteristics of the power amplifiers. Normally, the power amplifier is a block which generates harmonics as well as IM products due to its nonlinear operation. Furthermore, its gain is normally saturated by increasing the input amplitude. AM to AM conversion and AM to PM conversion are among the most important nonlinear characteristics of a power amplifier. Power amplifiers have different topologies and modes of operation. Class A, class B, class AB, class C, class D, class E, class F, and class S power amplifiers were introduced in this chapter along with their corresponding efficiencies and power capabilities.

Linearization techniques in power amplifiers are of great importance in the modern RF circuitry. Different linearization techniques were introduced in this chapter including predistortion, polar modulation feedback, Cartesian modulation feedback, feedforward, linear amplification with nonlinear components, envelope elimination and restoration, and switching parallel amplifiers. Each method has its own possibilities and complexities.

10.5 References and Further Reading

1. B. Razavi, *RF Microelectronics*, second edition, Castleton, NY: Prentice-Hall, 2011.
2. S. Cripps, *RF Power Amplifiers for Wireless Communications*, Nowrood, MA: Artech House, 1999.
3. A. Grebebbikov, *RF and Microwave Power Amplifiers Design*, Boston, MA: McGraw-Hill, 2005.
4. J. Everard, *Fundamentals of RF Circuit Design with Low Noise Oscillators*, United Kingdom: J. Wiley & Sons, Inc., 2000.
5. U.L. Rohde, A.M. Pavio, G.D. Vendelin, *Microwave Circuit Design Using Linear and Nonlinear Techniques*, Hoboken, NJ: J. Wiley & Sons, Inc., 2005.
6. J.R. Smith, *Modern Communication Circuits*, second edition, New York, NY: McGraw Hill, 1997.
7. K.K. Clarke, D.T. Hess, *Communication Circuits, Analysis and Design*, United States: Krieger Publishing Company, 1994.
8. R. Chi-Hsi Li, *RF Circuit Design*, Hoboken, NJ: J. Wiley & Sons, Inc., 2009.
9. N. O. Sokal, A. D. Sokal, "Class E-A New Single-Ended Class of High-Efficiency Tuned Switching Power Amplifiers," *IEEE J. Solid-State Circuits*, Vol. 10, pp. 168–176, 1975.
10. J.Y. Hasani, M. Kamarei, "Analysis and Optimum Design of a Class E RF Power Amplifier," *IEEE Transactions on Circuits and Systems I: Regular Papers,* Vol. 55, Issue 6, pp. 1759–1768, July 2008.
11. A. Shirvani, D. K. Su, and B. A.Wooley, "A CMOS RF power amplifier with parallel amplification for efficient power control," *IEEE J. Solid-State Circuits*, vol. 37, no. 7, pp. 684–693, June 2002.
12. Hae-Seung Lee and Michael H. Perrott, course materials for 6.776 *High Speed Communication Circuits*, Spring 2005. MIT OpenCourseWare (http://ocw.mit.edu/), Massachusetts Institute of Technology.

10.6 **Problems**

Problem 10.1 In a handheld transceiver as shown in Figure 10.50, the power amplifier has a maximum RF output power of 20 dBm, the transmitter bandpass filter has a 2 dB insertion loss, and the duplexer's isolation is about 30 dB. If the receiver has a bandpass filter with 1 dB insertion loss and 90 dB out-of-band rejection at f_{TX}, determine the minimum possible sensitivity of the receiver for a required C/I of 8 dB.

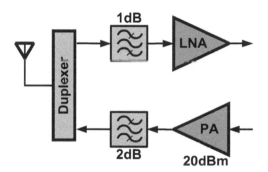

Figure 10.50: Block diagram of the handheld transceiver RF front-end.

Problem 10.2 An active device has a nonlinear transconductance which can be described by the following equation

$$i = \alpha v + \beta v^2 + \delta v^5 \tag{10.70}$$

For an input with the following form, determine the third-order IM products, furthermore, show that if the modulating signals $a(t)$ and $b(t)$ are band limited to W, the bandwidth of the third-order IM product components would be 5 W:

$$v_1 = a(t) V_1 \cos(\omega_1 t) \tag{10.71}$$
$$v_2 = b(t) V_2 \cos(\omega_2 t) \tag{10.72}$$

Problem 10.3 In a MOS transistor, the nonlinear gate–source capacitance is modeled as the following

$$C_{gs}(V_{gs}) = \frac{C_0}{\sqrt{1 + \frac{V_{gs}}{V_0}}} \tag{10.73}$$

First, determine the nonlinear $q - V$ characteristics of the junction capacitance, by integrating the above equation with respect to V_{gs}. Then, develop the Taylor's series expansion of this characteristics up to V_{gs}^3 term. Now, considering a large-signal input voltage of the form $V_{gs} = V_1 \cos(\omega t)$, determine the large-signal input capacitance of the gate–source junction. For the given values of the transistor's equivalent circuit model, compute the AM-PM characteristics (ϕ_{out} versus V_S) for the given stage at 5 GHz. For this purpose, assume that the value of V_S varies between 200 mV and 2 V.

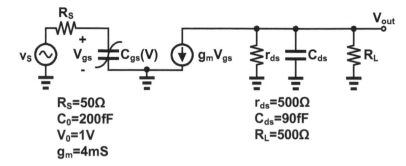

$$R_S=50\Omega$$
$$C_0=200fF$$
$$V_0=1V$$
$$g_m=4mS$$

$$r_{ds}=500\Omega$$
$$C_{ds}=90fF$$
$$R_L=500\Omega$$

Figure 10.51: The equivalent circuit of the MOS stage with a nonlinear gate–source capacitance.

Problem 10.4 A power MOSFET has the following square-law transfer characteristics $I_{DS} = k(V_{GS} - V_{TH})^2$, with a threshold voltage of $0.2V$, and $k = 50\frac{mA}{V^2}$. We intend to design a class A power amplifier with $V_{DD} = 1V$ and $R_L = 100\Omega$. First, determine the required bias point of the transistor, and the maximum drain current and the maximum drain voltage. Secondly, determine the gain of the stage, and the required input voltage swing to achieve the maximum output power. What would be the maximum output power in this case?

Problem 10.5 Consider the transfer characteristics of a power MOSFET as

$$I_{DS} = k(v_{GS} - V_{TH})^2 \quad v_{GS} > V_{TH} \tag{10.74}$$
$$I_{DS} = 0 \quad v_{GS} \le V_{TH} \tag{10.75}$$

Now, consider the input gate–source voltage as

$$v_{GS} = V_{GS_0} + V_1 \cos(\omega_0 t) \tag{10.76}$$

where $V_{GS_0} \le V_{TH}$. Determine the output current conduction angle as a function of V_1 and V_{GS_0}, and the output current waveform in this case. Now, compute the DC component and the first harmonic component of the output current, and consequently, compute the AC and the DC powers and the efficiency of this amplifier for class B and class C operation.

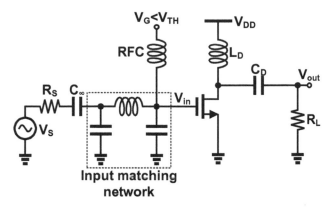

Figure 10.52: A MOSFET class B or class C power amplifier with bias and matching circuits.

Index

469

Authors Biographies

Forouhar Farzaneh was born in Tehran, Iran in 1957. He received his B.S. in Electrical Engineering from the University of Shiraz, in 1980, Master degree from E.N.S.T., Paris in1981, DEA and Doctorate from University of Limoges, France in 1982 and 1985, respectively. He was with Tehran Polytechnic from 1985 to 1989. Since 1989 he has been with the Department of Electrical Engineering, Sharif University of Technology where he is a Professor. He was the Chairman of the Department of Electrical Engineering, from 1992 till 1995.

His main areas of interest are Nonlinear RF Circuits, Microwave and Millimeter wave systems, Antenna Arrays and Wireless Communications. He is the author of a book in the field of Communication Circuits in Persian, published by Sharif University Press.

He has been a Senior Member of IEEE since 1997. He was a co-recipient of the Microwave Prize-European Microwave Conference in 1985, a recipient of the Maxwell Premium of IEE, U.K. in 2001, and co-recipient Mojtahedi Innovation Award (Sharif University of Technology) in 2010. He was also the recipient of the 2015 Hakkak award, in recognition of his tremendous life-time contribution to national development and propagation of research in Communications Engineering presented by IEEE-Iran Section.

Ali Fotowat was born in Tehran, Iran, in 1958. He received the B.S. degree in Electrical Engineering from the California Institute of Technology, Pasadena, CA, USA, in 1980, and the M.S. and Ph.D. degrees in Electrical Engineering from Stanford University, Stanford, CA, USA, in 1982 and 1991, respectively.

He started his career at Philips Semiconductor in Sunnyvale, CA, USA, in 1987 where he developed several integrated circuits for mobile phones. In 1991 he joined the Electrical Engineering Department of Sharif University of Technology, Tehran, Iran. His research interests include advanced integrated circuits for energy savings and communication/positioning applications. Due to his interests in entrepreneurial engineering, he has been the co-founder of several companies, including KavoshCom Asia R&D Company, alongside advising his students in the field.

Dr. Fotowat-Ahmady is a three times recipient of the Khwarizmi Science and Engineering Award for his work on low-power microelectronics and communication ICs. He is a member of the IEEE Solid-State Society and has been the adviser of the society's Sharif Electrical Engineering student chapter.

Mahmoud Kamarei received M.S. in Electrical Engineering from University of Tehran in 1979, M.S. in Communications Engineering from E.N.S.T., Paris, France, in 1981, and Ph.D. degree in electronics from the Institute National Polytechnique de Grenoble (INPG), Grenoble, France, in 1985.

He was a Researcher at the Laboratoire d'Electromagnétisme Micro-ondes et Optoélectronique, INPG, from 1982 till 1985. He was a Master of Conferences at the J. Fourier University of Grenoble from 1985 till 1991. He has been with the University College of Engineering, School of Electrical and Computer Engineering, University of Tehran since 1991, where he is currently a Professor of Electrical Engineering. He was the Dean of the University College of Engineering, University of Tehran for 8 years, 2009-2017. He is also a faculty member at the Center of Excellence on Applied Electromagnetic Systems, School of Electrical and Computer Engineering, University of Tehran.

His main research interest areas include RF CMOS IC Design, Design and Linearization of the RF Power amplifiers, Low-phase-noise oscillator design, PLLs and Injection Locked Oscillators.

Ali Nikoofard was born in Tehran, Iran, in 1990. He received his B.S. in Electrical Engineering from Shahed University, Tehran, in 2012, M.S. in Electronics from Sharif University of Technology, Tehran, in 2014, and M.S. from Case Western Reserve University, Cleveland, OH, USA, in 2017 in integrated circuit design. He was with KavoshCom Asia R&D Company from 2014 to 2015. He is now working toward Ph.D. at University of California at San Diego, La Jolla, CA, USA, on ultra-low-power transceiver design with new power efficient modulation schemes. He is a member of the IEEE Solid-State Circuits Society since 2013. His current research interests include wireless transceivers, frequency synthesizers, phase-locked loops and ultra-low-power circuit.

Mohammad Elmi was born in Tehran, Iran, in 1988. He received his B.S. degree in Electrical Engineering from Noshirvani University of Technology, Babol, Iran, in 2013, and M.S. degree in Electrical Engineering (Analog Electronics) from Shahid Beheshti University, Tehran, Iran, in 2015. He has been with KavoshCom Asia R&D Company since 2015. He is now working on a low-power wireless heart monitoring system as a member of research team at KavoshCom Asia. His research interests include RF integrated circuits design, low-power analog circuit design, wireless communication transceivers, and mm-wave integrated circuits.

Lightning Source UK Ltd.
Milton Keynes UK
UKHW021130240520
363760UK00001B/9